Connective Tissues

Biochemistry and Pathophysiology

Edited by R. Fricke and F. Hartmann

With 121 Figures

Springer-Verlag Berlin Heidelberg New York 1974

Professor Dr. Reinhard Fricke, Medizinische Poliklinik,
Hannover-Kleefeld

Professor Dr. Fritz Hartmann, Medizinische Poliklinik,
Hannover-Kleefeld

Illustration on the cover designed by Leonhard Fricke by modification of an electron microscopic picture (Wellauer, P., Wyler, T., Buddecke, E.: Electron microscopic and physicochemical studies on bovine nasal cartilage proteoglycan. Hoppe-Seyler's Z. Physiol. Chem. 353, 1043-1052 (1972)) of a proteoglycan molecule by Dr. P. Wellauer, Institute of Zoology, University of Bern, and by incorporating a figure of R. I. G. Morrison, this book, page 151.

ISBN-13: 978-3-642-61934-2 e-ISBN-13: 978-3-642-61932-8
DOI: 10.1007/ 978-3-642-61932-8

Softcover reprint of the hardcover 1st edition 1974
Offsetprinting: Julius Beltz, Hemsbach/Bergstr.

Preface

Connective tissues are essential for the physical functioning of the animals's body. The condition of the various connective tissues is governed by biochemical factors, anabolism and catabolism, that are controlled by specific enzymes. Any change outside the normal range of metabolism, for instance induced by immunological reactions, may induce a pathological disturbance. The result can be acute or chronic inflammation, or loss of normal function, expressed in loosening, dilatation, breaking, wear, stiffness, shrinking, scars, stenosis, and cirrhosis or any other kind of fibrosis.

A first step toward improving our understanding of the feedback mechanism that maintains the biological status and texture of a given connective tissue is to combine what is known about synthesis and enzymatic degradation of the components of fibers and ground substance. Common pathological phenomena like chronic inflammation of immune reactions can be either the result of the cause of disturbances in the sensitive balance of connective tissue metabolism. Nowadays connective tissues are less and less regarded as bradytrophic tissue but rather as a stimulating and many-sided problem of research.

Before we can understand the pathogenesis of the connective tissue diseases that result in the destructive processes mentioned above, basic research will be necessary. This research will be furthered by a constant exchange of information and the results of observations.

To promote this exchange of information between scientists, symposia on connective tissue research are organized at regular intervals. However, a symposium is not the place to present surveys of all the themes discussed. Therefore, at the symposium on connective tissue research held in Hannover, Germany, some time ago we seized the opportunity to invite specialists to write review articles on the field of research on which they presented papers at the symposium. A synopsis of the present state of knowledge in the field is a necessity for the promotion of research. We are therefore glad to be able to present review articles and reports of new data on selected topics in connective tissue.

Spring 1974 R. Fricke and F. Hartmann

Table of Contents

II. Synthesis

III. Degradation

Contributors

Adam, M., Research Institute for Rheumatic Diseases, Prague/Czechoslovakia
Anderson, A.J., Lower Flat, 9 Welbeck Avenue, Hove, Sussex/Great Britain
Barnes, M.J., Dunn Nutritional Laboratory, University of Cambridge and Medical Research Council, Milton Road, Cambridge CB4 1XY/Great Britain
Bates, C.J., Dunn Nutritional Laboratory, University of Cambridge, Milton Road, Cambridge CB4 1XJ/Great Britain
Baxter, Elsmaree, Department of Medicine, Royal Melbourne Hospital, University of Melbourne, P.O. Victoria 3050/Australia
Bazin, S., Service de Pathologie Expérimentale, Institut Pasteur, 92380 Garches/France
Bollwahn, W., Tierärztliche Hochschule Hannover, Klinik für kleine Klauentiere, Bischofsholer Damm 15, 3000 Hannover/Federal Republic of Germany
Bornstein, P., Departments of Biochemistry and Medicine, University of Washington, School of Medicine, Seattle, WA 98105/USA
Buddecke, E., Physiologisch-Chemisches Institut der Universität, Waldeyerstr. 15, 4400 Münster/Federal Republic of Germany
Codington, J.F., Massachusetts General Hospital, Laboratory for Carbohydrate Research, Boston, MA 02114/USA
Dabew, D., Zentrallaboratorium, Städtisches Krankenhaus Holweide, Neufelderstr. 32, 5000 Köln 80/Federal Republic of Germany
Delaunay, A., Institut Pasteur, Service de Pathologic Expérimentale, Annexe de Garches 92 (Hauts-De-Seine)/France
Deyl, Z., Institute of Physiology, Czechoslovak Academy of Science, Prague/Czechoslovakia
Eberhard, A., Klinisch-Chemisches Zentrallaboratorium an der Medizinischen Fakultät der RWTH Aachen, Goethestr. 27/29, 5100 Aachen/Federal Republic of Germany
Erl, D., Pathologisches Institut der Universität Hamburg, Martinistr. 52, 2000 Hamburg 20/Federal Republic of Germany
Evanson, J.M., Department of Medicine, University Hospital of South Manchester, Nell Lane, Manchester M20 8LR/Great Britain
Figura, K.v., Physiologisch-Chemisches Institut der Universität, Waldeyerstr. 15, 4400 Münster/Federal Republic of Germany
Fiszer-Szafarz, B., Fondation Curie, Institut du Radium, 15, Rue G.-Clémenceau, Bâtiment 111-91405 Orsay/France
Fricke, R., Medizinische Klinik, Medizinische Hochschule, Karl-Wiechert-Allee 9, 3000 Hannover-Kleefeld/Federal Republic of Germany
Gerlach, U., Medizinische Universitäts-Klinik, Westring 3, 4400 Münster/Federal Republic of Germany
Gotte, L., Istituto di Istologia-Embriologia, Università di Padova, Via Falloppio 16, 35100 Padova/Italy
Gottschalk, A.†, Max-Planck-Institut für Virusforschung, Spemannstr. 35, 7400 Tübingen/Federal Republic of Germany
Grade, J., Pathologisches Institut, Universitäts-Krankenhaus Eppendorf, Martinistr. 52, 2000 Hamburg 20/Federal Republic of Germany

Grasedyck, K., I. Medizinische Klinik, Universitäts-Krankenhaus Eppendorf, Martinistr. 52, 2000 Hamburg 20/Federal Republic of Germany
Greiling, H., Klinisch-Chemisches Zentrallaboratorium der Medizinischen Fakultät an der RWTH Aachen, Goethestr. 27/29, 5100 Aachen/Federal Republic of Germany
Gries, G., Perfallstr. 1, 8000 München 80/Federal Republic of Germany
Hall, D.A., The Department of Medicine, University of Leeds, The General Infirmary, Leeds LS1 3EX/Great Britain
Happey, F., Postgraduate School of Studies in Textiles, University of Bradford, Bradford, Yorkshire BD7 1DP/Great Britain
Hardingham, T.E., The Mathilda and Terence Kennedy Institute of Rheumatology, Hammersmith, London W6 7DW/Great Britain
Hartmann, F., Medizinische Klinik, Medizinische Hochschule, Karl-Wiechert-Allee 9, 3000 Hannover-Kleefeld/Federal Republic of Germany
Hauss, W.H., Medizinische Universitätsklinik und Poliklinik, Westring 3, 4400 Münster/Federal Republic of Germany
Hörmann, H, Max-Planck-Institut für Biochemie, 8033 Martinsried bei München/Federal Republic of Germany
Jeanloz, R.W., Laboratory for Carbohydrate Research, Massachusetts General Hospital, Boston MA 02114/USA
Kennedy, J.F., Department of Chemistry, University of Birmingham, Edgbaston, Birmingham B15 2TT/Great Britain
Kisters, R., Klinisch-Chemisches Zentrallaboratorium der Medizinischen Fakultät an der RWTH Aachen, Goethestr. 27/29, 5100 Aachen/Federal Republic of Germany
Kivirikko, K.I., Department of Medical Chemistry, University of Oulu, Kajaanintie 52 A, 90220 Oulu 22/Finnland
Kleine, T.O., Universitäts-Nervenklinik, Ortenbergstr. 48, 3550 Marburg/Federal Republic of Germany
Kölln, H., Pathologisches Institut, Universitäts-Krankenhaus Eppendorf, Martinistr. 52, 2000 Hamburg 20/Federal Republic of Germany
Kresse, H., Physiologisch-Chemisches Institut der Universität, Waldeyerstr. 15, 4400 Münster/Federal Republic of Germany
Leibovich, S.J., School of Medicine, University of Washington, Seattle WA/USA
Levene, C.I., Dunn Nutritional Laboratory, University of Cambridge, Milton Road, Cambridge CB4 1XJ/Great Britain
Lindner, J., Pathologisches Institut, Universitäts-Krankenhaus Eppendorf, Martinistr. 52, 2000 Hamburg 20/Federal Republic of Germany
Lindsley, H., Department of Medical Zoology, Walter Reed Army Institute of Research, Walter Reed Army Medical Center, Washington, DC 20012/USA
Lindy, S., Department of Medical Chemistry, University of Helsinki, Siltavuorenpenger 10, 00170 Helsinki 17/Finnland
Mannik, M., Department of Medicine, University of Washington, Seattle, WA 98195/USA
Merkel, Mathilde, Institut für Pharmakologie der Ruhr-Universität, Im Lottental, 4630 Bochum/Federal Republic of Germany
Messow, C., Pathologisches Institut, Tierärztliche Hochschule Hannover, Bischofsholer Damm 15, 3000 Hannover/Federal Republic of Germany
Moczar, E., Laboratoire de Biochimie du Tissu Conjonctif, Université de Paris-Val de Marne, Faculté de Médecine, 6 Rue du Général Sarrail, 94-Créteil/France
Moczar, M., Laboratoire de Biochimie du Tissu Conjonctif, Université de Paris-Val de Marne, Faculté de Médecine, 6 Rue du Général Sarrail, 94-Créteil/France
Morrison, R.I.G., Strangeways Research Laboratory, Wort's Causeway, Cambridge CB1 4RN/Great Britain
Muir, Helen, The Mathilda and Terence Kennedy Institute of Rheumatology, Hammersmith, London W6 7DW/Great Britain

Naylor, A., Bradford Royal Infirmary, Bradford, Yorkshire BD7 1DP/ Great Britain
Németh-Csóka, M., Zentral Laboratorium Fövárosı, Tétényi-uti-Kórház, tétényi-u 14 - 16, Budapest XI/Hungary
Osborn, J.M., Postgraduate School of Studies in Textiles, University of Bradford, Bradford, Yorkshire BD7 1DP/Great Britain
Palframan, J., Postgraduate School of Studies in Textiles, University of Bradford, Bradford, Yorkshire BD7 1DP/Great Britain
Pearson, C.H., Department of Oral Biology, Faculty of Dentistry, University of Alberta, Edmonton, Alberta/Canada
Prinz, G., Pathologisches Institut, Universitäts-Krankenhaus Eppendorf, Martinistr. 52, 2000 Hamburg 20/Federal Republic of Germany
Prokop, O., Institut für Gerichtliche Medizin der Humboldt-Universität, Hannoversche Str. 6, X104 Berlin/German Democratic Republic
Reed, F.B., Department of Biochemistry, The Royal Dental Hospital, University of London, School of Dental Surgery, 32 Leicester Square, London WC2H 7LJ/Great Britain
Render, R.M., Postgraduate School of Studies in Textiles, University of Bradford, Bradford, Yorkshire BD7 1DP/Great Britain
Robert, B. †, Laboratoire de Biochimie du Conjonctif, Faculté de Médecine, Université de Paris-Val de Marne, 6 Rue de Gal-Sarrail, 94 Crêteil/France
Robert, L., Laboratoire de Biochimie du Conjonctif, Faculté de Médecine, Université de Paris-Val de Marne, 6 Rue du Gal-Sarrail, 94 Crêteil/ France
Rodén, L., University of Alabama in Birmingham, University Station, OBSB, Birmingham, AL 35294/USA
Rosmus, J. †, Research Institute of Food Industry, Czech Agricultural Academy, Prague/Czechoslovakia
Rothe, A., Abteilung für Experimentelle Innere Medizin der Medizinischen Universitäts-Klinik, Kerpenerstr. 15, 5000 Köln 41/Federal Republic of Germany
Schauer, R., Institut für Physiologische Chemie, Ruhr-Universität Bochum, Buscheystraße, 4630 Bochum-Querenburg/Federal Republic of Germany
Schoop, H.J., Zentrallabor der Krankenanstalten, Bogenstraße, 2850 Bremerhaven/Federal Republic of Germany
Schwartz, Nancy B., Department of Pediatrics, University of Chicago, 5825 Maryland Avenue, Chicago, IL 60637/USA
Segeth, G., Medizinische Klinik der Universität, Westring, 4400 Münster/ Federal Republic of Germany
Stockhorst, Eva H., Sixth Street 554, Ann Arbor, MI 48103/USA
Struck, H., II. Chirurgischer Lehrstuhl, Biochemische Abteilung an der Chirurgischen Universitäts-Klinik, Ostmerheimerstr. 200, 5000 Köln-Merheim/Federal Republic of Germany
Stuhlsatz, H.W., Klinisch-Chemisches Zentrallaboratorium der Medizinischen Fakultät an der RWTH Aachen, Goethestr. 27/29, 5100 Aachen/ Federal Republic of Germany
Szafarz, D., Fondation Curie, Institut du Radium, 15 Rue G.-Clemenceau, Bâtiment, 111-91405 Orsay/France
Turner, R.L., Postgraduate School of Studies in Medical and Surgical Sciences, University of Bradford, Bradford, Yorkshire BD7 1DP/ Great Britain
Turto, H., Department of Medical Chemistry, University of Helsinki, Siltavuorenpenger 10, 00170 Helsinki/Finnland
Uhlenbruck, G., Abteilung für Experimentelle Innere Medizin der Medizinischen Universitäts-Klinik, Kerpenerstr. 15, 5000 Köln 41/ Federal Republic of Germany
Uitto, J., Department of Biochemistry, Rutgers Medical School, College of Medicine and Dentistry of New Jersey, Piscataway, NJ 08854/USA
Vainio, K., Department of Surgery, Rheumatism Foundation Hospital, Heinola/Finnland

Weiss, Jacqueline B., Department of Medical Biochemistry, University of Manchester, York Place, Manchester M13 9PT/Great Britain
Wollmer, A., Abteilung Physiologische Chemie der Medizinischen Fakultät an der RWTH Aachen, Melatener Str. 211/213, 5100 Aachen/Federal Republic of Germany
Wulff, U., Kinderklinik der Medizinischen Akademie, Ratzeburger Allee 160, 2400 Lübeck/Federal Republic of Germany

I. Structure

Metabolic Heterogeneity of Acidic Glycosaminoglycans in Bovine Arterial Tissue

E. Buddecke, H. Kresse, G. Segeth, and K. v. Figura

Mammalian arterial tissue is known to contain hyaluronate (HA), chondroitin 4(6)-sulfate (CS) dermatan sulfate (DS) heparan sulfate (HS) (Kaplan and Meyer, 1960; Antonopoulos et al., 1965; Dunstone, 1967; Nakamura et al., 1968; Buddecke and Kresse, 1960; Kresse and Buddecke, 1970) and their protein complexes respectively (Jacobs and Muir, 1963; Buddecke et al., 1963; Fransson and Havsmark, 1970) in a total amount of 1 - 2% of the dry weight of tissue.

Studies on age dependent and pathological changes in the distribution pattern of acidic glycosaminoglycans (GAG) in human arterial tissue gave inconsistent results (Kaplan and Meyer, 1960; Manley et al., 1969 Dalferes et al., 1971). However, the distribution pattern will be of particular interest, as an increased incorporation of ^{35}S-sulfate into the sulfated GAG has been described to be one of the first metabolic alterations in spontaneous and experimental arteriosclerosis (Hauss et al., 1968; Sanwald et al., 1971).

Since difficulties in distinguishing between normal human arterial tissue and barely detectable arteriosclerotic lesions prevented a systematic investigation of the metabolic feature of GAG, we studied the metabolism of the different types of GAG and proteoglycans in bovine aorta as a model for the normal.

1. Metabolic Heterogeneity of Aortic GAG

When segments of calf aorta were incubated under appropriate conditions in vitro in the presence of U-^{14}C-glucose or ^{35}S-sulfate over a period of 12 h, significant differences in the rate of ^{35}S- and ^{14}C-incorporation into the GAG are observed. The relative rates of biosynthesis of the individual GAG were calculated on the basis of relative proportions of HA : CS : DS : HS and their specific radioactivities. The highest rate of synthesis was found for HA and

Table. Relative rates of biosynthesis of arterial tissue GAG following in vitro incubation in the presence of U-^{14}C-glucose and ^{35}S-sulfate resp. (Kresse and Buddecke, 1970)

Type of GAG	mg/100 mg total GAG	Relative rate of synthesis following incorporation of ^{14}C	^{35}S	Relative turnover rate following incorporation of ^{14}C	^{35}S
HA	30	58		1.9	
CS	49	21	41	0.4	0.3
DS	9.5	10	29	1.0	1.0
HS	11.5	11	30	0.9	0.9

decreasing amounts for the sulfated GAG in the order CS, HS, and DS. The Table summarizes the results (Kresse and Buddecke, 1970). HA and HS were found to be synthesized from two differently labelled glucosamine precursor-pooles (Figura et al, 1973).

2. Age Dependent Changes in Specific Radioactivity of GAG

A study of age dependent changes of GAG metabolism revealed the following results: in bovine aortae the total GAG content increases from 1.19% to 1.70% over the first 3 years and decreases subsequently to 1.2% up to the 10th year, all values referring to the dry weight of tissue. No major changes in the distribution pattern of GAG are observed after this period. The specific radioactivity for all sulfated GAG decreases exponentially during the first 2 years but remains nearly constant for the following years, identical results being obtained for ^{14}C- and ^{35}S-GAG. Surprisingly, however, it was found that the specific ^{14}C-radioactivity of the HA increases twofold during the first 3 years reaching a plateau for the following years (Buddecke and Segeth, in press). The figure shows the age profile of ^{14}C-labelled HA and DS, the latter being representative for all sulfated GAG.

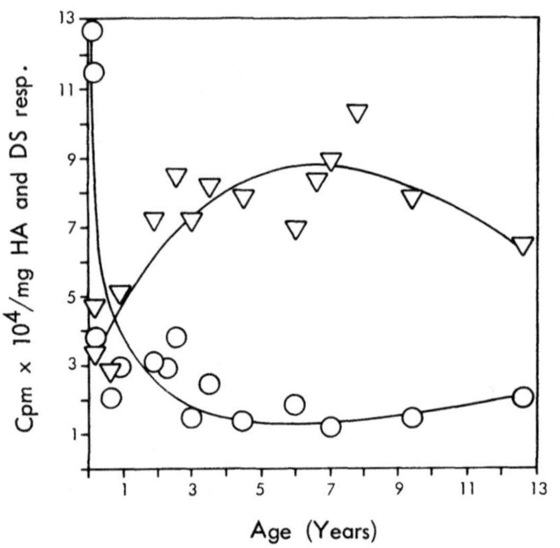

Fig. Age dependent changes in specific radioactivity of ^{14}C-hyaluronic acid(▽) and ^{14}C-dermatansulfate (O) after in vitro incubation in the presence of U-^{14}C-glucose

The reason for the higher rate of biosynthesis of HA and its further age dependent increase is still obscure, but a selective regulation of the biosynthesis of the individual GAG must be concluded from these results. An independent control of biosynthesis of the different GAG in arterial tissue is also reflected by the fact that a selective increase in the rate of synthesis of DS and HS of arterial tissue during genetic and experimental hypertonus in the rat has been observed (Kresse et al., 1971b).

3. Metabolic Heterogeneity of Bovine Aorta CS-DS-Proteoglycans

The observed differences between CS and DS in the rates of biosynthesis are especially striking with regard to the well established finding that in the aorta a large proportion of DS contains CS as well as DS disaccharide units in a copolymeric structure (Fransson and Roden, 1967; Fransson, 1968; Fransson and Havsmark, 1970).

Studying the chemical and metabolic features of bovine arterial tissue proteoglycans, we isolated and purified a CS-DS-proteoglycan containing CS as well as DS, both attached to the same protein core. This CS-DS-proteoglycan behaved as a single compound by several criteria of homogeneity. On incubation in vitro in the presence of ^{35}S-sulfate the CS-DS-proteoglycan was obtained in a ^{35}S-sulfate labelled form. From light scattering measurements a molecular weight of 2.0×10^6 and a random coil shape were calculated, chemical analysis indicated a protein content of 20.5% and a polysaccharide content of 79.5%, the latter consisting of 73.0% CS and 27.0% DS.

An exhaustive degradation of the ^{35}S-SD-DS-proteoglycan by testes hyaluronidase and pronase produces ^{35}S-DS and a series of ^{35}S-saccharide fragments ranging from disaccharides to decasaccharides. The tetrasaccharide fraction was found to contain iduronic acid (IdUA) besides glucuronic acid (GlcUA) indicating a copolymeric structure of CS and DS units. The occurence of IdUA in the tetrasaccharide fraction (as referred to the total uronic acid content) is conclusive for the presence of tetrasaccharide molecules having the structure

```
       SO3-        SO3-
        |           |
GlcUA-GalNAc-IdUA-GalNAc
```

From these data and from the results of alkali degradation the macromolecular structure of the CS-DS-proteoglycan may be summarized as follows: about 80 polysaccharide chains at a molecular weight between 15,000 and 20,000 are attached to a protein core by alkali sensitive O-seryl (threonyl) bonds giving a random coiled shaped macromolekule with a molecular weight of 2.0×10^6. It appears that (a) the CS and DS polymer chains are attached to a single protein core and (b) that CS and DS chains occur partly as copolymers.

Determinations of the specific radioactivity and the iduronic acid content of the DS fraction and the DS and CS units containing saccharide fragments obtained from testes hyaluronidase degradation showed that the specific radioactivity of the sulfate ester group of the DS units was about three times higher than those of the CS units (Kresse et al., 1971).

Further studies on ^{14}C-GlcN-labelled CS-DS-preparations revealed a varying degree of hybridization between CS and DS. A highly labelled ^{14}C-DS fraction hybridized with 19 % of a likewise highly labelled CS und a lower labelled CS fraction containing 20 % of a lower labelled DS as copolymer constituent were isolated by subfractionation of the aortic tissue GAG. These results are conclusive for the assumption that two topographically different sites exist for the biosynthesis of DS and CS, one delivering a highly labelled DS(CS)-hybrid, the other a lower labelled CS(DS)-hybrid (v. Figura et al., unpublished).

References

Antonopoulos, C.A., Gardell, S., Hamnström, B.J.: Separation of the glycosaminoglycans (mucopolysaccharides) from aorta by a column procedure using quaternary ammonium compounds. J. Atheroscler. Res. 5, 9 (1965).

Buddecke E., Kresse, H.: Mucopolysaccharide und Enzyme des Mucopolysaccharidstoffwechsels im Arterien- und Venengewebe. Angiologica 6, 89 (1969).

Buddecke, E., Kröz, W., Lanka, E.: Chemische Zusammensetzung und makromolekulare Struktur von Chondroitinsulfat-Proteinen. Hoppe-Seylers Z. physiol. Chem. 331, 196 (1963).

Buddecke, E., Segeth, G.: Altersabhängige Veränderungen im Stoffwechsel ^{14}C- und ^{35}S-markierter Glycosaminoglykane des Arteriengewebes. Z. Altersforsch. In press.

Dalferes, E.R.,jr., Ruiz, H., Kumar, V., Radhakrishnamurthy, B., Berenson, G.S.: Acid mucopolysaccharides of fatty streaks in young, human male aortas. Atherosclerosis 13, 121 (1971).

Dunstone, J.R.: Acid glycosaminoglycans of pig aorta. Experientia (Basel) 23, 344 (1967).

Figura, K.v., Kiowski, W., Buddecke, E.: Differently labelled glucosamine precursor pooles for the biosynthesis of hyaluronate and heparan sulfate. Europ. J. Biochem 40, 89 (1973).

Fransson, F.A.: Structure of dermatan sulfate III. The hybrid structur of dermatan sulfate from umbilical cord. J. biol. Chem. 243, 1504 (1968).

Fransson, F.A., Havsmark, B.: Structure of dermatan sulfate VII. The copolymeric structure of dermatan sulfate from horse aorta. J. biol. Chem. 245, 4770 (1970).

Fransson, F.A., Roden, L.: Structure of dermatan sulfate I. Degradatic by testicular hyaluronidase. II. Characterisation of products obtained by hyaluronidase digestion of dermatan sulfate. J. biol. Chem. 242, 4161, 4170 (1967a, b).

Hauss, H.W., Junge-Hülsing, G., Gerlach, U.: Die unspezifische Mesenchymreaktion. Stuttgart: G. Thieme 1968.

Jacobs, S., Muir, H.: A heparan sulfate-peptide from human aorta. Biochem. J. 87, 38 (1963).

Kaplan, K., Meyer, K.: Mucopolysaccharides of aorta at various ages. Proc. Soc. exp. Biol. (N.Y.) 105, 78 (1960).

Kresse, H., Buddecke, E.: Stoffwechselheterogenität ^{14}C- und ^{35}S-markierter Glycosaminoglykane (saurer Mucopolysaccharide) bei In-vitro-Inkubation. Hoppe-Seylers Z. physiol. Chem. 351, 151 (1970).

Kresse, H., Filipovic, I., Wessels, F., Wessels, G., Buddecke, E.: Selektive Zunahme der Syntheserate von Dermatansulfat und Heparansulfat des Arteriengewebes bei genetischem bzw. experimentellem Hypertonus der Ratte. Z. klin. Chem. 9, 21 (1971a).

Kresse, H., Heidel, H., Buddecke, E.: Chemical and metabolic heterogeneity of a bovine aorta chondroitin sulfate-dermatan sulfate-proteoglycan. Europ. J. Biochem. 22, 557 (1971)

Manley, G., Mullinger, R.N., Lloyd, P.H.: Properties of heparan sulfate and chondroitin sulfate from young and old human aortae. Biochem. J. 114, 89 (1969).

Nakamura, T., Tokita, K., Tateno, S., Kotoko, T., Ohba, T.: Human aortic acid mucopolysaccharides and glycoproteins. J. Atheroscler. Res. 8, 891 (1968).

Sanwald, R., Ritz, E., Wiese, G.: Acid mucopolysaccharide metabolism in early atherosclerotic lesions. Atherosclerosis 13, 247 (1971).

Chemical and Molecular Heterogeneity of Cartilage Proteoglycans

T. E. Hardingham, Elsmaree Baxter, and Helen Muir

The extracellular phase of cartilage, as in all connective tissues, is composed of collagen fibers and a polysaccharide-rich ground substance. The polysaccharide constituents have been characterized as proteoglycans containing chains of chondroitin-4-sulfate, chondroitin-6-sulfate and keratan sulfate covalently linked to a central protein core (see Muir, 1969). Early analyses of proteoglycan preparations from various cartilages showed that the relative proportions of these three constituents and protein were different in different sites in the same animal (Rosenberg et al., 1965; Gower and Pedrini, 1969; Tsiganos and Muir, 1970). Moreover, the composition of proteoglycan from any one site also varied with age (Goh and Lowther, 1966; Gower and Pedrini, 1969; Kröz and Buddecke, 1967; Brandt and Muir, 1969b; Rosenberg et al., 1969). These findings outlined large variations in composition and suggested that cartilage proteoglycan was not a single molecular entity, but a varying population of several molecular types. Although techniques have been perfected for the complete fractionation of the constituent glycosaminoglycan chains after proteolysis, fractionation of the intact proteoglycans has proved much more difficult due to their polyanionic macromolecular nature. However, the advent of both mild methods of preparation and new techniques of fractionation has now made it possible to demonstrate the possible extent of heterogeneity of proteoglycans under conditions where the possibility of any degradative changes has been largely eliminated.

In solutions of low ionic strength proteoglycans do not readily diffuse out of cartilage (Shatton and Schubert, 1954) unless the collagen network is subjected to severe mechanical disruption (Malawista and Schubert, 1958). However, if the ionic strength of the extracting solution is increased, increased yields of proteoglycan are obtained (Tsiganos and Muir, 1969a). Thus, it would appear, that the proteoglycans were not all bound within the tissue to the same extent, some being more readily extracted than others. On this principle sequential extraction of cartilage with solutions of increasing ionic strength has been used to obtain partial fractionation of proteoglycan molecules, which differ in size and composition.

When pig laryngeal cartilage was extracted initially by low speed homogenization in 0.15M sodium acetate and then by suspension in 10% calcium chloride (w/v) the proteoglycans released by the former had a lower proportion of keratan sulfate and protein than the latter, and comparison of size by gel-chromatography on 6% agarose showed that the 0.15M sodium acetate fraction had a larger proportion of smaller molecules than the subsequent fraction (Tsiganos and Muir, 1969b). This was also found when sequential extractions were carried out with pig articular cartilage (Fig. 1) (Brandt and Muir, 1969a,b, 1971).

Successive extractions produced proteoglycans of increased average size and with increasing keratan sulfate and protein content. However, the increase in keratan sulfate and protein contents not only reflec-

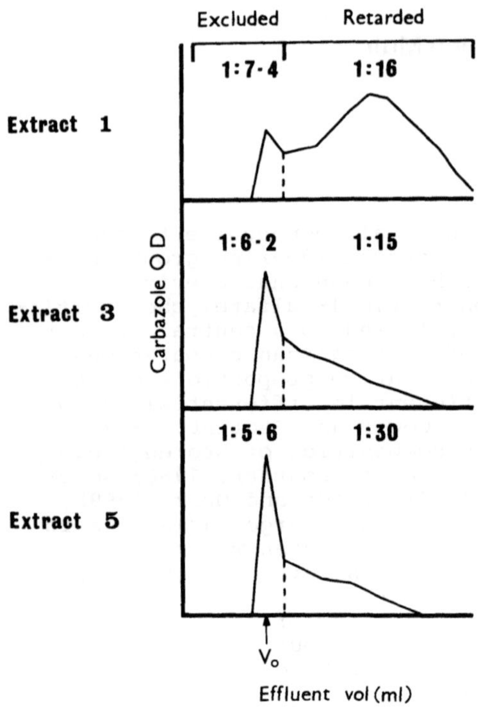

Fig. 1. Gel-chromatography on Sepharose 4B of proteoglycans from sequential extracts of pig articular cartilage from which the smallest molecules retarded by Sepharose 6B had been removed. The excluded and retarded fractions were isolated and their glucosamine/galactosamine molar ratios were determined. (From Brandt and Muir, 1971)

ted a higher proportion of large molecules, because in later extracts even the smallest molecules had a higher keratan sulfate and protein content than those in earlier extracts. Thus, even amongst molecules of the same size, it appeared that those with higher keratan sulfate and protein content were bound more firmly in the matrix. The small components in the initial extracts resembled fractions of high electrophoretic mobility observed in the proteoglycans from pig larynx (Muir and Jacobs, 1967),bovine nasal cartilage, human costal cartilage and nucleus pulposus (Pedrini, 1969), which likewise were rich in chondroitin sulfate and low in keratan sulfate and protein and which were of particularly low molecular weight.

Comparison of cartilage of different age revealed that the overall extractability of proteoglycans diminished with age. Thus, only 5 - 6% of the total uronic acid was extracted by 0.15M sodium acetate from articular cartilage of pigs 3 and 5 years old, whereas with fetal cartilage over half the total uronic acid was in this extract (Simunek and Muir, 1972). After extraction with 2M $CaCl_2$, the proportion of proteoglycan remaining firmly associated with the collagenous residue increased from a negligible amount in fetal tissue to one third at 3 and 5 years of age. The proportion of keratan sulfate in each extract increased with increasing age and the size of the readily extractable proteoglycans also varied. The proteoglycans extracted from fetal cartilage by 0.15M sodium acetate contained virtually no material retarded on Sepharose 6B, whereas 20% of that from 10-week-old-pigs and 12 - 13% from 25-weeks and 3-year-old animals was retarded.

The total amount of proteoglycan extracted from cartilage by solutions of different ionic strength revealed an optimal concentration of the extracting solution, considerably in excess of that used for initial sequential extractions. Both above and below the optimal concentration the yield of proteoglycan was considerably reduced; and the concentration of solution achieving the maximum yield of proteoglycan depended on the salt being used (Sajdera and Hascall, 1969).

When bovine nasal cartilage was extracted with the optimum concentration of guanidine HCl (4M) the largest (s_o = 40 S) proteoglycan species extracted were found to be aggregates of smaller molecules, (Average s_o = 25 S) (Hascall and Sajdera, 1969). In the presence of 4M guanidine HCl these aggregate molecules were dissociated into proteoglycan 'sub-units' and they were reformed upon lowering the concentration of guanidine HCl. The formation of aggregates required the interaction of proteoglycan sub-units with a 'glycoprotein-like' fraction which was separable by equilibrium density gradient centrifugation in the presence of 4M guanidine HCl. The original proteoglycan preparations contained both aggregate and monomer, their relative proportions depending on the pH of the extraction and the salt by which they were extracted.

Fractionation of proteoglycan preparations by density gradient centrifugation in the presence of 4M guanidine HCl separates molecules largely according to their relative content of protein and carbohydrate. In a preparation of proteoglycans which had been extracted from pig laryngeal cartilage in 4M guanidine HCl, approximately one-third of the protein was in a fraction at the top of a density gradient which contained very little uronic acid (Tsiganos et al, 1971). The uronic acid content of the fractions increased with increasing density, while the protein content decreased, producing a continuous distribution of molecules of varying composition. Most of the chondroitin sulfate was present in molecules of low protein content at the bottom of the gradient (Fig. 2, Table 1).

In general, throughout the gradient, the glucosamine content (reflecting primarily keratan sulfate) varied with the protein content.

In view of the possible presence of aggregates in proteoglycan preparations, it is necessary to reassess the heterogeneity observed under non-disaggregating conditions. Chromatography of the proteoglycan before disaggregation showed a large proportion of material to be excluded from Sepharose 2B. After disaggregation and separation of the 'glycoprotein-like' component, most of the proteoglycans were retarded by the gel, suggesting that aggregates were present in the original excluded fraction (Tsiganos and Muir, 1970) (Fig. 3).

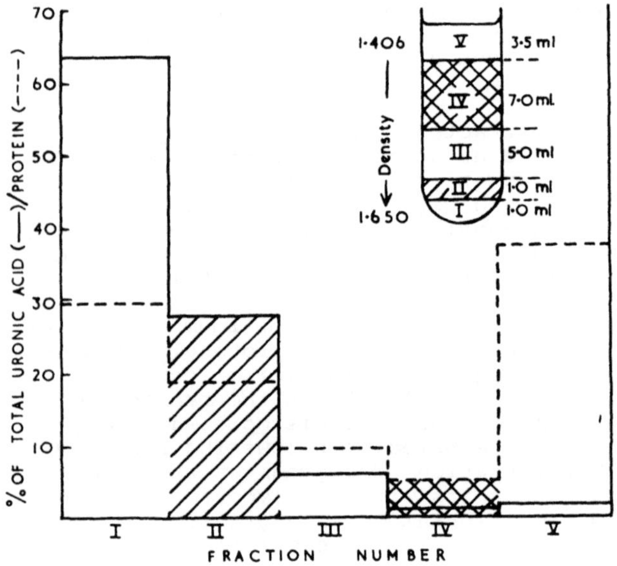

Fig. 2. Distribution of uronic acid and protein after equilibrium density gradient centrifugation in the presence of 4M guanidine HCl of proteoglycans from pig laryngeal cartilage. Insert shows the volume and relative position of the fractions. (From Tsiganos, Hardingham and Muir, 1971)

Table 1. Composition of caesium chloride density gradient fractions of proteoglycans of pig cartilage (Fig. 2) (Tsiganos et al., 1971)

Fraction No.		Weight ratios		Molar ratios
	% of total uronic acid	Uronic acid/ protein	Hexosamine/ protein	Glucosamine/ galactosamine
I	63.1	2.5	2.9	1:16
II	27.7	1.8	2.0	1:12
III	6.0	0.8	0.9	1:5.4
IV	0.9	0.25	0.3	1:3.5
V	1.6	0.05	0.09	1:2.3
starting material		1.74	1.9	1:11

In the initial preparation 40 - 50% of the total uronic acid was in molecules of smaller size, that were retarded by Sepharose 2B while the distribution of this fraction on gel chromatography was unaltered by disaggregation. This retarded component appeared unable to take part in the formation of aggregates, for when a radioactive retarded fraction was mixed under disaggregating conditions (in 4M guanidine HCl) with a non-radioactive excluded fraction, on reaggregation by dialysis to low ionic strength, gel chromatography showed no exchange

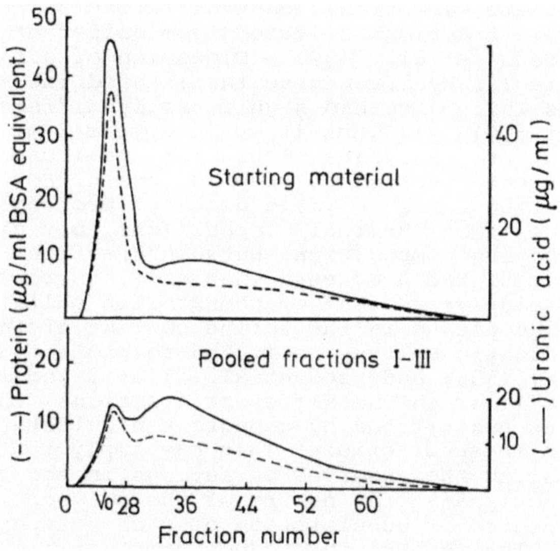

Fig. 3. Gel-chromatography on Sepharose 2B of isolated proteoglycans (starting material) from pig laryngeal cartilage and of pooled fractions I -III (Fig. 2) from density gradient fractionation of the starting material under disaggregating conditions. (From Tsiganos, Hardingham and Muir, 1971)

of radioactivity from the retarded to the excluded fraction. (Hardingham and Muir). Thus, in proteoglycans extracted from cartilage under disaggregating conditions there would appear to be two populations; those excluded from Sepharose 2B and disaggregated by 4M guanidine HCl and those mainly retarded by Sepharose 2B and left unaffected by 4M guanidine HCl.

The proteoglycans most readily extracted from cartilage by solutions of low ionic strength are mainly those retarded by gel chromatography on Sepharose 2B. Since solutions of low ionic strength do not dissociate proteoglycan aggregates, it is evident from their size that these molecules exist within the tissue in a non-aggregated form. In sequential extractions with solutions of increasing ionic strength, the proportion of proteoglycan retarded by Sepharose 2B steadily decreased until the final extracts only contained those excluded from the gel. Although the differences in composition between those fractions excluded and retarded by Sepharose 2B reflected to some extent the presence in the former of the 'glycoprotein-like' component that induces aggregation, the differences in composition between various fractions in sequential extracts of cartilage that were retarded by Sepharose 2B or gels of lower porosity indicates that there are non-aggregated proteoglycans of different composition.

If the variations in composition observed were the result of the attachment of different amounts of glycosaminoglycans to a single type of polypeptide chain, after removal of the glycoprotein-like component there should be negligible differences in the amino acid analysis of all proteoglycan fractions. However, proteoglycans isolated by density gradient centrifugation in the presence of 4M guanidine HCl contained several N-terminal amino acids (Baxter, 1972) and there were significant differences in amino acid compositions between fractions selected from different points in the gradient (Tsiganos et al., 1971). Digestion of disaggregated proteoglycans by testicular hyaluronidase facilitated the separation of several fractions which also had significantly different amino acid compositions (Baxter, 1972). Thus it would appear that the glycosaminoglycans in the proteoglycans of pig laryngeal cartilage are linked to several different types of polypeptide chain. However, density gradient fractionation fails to yield discrete proteoglycan fractions with distinct and constant chemical compositions, but gives a spectrum of molecules of varying composition. Between fractions from the bottom and the middle of the gradient, (Table 1), there was a ten-fold decrease in the relative content of chondroitin sulfate and protein, but only a small decrease in the serine content of the protein (Tsiganos). As chondroitin sulfate is linked to protein through serine residues (Muir, 1958; Anderson et al., 1965), and since its average chain length was similar in the different fractions, the proportion of serine residues substituted by chondroitin sulfate must decrease markedly between these fractions. This may imply that there are two factors influencing the observed variations in the composition of cartilage proteoglycans; the nature of the polypeptide core and variation in the degree of substitution of each polypeptide with chondroitin sulfate. The failure to produce distinct proteoglycan fractions upon density gradient fractionation may then be a consequence of the overlapping of the different polypeptide types, each with a variable number of chondroitin sulfate chains attached. It is of note that there is much less variation in the content of keratan sulfate as compared to protein throughout the density gradient.

The large changes in proteoglycan composition observed during ageing have not yet been fully characterized. They may result from changes in the relative synthesis of different protein cores or alterations at the level of attachment of glycosaminoglycan chains to the protein backbone or a combination of these two possibilities.

Having described the large degree of heterogeneity found in proteoglycan preparations from cartilage, it is necessary to assess its possible relevance to the function of cartilage. Studies on normal samples of human femoral articular cartilage have shown a positive correlation between the stiffness to compression and the content of chondroitin sulfate and keratan sulfate (Kempson et al., 1970). The correlation of stiffness with increased keratan sulfate content was somewhat greater than with chondroitin sulfate content. This may be related to the findings discussed above, that both increase in size and degree of aggregation of proteoglycans was paralleled by an increase in keratan sulfate content. The larger proteoglycan species may thus make a relatively greater contribution to the physical properties of cartilage.

When proteoglycans extracted from articular cartilage of lame and sound pigs were compared, in spite of similarities in overall composition, gel chromatography revealed the proteoglycans from lame animals to be of much smaller average size than those from normal animals (Simunek and Muir, 1971). This raised the prossibility of some correlation between abnormal proteoglycan populations and a pathological situation in cartilage.

When proteoglycans were selectively removed by the action of testicular hyaluronidase from the cartilage matrix of chick embryo limb rudiments in organ culture, followed by recovery in the absence of enzyme, the cartılage responded by rapidly replacing the lost proteoglycan (Fitton-Jackson, 1970). However, the newly synthesized material incorporated into the matrix was initially of smaller average size and higher glucosamine and protein content than proteoglycan in control rudiments and only slowly approached the composition of the control proteoglycans (Hardingham et al.). The synthesis of an apparently abnormal population of pıoteoglycans under these circumstances may reflect the specific synthesis of this type of proteoglycan in response to the depletion of the matrix or the failure of biosynthetic mechanisms to produce normal proteoglycans under conditions involving a much enhanced rate of synthesis (250%).

It is evident from these results that there is a feed-back control from the cartilage matrix to the chondrocyte that regulates the production of matrix constituents, and if increased degradation of proteoglycans ıllıcıts faster synthesis by the cells it is important that this should be controlled and normal.

Collagen and proteoglycans are the two main constituents of the cartilage matrix. Collagen is a structural component of very slow metabolic turnover (Gross, 1970). In cartilage the collagen network is maintained fully extended by a densely packed proteoglycan population which is of relatively fast turnover (Gross et al., 1960; Davidson and Small, 1963; Mankin and Lipiello, 1969). Fractionation of radioactive proteoglycans from cartilage, after pulse labelling in vitro with (^{35}S) sulfate and various times of non-radioactive chase incubation, could show no metabolic relationships between fractions of different size or composition (Hardingham and Muir) and, although there was evidence of metabolic heterogeneity, fractions of all sizes appeared to be formed at the same time and no fraction was found that was metabolically ıneıt. It is thus evident that the whole proteoglycan population may be rapidly changed by any factor that affects the control of their biosynthesis and/or degradation, and as proteoglycans make a large contribution to the mechanical properties of cartilage, such changes might be critical to its normal functioning. This would be particularly important for articular cartilage which is constantly subjected to mechanical stress.

<u>References</u>

Anderson, B., Hoffman, P., Meyer, K.: The O-serine linkage in peptides of Chondroitin 4- or 6-sulphates. J. biol. Chem. <u>240</u>, 156 (1965).
Baxter, E.: Proteoglycans of Cartilage: The Isolation and Study of Their Protein Moieties. PhD Thesis. University of London (1972).
Brandt, K.D., Muir, H.: Differences in composition and size of protein polysaccharide extracted from pig articular cartilage. FEBS Letters <u>4</u>, 16 (1969a).
Brandt, K.D., Muir, H.: Characterisation of protein-polysaccharides of articular cartilage from mature and immature pigs. Biochem. J. <u>114</u>, 871 (1969b).
Brandt, K.D., Muir, H.: Heterogeneity of protein-polysaccharides of porcine articular cartilage: the sequential extraction of chondoitin sulphate proteins with iso-osmotic neutral sodium acetate. Biochem. J. <u>121</u>, 261 (1971).
Davidson, E.A., Small, W.: Metabolism in vivo of connective tissue Mucopolysaccharides. III Chondroitin sulphate and Kerato sulphate of cartilage. Biochim. biophys. Acta (Amst.) <u>69</u>, 459 (1963).

Fitton-Jackson, S.: Environmental control of macromolecular synthesis in cartilage and bone: morphogenetic response to hyaluronidase. Proc. roy. Soc. B 175, 405 (1970).
Goh, A.T.W., Lowther, D.A.: Effect of age on the composition of bovine nasal cartilage. Nature 210, 1270 (1966).
Gower, W.E., Pedrini, V.: Age-related variations in protein-polysaccharides from human nucleus pulposus, annulus fibrosis and costal cartilage. J. Bone Surg. 51A, 1154 (1969).
Gross, J.I., Mathews, M.B., Dorfman, A.: Sodium chondroitin sulphate protein complexes of cartilage. J. biol. Chem. 235, 2889 (1960).
Gross, J.: The animal collagenases. In Chemistry and Molecular Biology of the Intercellular Matrix (E.A. Balazs, Ed.), Vol. 3, p.1623. London, New York: Academic Press 1970.
Hardingham, T.E., Fitton Jackson, S., Muir, H.: Replacement of proteoglycans in embryonic chick cartilage in organ culture after treatment with testicular hyaluronidase. Biochem. J. 126, 101 (1972).
Hardingham, T.E., Muir, H.: Biosynthesis of proteoglycans in cartilage slices, Biochem. J. 126, 791 (1972).
Hascall, V.C., Sajdera, S.W.: Proteinpolysaccharide complex from bovine nasal cartilage. The function of glycoprotein in the formation of aggregates. J. biol. Chem. 244, 2384 (1969).
Kempson, G.E., Muir, H., Swanson, S.A.V., Freeman, M.A.R.: Correlations between stiffness and the chemical constituents of cartilage on the human femoral head. Biochim. biophys. Acta (Amst.) 215, 70 (197o).
Kröz,W., Buddecke, E.: Chemische und markomolekulare Altersveränderungen von Polysaccharid-proteinen aus menschlichen Rippenknorpel. Z. physiol. Chem. 348, 665 (1967).
Malawista, I., Schubert, M.: Chondromucoprotein; new extraction method and alkaline degradation. J. biol. Chem. 230, 535 (1958).
Mankin, H.J., Lippiello, L.: The turnover of adult rabbit articular cartilage. J. Bone Joint Surg. 51A, 1591 (1969).
Muir, H.: The nature of the link between protein and carbohydrate of a chontroitin sulphate complex from hyaline cartilage. Biochem. J. 69, 195 (1958).
Muir, H.: The structure and metabolism of Mucopolysaccharides (glycosamineoglycans) and the problem of the Mucopolysaccharidoses. Amer. J. Med. 47, 673 (1969).
Muir, H., Jacobs, S.: Protein polysaccharides of pig laryngeal cartilage. Biochem. J. 103, 367 (1967).
Pedrini, V.: Electrophoretic heterogeneity of protein polysaccharides. J. biol. Chem. 244, 1540 (1969).
Rosenberg, L., Johnson, B., Schubert, M.: Protein-polysaccharides from human articular and costal cartilage. J. clin. Invest. 44, 1647 (1965).
Rosenberg, L., Johnson, B., Schubert, M.: The protein-polysaccharides of human costal cartilage. J. clin. Invest. 48, 543 (1969).
Sajdera, S.W., Hascall, V.C.: Proteinpolysaccharide complex from bovine nasal cartilage. Comparison of low and high shear extraction procedures. J. biol. Chem. 244, 77 (1969).
Shatton, J., Schubert, M.: Isolation of a mucoprotein from cartilage. J. biol. Chem. 211, 565 (1954).
Šimůnek, Z., Muir, H.: Macromolecular differences of articular cartilage of lame and sound pigs during rapid weight gain. Abst. 7th European Rheumatology Congress, 24. February, Brighton/England 1971a.
Šimůnek, Z., Muir, H.: Changes in the protein-polysaccharides of pig articular cartilage during prenatal life, development and old age. Biochem. J. 126, 515 (1972).
Tsiganos, C.P.: Personal communication.

Tsiganos, C.P., Hardingham, T.E., Muir, H.: Proteoglycans of cartilage: An assessment of their structure. Biochim. biophys. Acta (Amst.) 229, 529 (1971).
Tsiganos, C.P., Muir, H.: Studies on protein-polysaccharides from pig laryngeal cartilage. Extraction and purification. Biochem. J. 113, 879 (1969a).
Tsiganos, C.P., Muir, H.: Studies on protein polysaccharides from pig laryngeal cartilage. Heterogeneity, fractionation and characterisation. Biochem. J. 113, 885 (1969b).
Tsiganos, C.P., Muir, H.: The natural heterogeneity of proteoglycans of porcine and human cartilage. In: Chemistry and Molecular Biology of the Intercellular Matrix (Ed. E.A. Balazs), p. 859. London, New York: Academic Press 1970.

Proteochondroitin Sulfates, Chemical and Metabolic Heterogeneity, Immunology and Biosynthesis

T. O. Kleine

In vertebrates proteochondroitin sulfates represent the main component of the ground substance of connective tissues. They participate in growth, aging and in many pathological changes. Up to this day many details of structure and biosynthesis of chondroitin sulfate proteins are still unknown. Difficulties arising during the investigation of this macromolecule are caused by its complexity as well as by its chemical and metabolic heterogeneity.

Proteochondroitin sulfates consist of a carbohydrate component (80 to 90%) and a protein core (20 to 6%). 80 to 90% of the carbohydrate component is composed of three types of polymeric disaccharide units: chondroitin-4-sulfate (Ch-4-S), chondroitin-6-sulfate (Ch-6-S) or dermatan sulfate (DS). The rest consists of keratan sulfate and glycoproteins (Balazs, 1970; Barret, 1968; Boström and Roden, 1966; Brimacombe and Webber, 1964; Dodgson and Lloyd, 1968; Fitton Jackson, 1964; Kleine, 1972; Rodén, 1966; Schubert and Hamerman, 1968; Stoolmiller and Dorfman, 1969). Two or three of these constituents have been found in hybrid proteochondroitin sulfate molecules (Fransson, 1968; Fransson and Roden, 1967; Kresse and Buddecke, 1968).

The chondroitin sulfate chains [mol. weight 10,000 to 30,000 (Buddecke et al., 1963; Choi et al., 1971; Greiling et al., 1967; Kleine, 1972; Luscombe and Phelps, 1967a; Lyons and Singer, 1971; Marler and Davidson, 1965; Partridge et al., 1961; Tanford et al., 1964)] exhibit a considerable chemical and metabolic heterogeneity (see below). Generally, they contain 20 to 50 disaccharide units and are covalently bound by two molecules galactose and one molecule xylose to one molecule serine (Lindahl and Rodén, 1966; Rodén and Armand, 1966; Rodén and Smith, 1966) at a semirigid polypeptide (protein core or acceptor protein; mol.-weight between 110,000 and 140,000) (Buddecke et al., 1963; Fitton Jackson, 1964; Gregory et al., 1964; Luscombe and Phelps, 1967b; Partridge et al., 1961). The bonds of this linkage region are alkali-labile (Anderson et al., 1965; Muir, 1958). Some evidences exist for another type of linkage not susceptibel to alkali between hexosamine residues of polysaccharide chains and acid amino acids (glutamic or aspartic acid) of the protein core (Katsura and Davidson, 1966; Radhakrishnamurthy et al., 1970).

The number of sulfopolysaccharide side chains bound to one acceptor protein varies: In Chicken cartilage 2 to 4 chains are fixed to one protein core (Fitton Jackson, 1964), in bovine nasal cartilage 20 to 30 (Buddecke et al., 1967; Luscombe and Phelps, 1967a, b; Partridge et al., 1961; Rosenberg et al., 1970a; Schubert and Hamerman, 1968), and in calf rib cartilage about 60 Ch-4-S chains are bound to one protein core (Kleine, 1972). Thus the molecular weight of one monomeric proteochondroitin sulfate unit differs considerably between 0.2×10^6 to 2×10^6 (Buddecke et al., 1963; Eyring and Tsi Yang, 1968; Kröz and Buddecke, 1967; Muir and Jacobs, 1967; Partridge et al., 1961; Rosenberg et al., 1969; Schubert and Hamerman, 1968).

At least two kinds of monomers must be distinguished: One consisting of chondroitin sulfate side chains attached to a protein core, the other carrying chondroitin sulfate and keratan sulfate side chains at the same acceptor protein (Kröz and Buddecke, 1967; Muir and Jacobs, 1967; Partridge et al., 1961; see also below). Two monomeric molecules can aggregate to one dimeric unit (Rosenberg et al., 1970a). Monomers and dimers are thought to conglomerate with the aid of a linking glycoprotein (Hascall and Sajdera, 1969, 1970) to first order aggregates: In chicken cartilage 5 to 6 units aggregate having a mol.-weight of 1.16×10^6 (Fitton Jackson, 1964), in bovine nasal cartilage 4 to 6 (mol.-weight $2 - 4 \times 10^6$; Luscombe and Phelps, 1967a, b; Mathews and Lozaityte, 1958; Partridge et al., 1961) and in calf rib cartilage two units (mol.-weight 2.25×10^6; Kleine, 1972). Furthermore, the first order aggregates can again conglomerate to second order aggregates (Rosenberg et al., 1970a) of a molecular weight of 10 to 50×10^6 and more (Buddecke et al., 1963; Dunstone and Franek, 1969; Hascall and Sajdera, 1970; Mathews and Decker, 1968b).

Thus, a considerable heterogeneity of proteochondroitin sulfates must be taken into account (see below). In general the composition of these macromolecules extracted from connective tissues is either dependent on species, age and origin of the tissue examined, or on the extraction procedure applied.

1. Physico-Chemical Heterogeneity

a) Chondroitin Sulfate Side Chains

The greatest differences are thought to be found in growing tissue like calf rib cartilage. The total sulfopolysaccharide fraction of calf rib cartilage consists of 95% chondroitin-4-sulfate and 5% keratan sulfate (Kleine and Hilz, 1968a, b). After intensive digestion with papain, precipitation with cetylpyridinium chloride (Kleine et al., 1968) and chromatography on ECTEOLA cellulose (Kleine and Hilz, 1968a) it can be separated into six Ch-4-S peptide fractions differing in chain length, degree of sulfation and composition of peptide residues (Bubenzer et al., 1968; Kleine and Hilz, 1968b, 1969a, 1970; Kleine 1972). The shortest Ch-4-S side chains (about 30 disaccharide units) are undersulfated (SO_4/GlcA<0.9) (Table 1), the longer ones 30 to 40 disaccharide units exhibit a higher degree of sulfation (SO_4/GlcA>0.9<1.0), and the longest (40 to 50 disaccharide units) are slightly oversulfated (SO_4/GlcA>1.0). At present their structure is still unknown, although oversulfated ChS have already been characterized (Anno et al., 1971; Mathews and Decker, 1968a; Varadi and Griffiths, 1971). Electrophoretic migration of these Ch-4-S peptides partially conforms to their chain length and their degree of sulfation (Table 1). The calculated molecular weight of the Ch-4-S side chains of calf rib chondroitin sulfate proteins corresponds to that of ChS side chains extracted from other tissues (Buddecke et al., 1963; Fransson, 1968; Greiling et al., 1967; Luscombe and Phelps, 1967a; Partridge et al., 1961; Tanford et al., 1964). However, the differences between the calculated molecular weight and the molecular weight according to Scheraga and Mandelkern (1953, see Table 1) indicate the existence of two side chains at one peptide residue for longer and higher sulfated Ch-4-S side chains isolated by papain digestion from calf rib cartilage. Similar results have been described for ChS peptides after digestion with trypsin (Luscombe and

Table 1. Physico-chemical heterogeneity of chondroitin-4-sulfate peptides from calf rib cartilage (Separation on ECTEOLA cellulose)

ECTEOLA fractions eluted with 0.05 N HCl plus No.		Uronic acid (GlcA) %	Degree of sulfation SO_4^{2-}/GlcA	Electrophor. migration $cm^2 Volt^{-1} h^{-1}$	Chain length GlcA:Gal:Xyl	Molecular weight calculated	Molecular weight acc. to Scheraga and Mandelkern, 1953
I	-	2	0.78	-	(containing about 50% glucosamine and 50% galactosamine)		
II	0.10 N NaCl	6	0.87	0.30	32 : 2 : 1	16,900	22,000
III	0.25 N NaCl	25	0.84	0.43	31 : 2 : -	15,700	24,000
IV	0.50 N NaCl	30	0.89	0.46	39 : 2 : 1	20,000	30,000
V	0.75 N NaCl	22	0.93	0.52	38 : 2 : -	19,600	42,000
VI	1.00 N NaCl	13	1.08	0.47	44 : 2 : 1	24,100	53,000
VII	2.00 N NaCl	3	1.30	0.41	52 : 2 : -	26,200	-

Phelps, 1967a; Mathews and Sandson, 1969).

Table 2. Chemical heterogeneity of the peptide residue of Ch-4-S peptides from calf rib cartilage

	Starting fraction	ECTEOLA fractions eluted with 0.05 N HCl plus N NaCl						
		-	0.10	0.25	0.50	0.75	1.00	2.00
Amino acids/ 1/2 galactose	14	-	51	8	13	14	15	72
Amino acids/ serine	5	5	4	6	5	5	4.5	4.5
Serine/1/2 gal.	3	-	21	2	3	3.4	3	7
Serine	224	226	252	235	236	228	240	247
Threonine	35	47	48	31	31	38	43	54
Glutamic acid	114	131	144	110	112	112	135	164
Aspartic acid	52	67	69	51	49	53	69	68
Proline	64	76	53	61	69	62	69	20
Glycine	203	177	193	186	196	203	178	200
Leucine	58	32	28	57	67	60	43	25
Isoleucine	38	19	19	25	34	29	30	20
Valine	40	32	23	60	26	50	35	29
Alanine	70	89	80	83	73	78	70	87
Phenylalanine	17	23	16	18	18	17	15	18
Tyrosine	16	18	17	20	12	10	14	18
Lysine	30	27	20	32	39	36	26	30
Histidine	15	30	27	19	29	22	24	28
Arginine	24	12	7	26	17	14	20	12

Differences in the amino acid composition of peptide residues of these Ch-4-S peptides are shown in Table 2. The amino acids serine, glycine, glutamic acid (glutamine), proline, and alanine dominate strongly. Cystine and methionin are not detectable (see also Greiling and Stuhlsatz, 1969; Greiling et al., 1967; Muir, 1958). The molar relation of total amino acids to one-half galactose indicates the number of amino acid residues per one ChS side chain. It differs considerably: the shortest and longest ChS side chains are bound to the longest peptide residues (consisting of 50 to 70 amino acid residues; see Table 2). The molar relation of serine residues to one-half galactose indicates that not every serine residue carries a ChS side chain (Kleine and Hilz, 1969a). Furthermore, the molar relation of total amino acid residues to serine residues seems to indicate that the protein core contains uniformly composed subunits (doublets) consisting of at least two serine and glycine residues and one glutamic acid (glutamine) residue (see also Mathews and Sandson, 1969).

b) Proteochondroitin Sulfates

Proteochondroitin sulfates are bound in different ways in connective tissues. The following types of bonds are discussed: covalent, coordinative, metallic, sandwich and hydrogen bonds as well as ionic and topological ones (Schubert, 1966). Furthermore, the occurrence of different amounts of keratan sulfate (Kleine and Hilz, 1970; Kresse and Buddecke, 1970; Kröz and Buddecke, 1967; Tsiganos and Muir, 1967, 1969), glycoproteins (Hascall and Sajdera, 1969, 1970; Luscombe and Phelps, 1967a, b) and sialoproteins (Anderson, 1962) together with proteochondroitin sulfates causes difficulties in isolating pure proteochondroitin sulfates.

Schubert and co-workers (1966) were the first to develop a *disruptive extraction procedure* in high volumes of salt solutions exhibiting a low degree of ionization (Gerber et al., 1960; Malawista and Schubert, 1958). By this method they could isolate at least five proteochondroitin sulfate fractions from bovine nasal cartilage and nucleus pulposus, human rib and joint cartilage demonstrating a considerable physicochemical heterogeneity (Rosenberg and Schubert, 1967; Rosenberg et al., 1969, 1970a, b; Schubert and Hamerman, 1968).

In calf rib cartilage 75 to 80% of the proteochondroitin-4-sulfate together with glycoproteins and collagen can be dissolved by this procedure (see Diagram 1) (Kleine and Hilz, 1968a, 1969b, 1970). This soluble proteoglycan fraction ("PPL") contains the shortest, uniformly sulfated (SO_4/GlcA = 1.0) Ch-4-S side chains of 32 disaccharide units. About 20% of the proteochondroitin-4-sulfate ("PPH" and "PPR" fractions) are insoluble, obviously bound to collagen fibrils (Smith et al., 1969). These proteoglycans exhibit undersulfated (SO_4/GlcA = 0.7 - 0.8) longer Ch-4-S side chains of 38 and more disaccharide units (Kleine and Hilz, 1969b).

The "PPL" fraction proved to be extremely heterogeneous when two different fractionation methods were applied:
CsCl density gradient centrifugation of the partially purified "PPL" fraction (see Diagram 1) yields three proteoglycan fractions (Fig. 1).

However, chromatography of cetylpyridinium complexes of the "PPL" fraction (cellulose column, stepwise increasing $MgCl_2$ concentrations see Diagram 1) separates at least seven subfractions (Hilz et al., 1969; Kleine and Hilz, 1969b, 1970). They are distinguished by an increasing degree of sulfation (SO_4/HexN = 0,89 - 1.21), by increasing chain lengths (31 to 38 disaccharide units) and by different glucosamine/galactosamine and protein contents. About 75% of their amino acid composition consists of serine, glycine, glutamic acid (glutamine), aspartic acid (asparagine), proline, leucine, and alanine (Table 3). It resembles the amino acid composition of proteochondroitin sulfates isolated from bovine nasal cartilage (Dunstone and Franek, 1969; Gerber et al., 1960; Hascall and Sajdera, 1969, 1970; Pal et al., 1966; Rosenberg et al., 1970b; Sajdera and Hascall, 1969; Schubert and Hamerman, 1968). However, there exist some differences in the content of serine, glycine, proline, hydroxyproline, leucine, alanine, methionine, and cystine in these "PPL" subfractions (see Table 3). Moreover, similar results have been obtained by cellulose chromatography of proteochondroitin sulfate purified in CsCl density gradients (see Table 4).

The largest number of amino acid residues per ChS side chain is found in undersulfated (SO_4/HexN<0.9) and short-chained (about 32 disaccharide units) proteoglycans containing some collagen-like material (see molar ratio of total amino acid residues to one-half galactose in Table 3).

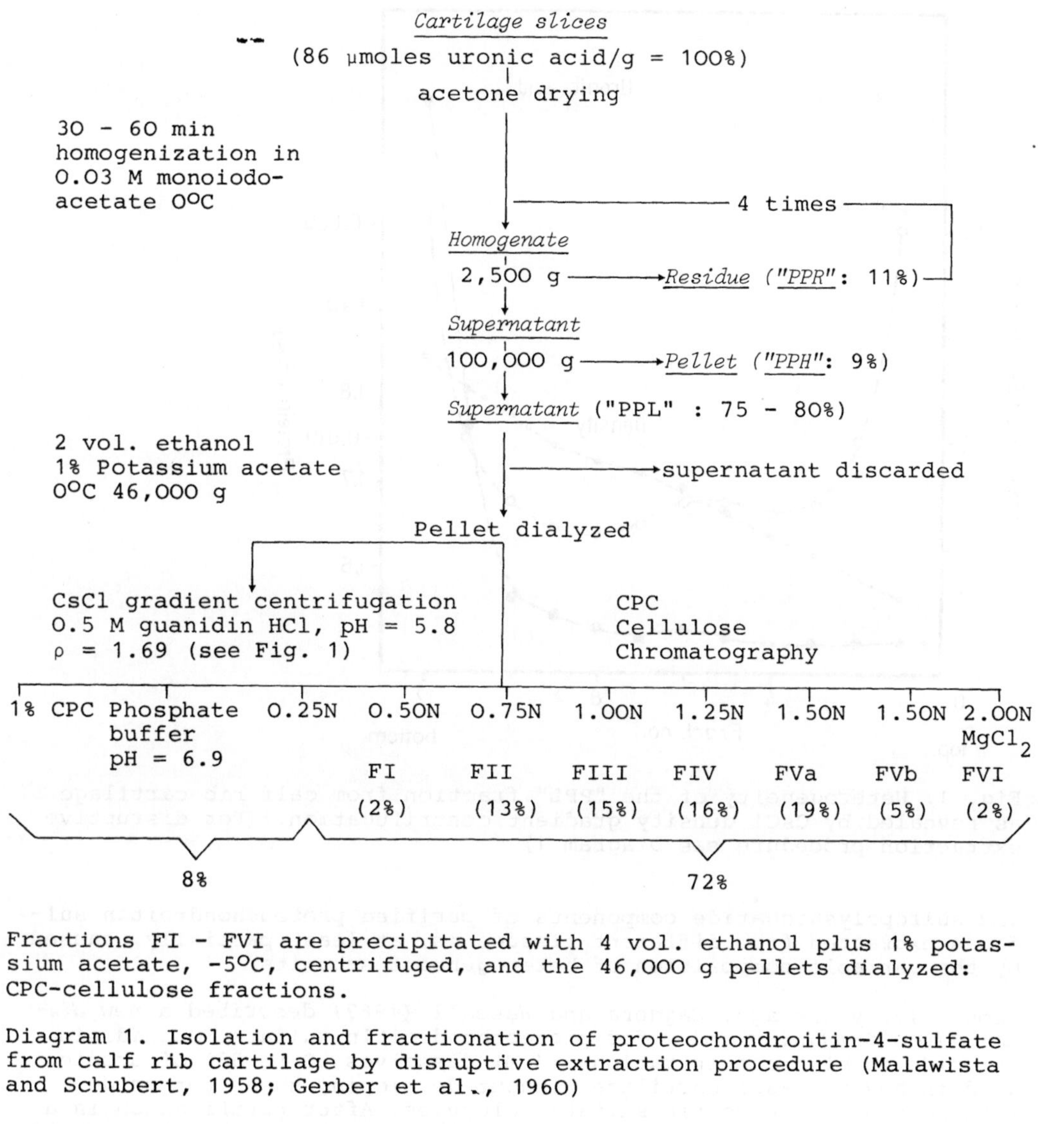

Fractions FI - FVI are precipitated with 4 vol. ethanol plus 1% potassium acetate, -5°C, centrifuged, and the 46,000 g pellets dialyzed: CPC-cellulose fractions.

Diagram 1. Isolation and fractionation of proteochondroitin-4-sulfate from calf rib cartilage by disruptive extraction procedure (Malawista and Schubert, 1958; Gerber et al., 1960)

The molar relation of total amino acid residues to serine residues and of total serine residues to one-half galactose vary considerably, indicating inhomogeneities in the amino acid sequence of the protein cores as well as variations in the number of ChS side chains bound to one protein core (see Table 3).

Similar physico-chemical heterogeneity of proteochondroitin sulfate from pig laryngeal (Muir and Jacobs, 1967; Tsiganos and Muir, 1967, 1969) and articular cartilage (Brand and Muir, 1969, 1971) has been demonstrated by using sequentially disruptive extraction procedures. By determination of N-terminal amino acid residues Tsiganos and Muir (1967, 1969) and Serafini-Fracassini et al. (1967) have found five to six differently terminated protein cores. At present the question cannot be answered whether chemical heterogeneity of protein cores

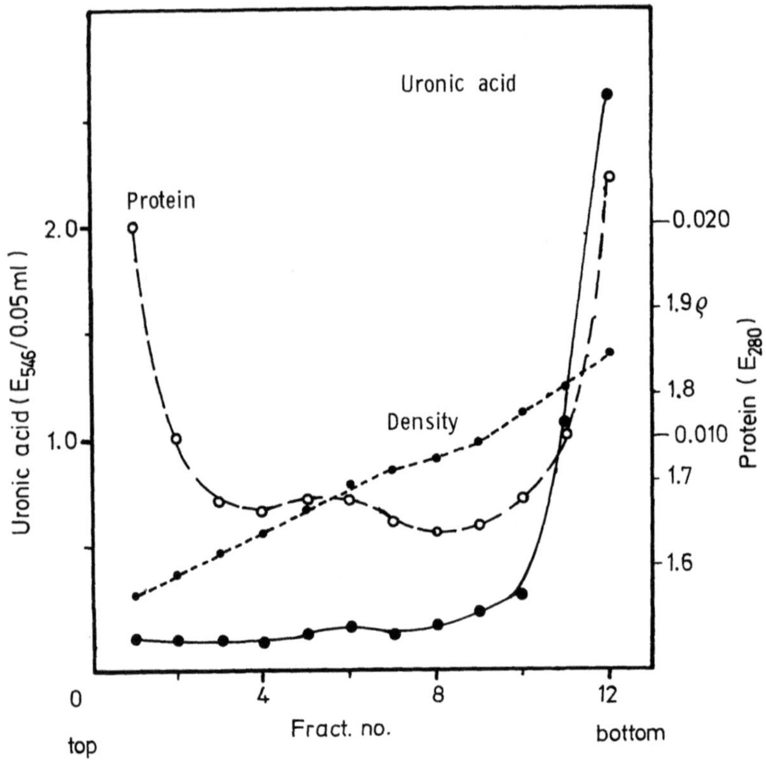

Fig. 1. Heterogeneity of the "PPL" fraction from calf rib cartilage as revealed by CsCl density gradient centrifugation. (For disruptive extraction procedure see Diagram 1)

and sulfopolysaccharide components of purified proteochondroitin sulfates extracted from different tissues is, at least partially, caused by the variable composition of heterogeneous subunits.

About five years ago, Sajdera and Hascall (1969) described a new *dissociative extraction procedure* for proteochondroitin sulfate. Guanidinium chloride in high concentrations (4 M) dissolves 80 to 85% of uronic acid in bovine nasal cartilage without developing shear forces deforming proteochondroitin sulfate molecules. After purification in a CsCl density gradient (Dunstone and Franek, 1969) a small glycoprotein fraction is separated from chondroitin sulfate protein exhibiting linkage tendencies to the bulk fraction by forming ionic, disulfide and hydrogen bonds (Hascall and Sajdera, 1969).

About 90% of proteochondroitin-4-sulfates together with soluble collagen and glycoproteins are extracted from calf rib cartilage by this procedure (Kleine et al., 1971a, b; Kleine, 1971a, b; see Diagram 2). The soluble proteochondroitin-4-sulfates demonstrate a physico-chemical heterogeneity similar to those extracted by the disruptive extraction procedure (see Diagram 1). They contain uniformly sulfated (SO_4/GlcA = 1.0) shorter Ch-4-S side chains (27 disaccharide units) than those of the insoluble undersulfated (SO_4/GlcA<1.0) proteochondroitin sulfates with 33 disaccharide units (Kleine et al.,

Table 3. Chemical heterogeneity of the protein component in soluble proteoglycans from calf rib cartilage (disruptive extraction procedure see Diagram 1)

	Starting fraction "PPL"	CPC-cellulose fractions eluted with N $MgCl_2$						
		0.50	0.75	1.00	1.25	1.50a	1.50b	2.00
Amino acids/ 1/2 galactose	56	108	29	19	22	17	32	27
Amino acids/ serine	14	6	7	10	8	8	6	6.5
Serine/ 1/2 galactose	4	21	5	2	3	2	5	4
Serine	82	213	187	131	132	139	174	156
Threonine	42	56	45	44	46	68	49	37
Glutamic acid	112	128	145	128	121	133	148	124
Aspartic acid	108	74	82	96	96	84	85	89
Proline	83	35	65	65	72	75	44	45
Hydroxyproline	39	23	6	1	8	4	3	12
Hydroxylysine	6	2	1	1	1	1	1	1
Glycine	178	179	175	155	161	144	190	203
Leucine	68	31	46	67	70	79	46	50
Isoleucine	19	24	30	30	29	40	21	20
Valine	28	33	25	25	39	45	28	58
Alanine	81	85	92	115	70	56	81	78
Methionine	5	8	4	2	2	8	7	11
Cystine	7	3	6	10	12	9	8	12
Phenylalanine	32	21	18	26	38	21	20	24
Tyrosine	22	20	15	21	21	15	22	21
Lysine	31	28	24	29	33	27	28	25
Histidine	19	25	18	32	23	26	28	18
Arginine	35	10	13	22	24	33	19	15

1971a, b). However, their amino acid composition (Table 4) does not resemble the one of "PPL" (see Table 3). This indicates that smaller amounts of soluble collagen and glycoproteins are extracted together with soluble proteoglycans by this dissociative extraction procedure than by the disruptive one.

Under "associative" conditions (Hascall and Sajdera, 1969; see Diagram 2) in a CsCl density gradient a light chondroitin sulfate protein fraction can be separated from the specific heavier bulk fraction besides two smaller proteoglycan fractions (Kleine et al., 1971b). The light chondroitin sulfate protein fraction contains very short and undersulfated (SO_4/HexN<0.8) ChS side chains of 21 disaccharide units and is associated with collagen-like material. Its amino acid compo-

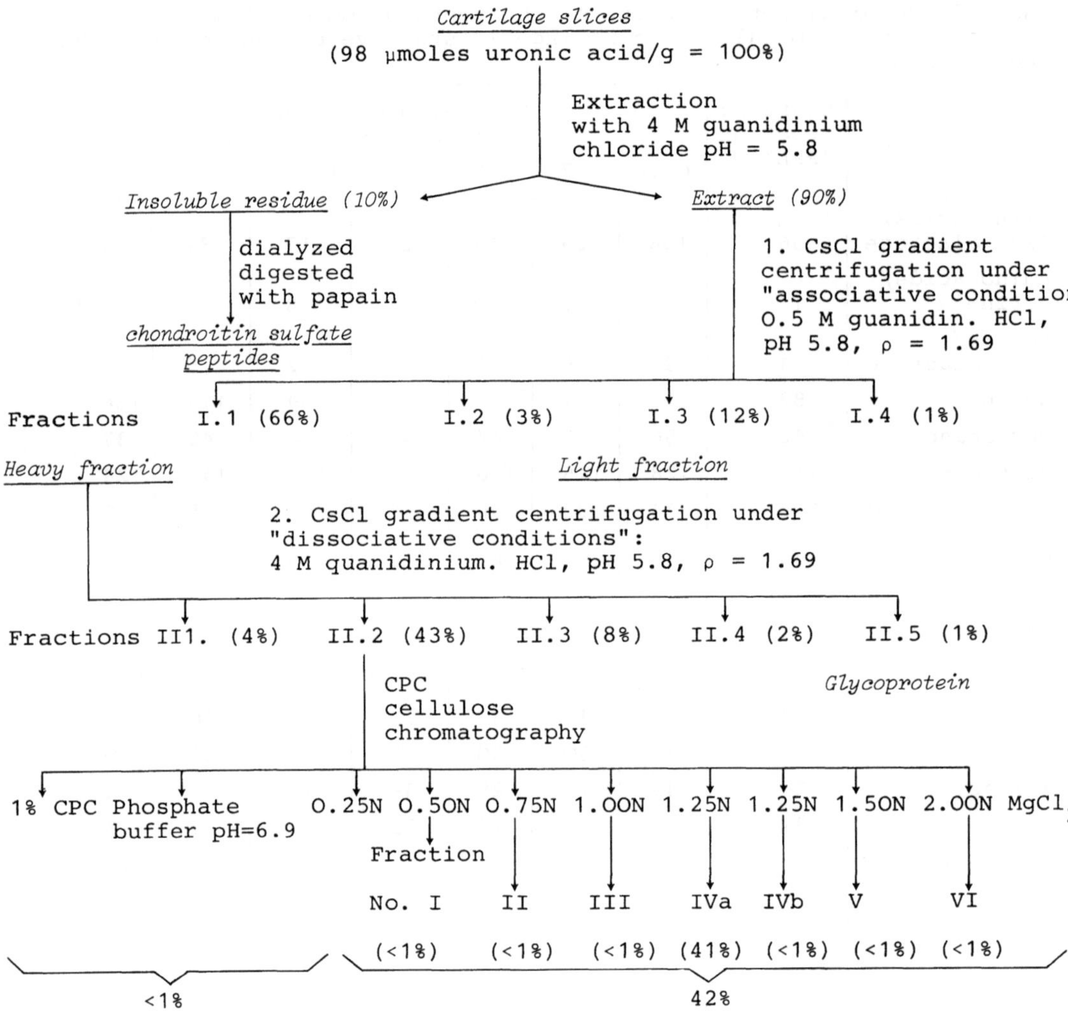

Diagram 2. Isolation and fractionation of proteochondroitin-4-sulfates from calf rib cartilage by dissociative extraction procedure (Hascall and Sajdera, 1969)

sition (Table 4) resembles the "PPL" one (see Table 3).

The partially purified proteochondroitin sulfate bulk fraction proves to be inhomogeneous. Under "dissociative" conditions (Hascall and Sajdera, 1969; see Diagram 2) three smaller proteoglycan fractions and a light glycoprotein one are separated from it (Kleine et al., 1971b). The glycoprotein fraction contains some chondroitin sulfate-like constituents, no soluble collagen and a specific amino acid composition (Table 4).

However, the purified proteochondroitin sulfate bulk fraction again proves to be heterogeneous: On a cellulose column (see Diagram 2) six smaller proteoglycan subfractions can be separated from it whose

Table 4. Chemical heterogeneity of protein component in soluble proteoglycans of calf rib cartilage (dissociative extraction procedure see Diagram 2)

	Extract	Light fraction I.3	Heavy fraction I.1	Glyco-protein fraction II.5	Heavy fraction without glycoprotein II.2
Amino acids/ 1/2 galactose	72	161	29	12	31
Amino acids/ serine	8	15	9	7	9
Serine/ 1/2 galactose	9	11	3	2	3
Serine	127	66	110	138	108
Threonine	79	43	63	31	64
Glutamic acid	121	106	122	103	118
Aspartic acid	78	119	87	105	87
Proline	51	98	70	79	86
Hydroxyproline	19	25	9	1	15
Hydroxylysine	10	2	6	3	11
Glycine	137	110	113	105	128
Leucine	70	89	62	96	58
Isoleucine	25	34	29	28	38
Valine	36	40	55	47	47
Alanine	70	70	73	75	60
Methionine	6	5	3	1	6
Cystine	13	20	9	1	13
Phenylalanine	55	33	55	43	53
Tyrosine	17	25	32	31	19
Lysine	28	44	35	36	34
Histidine	20	27	35	27	23
Arginine	37	43	32	52	32

amino acid composition exhibits a considerable resemblance to the "PPL" subfractions (see Tables 3, 4).

Thus, similar physico-chemical heterogeneity of sulfopolysaccharide constituents and of the protein core can be demonstrated for calf rib proteochondroitin sulfate by applying two different extraction procedures.

Table 4. Continued

CPC-cellulose fractions eluted with N $MgCl_2$						
0.50	0.75	1.00	1.25a	1.25b	1.50	2.00
-	-	-	-	-	-	-
4	4	4	6	6	11.5	7
-	-	-	-	-	-	-
233	258	250	159	176	87	145
54	55	42	71	60	80	35
139	143	129	128	154	114	116
83	76	96	82	91	115	99
28	27	32	73	53	44	60
22	3	6	6	1	5	17
11	21	6	1	29	6	24
154	161	160	108	158	79	149
26	29	30	81	33	79	51
26	21	29	46	19	29	21
34	-	-	51	34	46	53
87	101	91	52	80	96	73
6	1	1	4	7	18	13
4	1	8	10	9	7	26
12	15	-	19	15	32	22
16	19	20	23	16	26	24
28	28	33	27	22	60	24
29	31	54	31	30	20	32
6	12	13	26	14	56	15

2. Metabolic Heterogeneity

In most cases the physico-chemical heterogeneity corresponds to a metabolic one. In calf rib cartilage double labelled Ch-4-S peptides ($^{35}SO_4$ plus ^{3}H-glucosamine) can be separated on ECTEOLA cellulose into undersulfated and short ChS chains (see Table 1) having a small incorporation rate. The higher sulfated longer side chains exhibit higher rates of labelling. The highest labelling of both precursors is found in the longest, slightly oversulfated ChS side chains (Bubenzer et al., 1968; Hilz et al., 1969; Kleine and Hilz, 1968a, b, 1969a, 1970; Kleine et al., 1968, 1971b). Similar results have been found in in vitro systems of other kinds of cartilage and in leukemic myeloid cells (Bentley and Rokosova, 1970; Hardingham and Muir, 1970; Olsson, 1968).

Furthermore, in calf rib cartilage the incorporation rates of ^{3}H-leucine and ^{14}C-serine into the acceptor proteins of *soluble* and chemically heterogeneous proteochondroitin sulfates generally correspond to the synthesizing rate of their ChS side chains: Proteochondroitin sulfates carrying short, undersulfated and slowly labelled ChS side chains also possess a slowly labelled protein core, whereas proteoglycans with longer and more sulfated ChS side chains exhibit the highest incorporation rates of radioactive precursors both for the protein core and for the sulfopolysaccharide side chains (Kleine and Hilz, 1969b, 1970; Kleine et al., 1971b). Besides, *collagen bound* proteochondroitin sulfates with long, undersulfated ChS side chains exhibit higher labelling rates than soluble proteochondroitin sulfate with shorter, uniformly sulfated side chains (Kleine et al., 1971a, b; see Kresse and Buddecke, 1968).

3. Immunology

Further heterogeneity of the protein core from chondroitin sulfate proteins can be demonstrated by immunological techniques. In general, the whole macromolecule possesses weak (*in vitro*) antigenicity because all antigenic groups of the polypeptide are covered with strongly negatively charged ChS side chains which are not antigenic (Loewi and Muir, 1965; Sandson et al., 1966). Nevertheless, highly boostered rabbit antiserum against purified chondroitin sulfate protein precipitates this macromolecule in two lines in the OUCHTERLONY gel diffusion precipitin test (Fig. 2C) (Janis et al., 1967; Kleine and Hilz, 1968a; Loewi and Muir, 1965; Pankovich and Korngold, 1967; Sandson et al., 1966).

Two types of antibodies must be considered: One directed against determinants common to different species, the other directed against species-specific determinants (Di Ferrante and Pauling, 1964; Loewi and Muir, 1965; Pankovich and Korngold, 1967; Sandson et al., 1966; Tsiganos and Muir, 1969). The former determinants in the protein core very probably consist of the amino acid sequence of doublets common to different species. At least it is composed of two serine and glycine residues and one glutamic acid residue (see above and Mathews and Sandson, 1969). The specific antigenicity of glutamic acid residues is well known (Sela, 1964): however, that of the linkage region (glcA-gal-gal-xyl-ser) is uncertain. Relatively small chondroitin sulfate protein molecules with a short protein core consisting mostly of the amino acids named above could be isolated from pig laryngeal (Muir and Jacobs, 1967) and calf rib cartilage (Kleine, unpublished work). Antibodies against purified proteochondroitin sulfate absorbed with Ch-4-S peptides do not precipitate this small chondroitin sulfate protein diffusing more rapid in agar gel (Fig. 2C).

The species-specific determinants at longer protein cores very probably contain hydrophobic (tyrosine, phenylalanine) and basic amino acids (histidine, lysine and arginine) known to develop powerful antigenicity (Sela, 1964). But also the linkage glycoprotein possesses species-specific determinants (Di Ferrante et al., 1970). However,

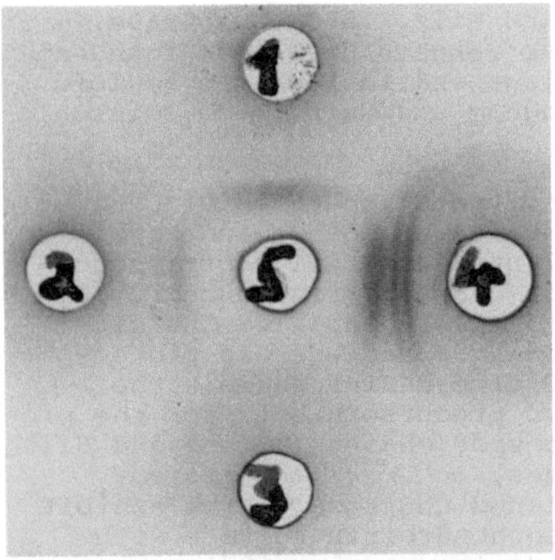

2A
No. 1 = antiserum to protein core fraction 1.0 M NaCl.
No. 2 = antiserum to protein core fraction 2.0 M NaCl.
No. 3 = serum No. 1 before application of antigen to the rabbit.
No. 4 = antiserum to purified Ch-4-S protein.
No. 5 = 4.5 µg chondroitin-4-sulfate protein digested with hyaluronidase

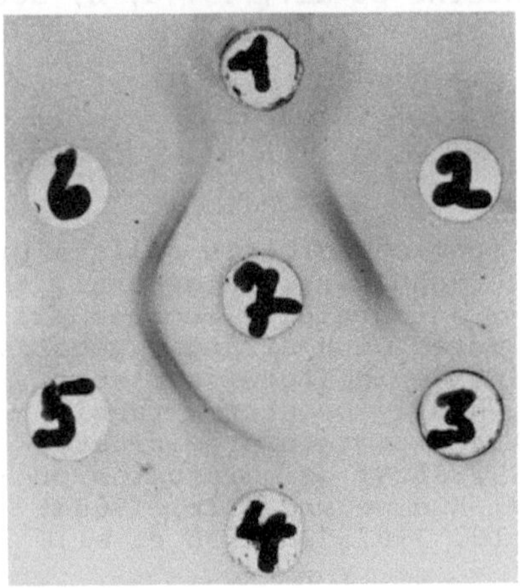

2B
No. 1 = antiserum to purified Ch-4-S protein absorbed with protein core fraction 1.0 M NaCl. No. 2 = antiserum to purified Ch-4-S protein untreated. No. 3 = antiserum of No. 2 absorbed with protein core fraction 2.0 M NaCl. No. 4 = antiserum of No. 2 partially absorbed with unpolar protein core fraction. No. 5 = see No. 2. No. 6 = antiserum of No. 2 treated with Ch-4-S peptides. No. 7 = 2.2 µg protein core fraction 2.0 M NaCl

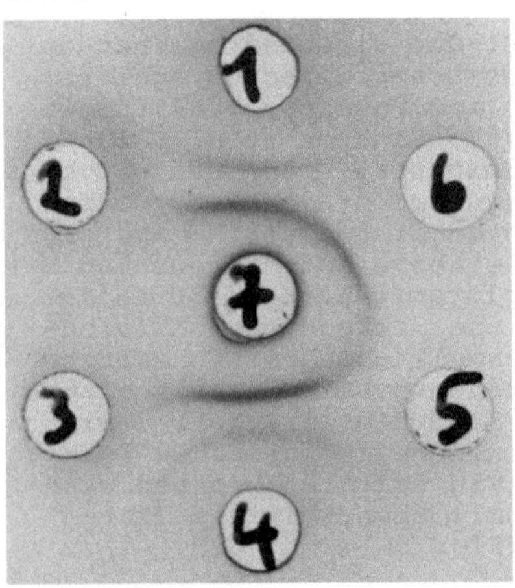

2C
No. 1 = antiserum to purified Ch-4-S protein untreated.
No. 2 = antiserum of No. 1 absorbed with protein core fraction 1.0 M NaCl. No. 3 = antiserum of No. 1 absorbed with protein core fraction 2.0 M NaCl. No. 4 = see No. 1.
No. 5 = antiserum of No. 1 partially absorbed with unpolar protein core fraction.
No. 6 = antiserum of No. 1 absorbed with Ch-4-S peptides.
No. 7 ="PPL" fraction (see Diagram 1) (50 µg protein)

Fig. 2A - C. *Characterization of protein core fractions from proteochondroitin-4-sulfates of calf rib cartilage by immunological techniques.* Ouchterlony double diffusion in 1% agar, 0.1 M sodium phosphate buffer pH 6.5 and 0.5 ‰ merthiolate (Lilly). Precipitation lines are stained with AmidoSchwarz 10 B. For details see text the distribution of these species-specific determinants in the longer protein cores is still unknown.

Purified proteochondroitin sulfate from calf rib cartilage is precipitated by its antiserum into two lines (Fig. 2C), and after degradation with hyaluronidase into four lines (Fig 2A). The constituents of the two faster migrating ones prove to be micromolecular on Sephadex G-75, those of the two slower migrating ones to be macromolecular. The slower migrating substances can be separated on Dowex (Cl^-form) with 1.0 and 2.0 M NaCl into two protein core fractions carrying residues of CH-4-S side chains or mostly keratan-6-sulfate side chains besides residues of ChS chains, respectively. Experiments with antibodies against these two protein core fractions alternatively absorbed by these fractions demonstrate a complete immunological identity of both protein cores (Fig. 2B). These results contrast with similar experiments of proteoglycans from pig laryngeal cartilage indicating no immunological identity (Tsiganos and Muir, 1967). Furthermore, differences in the amino acid composition between keratan sulfate and ChS chains carrying peptides are found in chondroitin sulfate protein from bovine nasal cartilage (Hoffman and Mashburn, 1967). Nevertheless, close immunological relationship exists between calf rib cartilage proteochondroitin sulfates and those of human cartilage, whereas proteochondroitin sulfate from bovine aorta resembles it weakly and proteochondroitin sulfate from human aorta not at all (Table 5). Other relationships between porcine and bovine proteochondroitin sulfates (Loewi and Muir, 1965), human, bovine and pig ones (Sandson et al., 1966) and human, steer and dog proteoglycans (Pankovich and Korngold, 1967) have been described.

Table 5. Immunological relationship of bovine and human proteochondroitin sulfates (PCS) and their antisera (double diffusion in agar)

PCS From Calf Rib Cartilage	PCS from bovine aorta: areciprocal (one line) PCS from human rib cartilage: reciprocal (two lines) PCS from human aorta: negativ

4. Biosynthesis

Relatively little is known of the biosynthesis of proteochondroitin sulfates to explain the considerable physico-chemical heterogeneity. The biosynthesis of the linkage protein seems to take place independently from the proteochondroitin sulfate: In calf rib cartilage the turnover rate of this glycoprotein is obviously slower than that of the bulk proteochondroitin-4-sulfate (Kleine, unpublished experiments).

In the biosynthesis of proteochondroitin sulfate four steps can be distinguished; they are reported in detail elsewhere (Kleine, 1972):

1. The formation of acceptor proteins takes place at the polysomes of the connective tissue cells (Stoolmiller and Dorfman, 1969). Experiments with puromycin and chemical and immunological analysis of the protein core of proteochondroitin sulfates (see above) indicate several, at least two, pools of acceptor proteins with different composition and turnover rates (Kleine and Hilz, 1969a, 1970; Kleine et al., 1968) (see also experiments with cycloheximide, Colf and Lowther, 1969).

2. The linkage region glcA-gal-gal-xyl-ser is built up by successively glycosidic binding of one activated molecule xylose, two molecules galactose and one molecule glucuronic acid at serine residues of the protein core (Helting and Rodén, 1969a, b). Four specific transferases localized at the rough membranes of the endoplasmatic reticulum (Stoolmiller and Dorfman, 1969; Horowitz and Dorfman, 1968) catalyze these reactions.

3. The polysaccharide chains are formed by the activated precursors UDP-N-acetyl-galactosamine and UDP-glucuronic acid. A polymerase carrying two active centers catalyzes these reactions at both smooth and rough membranes (Olsson and Gardell, 1971; Perlman et al., 1964; Silbert, 1964; Telser et al., 1965). The polysaccharide side chains grow by successive transfer of two different monosaccharide subunits from their appropriate nucleotide sugar precursors to the non-reducing and unsulfated end of the chain. However, the determination of the chain length is not yet clear. The polymerase is inhibited by ChS chains carrying a sulfate ester group at the C_4 atom (but not at the C_6 atom) of the free non-reducing N-acetyl-galactosamine residue (Telser et al., 1966). The biosynthesis of hybride molecules is also not clear.

4. The sulfation of the plysaccharide chain happens with 3'phospho-adenosine-5'-phosphosulfate by partially soluble C_4- and C_6sulfotransferases (Davidson and Riley, 1960; DeLuca and Silbert, 1968; George et al., 1970; Meezan and Davidson, 1967; Robbins and Lipman, 1958; Robinson and Dorfman, 1969; Silbert and DeLuca, 1969; Suzuki and Strominger, 1960; Wilson and Bandurski, 1958) at the smooth memebranes (Horowitz and Dorfman, 1968; Stoolmiller and Dorfman, 1969) of the cell, respectively in the golgi zone. (Neutra and Leblond, 1966; Peterson and Leblond, 1964).

The completed proteochondroitin sulfate molecule is stored in vesicles and probably leaves the cell by retrograde pinocytosis (see Kleine, 1972). Up to date it is not clear what kind of mechanisms release the completed and strongly negatively loaded proteochondroitin sulfate molecules from the cell membranes. Moreover, it is not known what kind of enzymes fix metabollically active proteochondroitin sulfate molecules (Kleine et al., 1971b) at extracellular collagen fibrils.

5. Regulation of Proteochondroitin Sulfate Biosynthesis

An extracellular control of the proteochondroitin sulfate synthesis by feed-back mechanisms is unknown. However, some intracellular regulating systems are found in cartilage cells indicating a strong coupling between the biosynthesis of the polysaccharide chain and its sulfation (Hilz et al., 1969; Kleine and Dapper, 1971; Kleine, 1972; Kleine et al., 1968; Meezan and Davidson, 1967). This phenomenon is only found in intact cells but not in cell extracts (Robinson et al., 1966; Telser et al., 1966). In intact cells of calf rib cartilage, elongation of polysaccharide chains and their sulfation must occur simultaneously and concomitantly, because after the specific inhibition of acceptor protein synthesis by puromycin, of sulfation by selenate and of polysaccharide formation by 6-diazo-5-oxo-norleucine (Ghosh et al., 1960) the relative diminuation of $^{35}SO_4$-incorporation in Ch-4-S peptides is usually higher than that of the ^3H-glucosamine one (Bubenzer et al., 1968; Hilz et al., 1969; Kleine and Hilz, 1968b, 1969a, 1970; Kleine et al., 1967, 1968). Furthermore, the synthesis of the linkage region ser-xyl-gal-gal-glcA must also be controlled by feed-back mechanisms because the incorporation rate of labelled serine into Ch-4-S peptides decreases under the same conditions where selenate (Kleine et al., 1967, 1968) and 6-diazo-5-oxo-norleucine (Telser et al., 1965) specifically inhibit sulfation or polysaccharide chain formation, respectively. The biosynthesis of total proteochondroitin sulfate, however, is controlled by the formation of ChS side chains (Kleine et al., 1968).

At present, chemical and metabolic heterogeneity of proteochondroitin sulfate can be explained by the different activity, specification and localization of transferases and kinases building up ChS side chains, as well as by some acceptor protein pools differing in turnover rates and composition (Kleine, 1972).

Acknowledgements: This work was supported by the Deutsche Forschungsgemeinschaft. This script is dedicated to Prof. H. Hilz, Hamburg, and Prof. F. Hartmann, Hannover.

References

Adams, J.B., Rienitz, K.G.: The biosynthesis of chondroitin sulphates. Influence of nucleotides and hexosamines on sulphate incorporation. Biochim. biophys. Acta (Amst.) 51, 567 (1961).
Anderson, A.J.: Some studies on the relationship between sialic acid and the mucopolysaccharide-protein complexes in human cartilage. Biochem. J. 82, 372 (1962).
Anderson, B., Hoffman, P., Meyer, K.: The O-serine linkage in peptide of chondroitin-4- or -6-sulfate. J. biol. Chem. 240, 156 (1965).
Anno, K., Seno, N., Mathews, M.B., Yamagata, T., Suzuki, S.: A new dermatan polysulfate, chondroitin sulfate H, from hagfish notochord. Biochim. biophys. Acta (Amst.) 237, 173 (1971).
Balazs, E.A.: Chemistry and Molecular Biology of the Intercellular Matrix. New York: Academic Press 1970.
Barrett, A.J.: In: Comprehensive Biochemistry (Florkin, M., E.H. Stotz, Eds.), p. 425. Amsterdam: Elsevier 1968.
Bentley, J.P., Rikosovà, B.: The metabolic heterogeneity of rabbit ear cartilage chondroitin sulfate. Biochem. J. 116, 329 (1970).
Boström, H., Rodén, L.: In: The Amino Sugars, Vol. IIb (Balazs, E.A., W.R. Jeanloz, Eds.), p. 45. New York: Academic Press 1966.
Brandt, K.D., Muir, H.: Characterization of protein-poly-saccharides of articular cartilage from mature and immature pigs. Biochem. J. 114, 871 (1969).
Brandt, K.D., Muir, H.: Heterogeneity of protein-polysaccharides of porcine articular cartilage. Biochem. J. 121, 261 (1971).
Brimacombe, J.S., Webber, J.M.: Mucopolysaccharides. Amsterdam: Elsevier 1964.
Bubenzer, H.J., Kleine, T.O., Hilz, H.: Chemical and metabolic heterogeneity of chondroitin-4-sulfate peptides from calf rib cartilage. Fed. Exp. Biol. Soc. (Prague), Abstr. V, p. 124 (1968).
Buddecke, E., Kröz, W., Lanka, E.: Chemische Zusammensetzung und makromolekulare Struktur von Chondroitinsulfat-Proteinen. Hoppe-Seylers Z. physiol. Chem. 331, 196 (1963).
Buddecke, E., Kröz, W., Tittor, W.: Chondroitinsulfat-Protein aus Rindernasenknorpel - Beziehungen zwischen makromolekularen Eigenschaften und Funktion. Hoppe-Seylers Z. physiol. Chem. 348, 651 (1967).
Choi, H.U., Meyer, K., Swarm, R.: Mucopolysaccharide and proteinpolysaccharide of a transplantable rat chondrosarcoma. Proc. nat. Acad. Sci. (Wash.) 68, 887 (1971).
Colf, N.N., Lowther, D.A.: The inhibition of chondroitin sulphate protein synthesis by cycloheximide. FEBS Letters 2, 351 (1969).

Davidson, E.A., Riley, J.G.: Enzymatic sulfation of chondroitin B. J. biol. Chem. 235, 3367 (1960).

DeLuca, S., Silbert, J.E.: Biosynthesis of chondroitin sulfate. J. biol. Chem. 243, 2725 (1968).

Di Ferrante, N.: Precipitins in the rabbit produced by protein-polysaccharide from bovine nasal cartilage. Science 143, 250 (1964).

Di Ferrante, N., Pauling, M.: Properties of antibodies to a protein-polysaccharide from bovine nasal cartilage. J. Lab. clin. Med. 63, 945 (1964).

Di Ferrante, N., Donnelly, P.V., Gregory, J.D., Sajdera, S.W.: Antigenic determinants of two components of proteoglycan complex from bovine cartilage. FEBS Letters 9, 149 (1970).

Dodgson, K.S., Lloyd, A.: In: Carbohydrate Metabolism and its Disorders, Vol. I (Dickens, F. et al., Eds.), p. 169. New York: Academic Press 1968.

Dunstone, J.R., Franek, M.D.: Connective tissue protein-polysaccharides. J. biol. Chem. 242, 3460 (1967); 244, 3654 (1969).

Eyring, J.E., Tsi Yang, J.: Conformation of protein-polysaccharide complex from bovine nasal septum. J. biol. Chem. 243, 1306 (1968).

Fitton Jackson, S.: In: The Cell, Vol. VI (Brachet, J., A.E. Mirsky, Eds.), p. 387. New York: Academic Press 1964.

Fransson, L.A.: Structure of dermatan sulfate III. J. biol. Chem. 243, 1504 (1968).

Fransson, L.A., Rodén, L.: Structure of dermatan sulfate I, II. J. biol. Chem. 242, 4161, 4170 (1967).

George, E., Singh, M., Bachhawat, B.K.: The nature of sulfation of uronic acid - containing glycosaminoglycans catalyzed by brain sulphotransferase. J. Neurochem. 17, 189 (1970).

Gerber, B.R., Franklin, E.C., Schubert, M.: Ultracentrifugal fractionation of bovine nasal chondromucoprotein. J. biol. Chem. 235, 2870 (1960).

Ghosh, S., Blumenthal, H.J., Davidson, E., Roseman, S.: Glucosamine metabolism. J. biol. Chem. 235, 1265 (1960).

Gregory, J.D., Laurent, T.C., Rodén, L.: Enzymatic degradation of chondromucoprotein. J. biol. Chem. 239, 3312 (1964).

Greiling, H., Stuhlsatz, H.W.: Glycosaminoglycan-Peptide aus dem humanen Kniegelenkknorpel. Hoppe-Seylers Z. physiol. Chem. 350, 449 (1969).

Greiling, H., Stuhlsatz, H.W., Plagemann, L.: Struktur und Stoffwechsel von Glykosaminoglycan-Proteinen. II. Die Chondroitin-4-sulfat-Peptide der Rinder-Cornea. Hoppe-Seylers Z. physiol. Chem. 348, 121 (1967).

Hardingham, T.E., Muir, H.: The effect of temperature on the biosynthesis of chondroitin-4-sulfate in cartilage slices in vitro. FEBS Letters 9, 145 (1970).

Hascall, V.C., Sajdera, S.W.: Proteinpolysaccharide complex from bovine nasal cartilage. The function of glycoprotein in the formation of aggregates. J. biol. Chem. 244, 2384 (1969).

Hascall, V.C., Sajdera, S.W.: Physical properties and polydispersity of proteoglycan from bovine nasal cartilage. J. biol. Chem. 245, 4920 (1970).

Helting, T., Rodén, L.: Biosynthesis of chondroitin sulfate I. Galactosyl transfer in the formation of the carbohydrate-protein linkage region. J. biol. Chem. 244, 2790 (1969a).

Helting, T., Rodén, L.: Biosynthesis of chondroitin sulfate II. Glucuronosyl transfer in the formation of the carbohydrate linkage region. J. biol. Chem. 244, 2799 (1969b).

Hilz, H., Kleine, T.O., Bubenzer, H.J.: Biosynthese und metabolische Heterogenität von Chondroitin-4-sulfat-Protein aus Kälber-Rippenknorpel. Hoppe-Seylers Z. physiol. Chem. 350, 669 (1969).

Hoffman, P., Mashburn, T.O.: Proteinpolysaccharide of bovine cartilage. The relation of keratan sulfate and chondroitin sulfate. J. biol. Chem. 242, 3805 (1967).

Horowitz, A.L., Dorfman, A.: Subcellular sites for synthesis of chondromucoprotein of cartilage. J. Cell Biol. 38, 358 (1968).
Janis, R., Sandson, J., Smith, S., Hamerman, D.: Synovial cell synthesis of a substance immunologically like cartilage proteinpolysaccharide. Science 158, 1464 (1967).
Katsura, N., Davidson, E.A.: Studies on porcine costal cartilage protein-polysaccharide complex. II. Chemical studies. Biochim. biophys. Acta (Amst.) 121, 128 (1966).
Kleine, T.O.: Metabolic pattern of proteoglycans in calf rib cartilage extracted by 0.4 and 4M guanidinium chloride. VII. Europ. Rheumatology Congress, Brighton 1971a, Abstract Nr. 24.4.
Kleine, T.O.: Biosynthesis of metabolically different chondroitin-4-sulfate proteins in calf rib cartilage. Fed. Exp. Biol. Soc. (Varna), Abstr. VII, p. 221 (1971b).
Kleine, T.O.: Struktur, Biosynthese und Heterogenität der Chondroitinsulfatproteine. Naturwissenschaften 59, 64 (1972).
Kleine, T.O., Dapper, F.: Different influence of antirheumatica on the biosynthesis of chondroitin-4-sulfate peptides in calf rib cartilage. Fed. Exp. Biol. Soc. (Varna), Abstr. VII, p. 224 (1971).
Kleine, T.O., Hilz, H.: Charakterisierung der Protein-Polysaccharide aus Kälberrippenknorpel und ihre In-vitro-Markierung mit $^{35}SO_4^{2-}$. Hoppe-Seylers Z. physiol. Chem. 349, 1027 (1968a).
Kleine, T.O., Hilz, H.: Unterschiedliche Stoffwechselaktivität der Chondroitin-4-sulfat-Peptide aus Kälberrippenknorpel. Hoppe-Seylers Z. physiol. Chem. 349, 1250 (1968b).
Kleine, T.O., Hilz, H.: Indications for two metabolically different chondroitin-4-sulfate proteins in calf rib cartilage. Fed. Exp. Biol. Soc. (Madrid), Abstr. VI, p. 258 (1969a).
Kleine, T.O., Hilz, H.: Isolation of several chondroitin-4-sulfate proteins from calf rib cartilage with different metabolic activities. XII. Congressus Rheumatologicus int. Prague 1969b, Abstr. Nr. 95.
Kleine, T.O., Hilz, H.: Metabolically different proteoglycans in calf rib cartilage. In: Chemistry and Molecular Biology of the Intercellular Matrix Vol. II (Balazs, E.A., Ed.), p. 907. New York: Academic Press 1970.
Kleine, T.O., Kirsig, H.J., Hilz, H.: On the regulation of chondromucoprotein synthesis. Fed. Exp. Biol. Soc. (Oslo), Abstr. IV, p. 71 (1967).
Kleine, T.O., Kirsig, H.J., Hilz, H.: Über die Koordinierung von Protein-, Polysaccharid- und Sulfatester-Synthese im Rippenknorpel von Kälbern. Hoppe-Seylers Z. physiol. Chem. 349, 1037 (1968).
Kleine, T.O., Kirsig, H.J., Hilz, H.: Metabolic pattern of calf cartilage proteoglycans separated by CsCl-gradient centrifugation. Z. klin. Chem. 9, 74 (1971a).
Kleine, T.O., Kirsig, H.J., Hilz, H.: Metabolic and chemical heterogeneity in calf rib cartilage as revealed by CsCl gradient centrifugation and CPC-cellulose chromatography. Hoppe-Seylers Z. physiol. Chem. 353, 479 (1971b).
Kresse, H., Buddecke, E.: Stoffwechselheterogenität von Chondroitin-4-sulfat-Proteinen und Kollagenstoffwechsel in Rindernasenknorpel. Hoppe-Seylers Z. physiol. Chem. 349, 1497 (1968).
Kresse, H., Buddecke, E.: Stoffwechselheterogenität ^{14}C- und ^{35}S-markierter Glycosaminoglycane (saurer Mucopolysaccharide) bei In-vitro-Inkubation. Hoppe-Seylers Z. physiol. Chem. 351, 151 (1970).
Kröz, W., Buddecke, E.: Chemische und makromolekulare Altersveränderungen von Polysaccharid-Proteinen aus menschlichem Rippenknorpel. Hoppe-Seylers Z. physiol. Chem. 348, 665 (1967).
Lindahl, U., Rodén, L.: The chondroitin-4-sulfate protein linkage. J. biol. Chem. 241, 2113 (1966).
Loewi, G., Muir, H.: The antigenicity of chondromucoprotein. J. Immunol. 9, 119 (1965).

Lowther, D.A., Preston, B.N., Meyer, F.A.: Isolation and properties of chondroitin sulphates from bovine heart valves. Biochem. J. 118, 595 (1970).
Luscombe, M., Phelps, C.F.: Action of degradative enzymes on the light fraction of bovine nasal-septa protein polysaccharide. Biochem. J. 103, 103 (1967a).
Luscombe, M., Phelps, C.F.: The composition and physicochemical properties of bovine nasal-septa protein polysaccharide complex. Biochem. J. 102, 110 (1967b).
Lyons, H., Singer, J.A.: The effect of mucopolysaccharide linkage region and intrachain alkali labilities on chain weight determination. J. biol. Chem. 246, 277 (1971).
Malawista, I., Schubert, M.: Chondromucoprotein; new extraction method and alkali degradation. J. biol. Chem. 230, 535 (1958).
Marler, E., Davidson, E.A.: Structure of a polysaccharide protein complex. Proc. nat. Acad. Sci. (Wash.) 54, 648 (1965).
Mathews, M.B., Decker, L.: The distribution of sulfate groups in chondroitin sulfate chains. Detection of "oversulfated" and "undersulfated" repeating periods. Biochim. biophys. Acta (Amst.) 156, 419 (1968a).
Mathews, M.B., Decker, L.: The effect of acid mucopolysaccharides and acid mucopolysaccharide-proteins on fibril formation from collagen solutions. Biochem. J. 109, 517 (1968b).
Mathews, M.B., Lozaityte, I.: Sodium chondroitin sulfate-protein complexes of cartilage. I. Molecular weight and shape. Arch. Biochem. 74, 158 (1958).
Mathews, M.B., Sandson, J.: Molecular structures and antigenic properties of chondroitin sulfate-proteins (CSP) of vertebrate cartilage. Proc. Symp. connective tissue, p. 186. XII. Congressus Rheumatologicus int. at Prague 1969.
Meezan, E., Davidson, E.A.: Mucopolysaccharide sulfation in chick embryo cartilage. J. biol. Chem. 242, 1685, 4956 (1967).
Muir, H.: The nature of the link between protein and carbohydrate of a chondroitin sulphate complex from hyaline cartilage. Biochem. J. 69, 195 (1958).
Muir, H., Jacobs, S.: Protein-polysaccharides of pig laryngeal cartilage. Biochem. J. 103, 367 (1967).
Neutra, M., Leblond, C.P.: Radioautographic comparison of the uptake of galactose - ^{3}H and glucose - ^{3}H in the golgi region of various cells secreting glycoproteins or mucopolysaccharides. J. Cell Biol. 30, 137 (1966).
Olsson, I.: Biosynthesis of glycosaminoglycans (mucopolysaccharides) in leucemic myeloid cells. Biochim. biophys. Acta (Amst.) 165, 324 (1968).
Olsson, I., Gardell, S.: Biosynthesis of glycosaminglycans in leucocytes. Biochim. biophys. Acta (Amst.) 237, 203 (1971).
Pal, S., Doganges, P.T., Schubert, M.: The sparation of new forms of the proteinpolysaccharides of bovine nasal cartilage. J. biol. Chem. 241, 4261 (1966).
Pankovich, A.M., Korngold, L.: A comparison of the antigenic properties of nucleus pulposus and cartilage protein polysaccharide complexes. J. Immunol. 99, 431 (1967).
Partridge, S.M., Davis, H.F., Adair, G.S.: The chemistry of connective tissue and the constitution of the chondroitin sulfate protein complex in cartilage. Biochem. J. 79, 15 (1961).
Perlman, R.F., Telser, A., Dorfman, A.: The biosynthesis of chondroitin sulfate by a cell-free preparation. J. biol. Chem. 239, 3623 (1964).
Peterson, M., Leblond, C.P.: Synthesis of complex carbohydrates in the golgi region as shown by radioautography after injection of labeled glucose. J. Cell Biol. 21, 143 (1964).

Radhakrishnamurthy, B., Skinivasan, S.R., Dalferes, E.R., Berenson, G.S.: Composition of glycopeptides from chondroitin sulfate protein complex from squid skin. Comp. Biochem. Physiol. 36, 107 (1970).
Robbins, P.W., Lipmann, F.: Separation of two enzymatic phases in active sulfate synthesis. J. biol. Chem. 233, 681 (1958).
Robinson, H.C., Dorfman, A.: The sulfation of chondroitin sulphate in embryonic chick cartilage epiphyses. J. biol. Chem. 244, 348 (1969).
Robinson, H.C., Telser, A., Dorfman, A.: Studies on biosynthesis of the linkage region of chondroitin sulfate protein complex. Proc. nat. Acad. Sci. (Wash.) 56, 1859 (1966).
Rodén, L.: In: 4th Intern. Conf. Cystic Fibrosis of the Pancreas (Mucoviscidosis), part 11. Bern/Grindelwald 1966.
Rodén, L., Armand, G.: Structure of chondroitin-4-sulfate-protein linkage region: Isolation and characterization of the disaccharide. J. biol. Chem. 241, 65 (1966).
Rodén, L., Smith, R.: Structure of the neutral disaccharide of the chondroitin-4-sulfate protein linkage region. J. biol. Chem. 241, 5949 (1966).
Rosenberg, L., Hellmann, W., Kleinschmidt, A.K.: Macromolecular models of proteinpolysaccharides from bovine nasal cartilage based on electron microscopic studies. J. biol. Chem. 245, 4123 (1970a).
Rosenberg, L., Johnson, B., Schubert, M.: The protein-polysaccharides of human costal cartilage. J. clin. Invest. 48, 543 (1969).
Rosenberg, L., Pal, S., Beale, R., Schubert, M.: A comparison of proteinpolysaccharides of bovine nasal cartilage isolated and fractionated by different methods. J. biol. Chem. 245, 4112 (1970b).
Rosenberg, L., Schubert, M.: The proteinpolysaccharides of bovine nucleus pulposus. J. biol. Chem. 242, 4691 (1967).
Sajdera, S.W., Hascall, V.C.: Proteinpolysaccharide complex from bovine nasal cartilage. A comparison of low and high shear extraction procedures. J. biol. Chem. 244, 77 (1969).
Sandson, J., Rosenberg, L., White, D.: The antigenic determinants of protein-polysaccharides of cartilage. J. exp. Med. 123, 817 (1966).
Scheraga, H.A., Mandelkern, L.: Consideration of the hydrodynamic properties of proteins. J. Amer. chem. Soc. 75, 179 (1953).
Schubert, M.: Structure of connective tissues, a chemical point of view. Fed. Proc. 25, 1047 (1966).
Schubert, M., Hamerman, D.: A Primer on Connective Tissue Biochemistry. Philadelphia: Lea & Febiger 1968.
Sela, M.: In: New Perspectives in Biology (Sela, M., Ed.), Vol. 4, p. 225. Amsterdam: Elsevier 1964.
Serafini-Fracassini, A., Peters, T.J., Floreani, L.: The protein-polysaccharide complex of bovine nasal cartilage. Studies on the protein core. Biochem. J. 105, 569 (1967).
Silbert, J.E.: Incorporation of ^{14}C and ^{3}H from labeled nucleotide sugars into a polysaccharide in the presence of a cell free preparation from cartilage. J. biol. Chem. 239, 1310 (1964).
Silbert, J.E., DeLuca, S.: Biosynthesis of chondroitin sulfate. J. biol. Chem. 244, 876 (1969).
Smith, J.W., Peters, T.J., Serafini-Fracassini, A.: Observations on the distribution of the proteinpolysaccharide complex and collagen in bovine articular cartilage. J. Cell Sci. 2, 129 (1969).
Stoolmiller, A.C., Dorfman, A.: In: Comprehensive Biochemistry (Florkin, M, E.H. Stotz, Eds.), p. 241. Amsterdam: Elsevier 1969.
Suzuki, S., Strominger, J.L.: Enzymatic sulfation of mucopolysaccharides of hen oviduct. J. biol. Chem. 235, 257 (1960).
Tanford, C., Marler, E., Jury, E., Davidson, E.A.: Determination of the molecular weight distribution of chondroitin sulfate B by sedimentation equilibrium. J. biol. Chem. 239, 4034 (1964).
Telser, A., Robinson, H.C., Dorfman, A.,: The biosynthesis of chondroitin sulfate-protein complex. Proc. nat. Acad. Sci. (Wash.) 54, 371 (1965).

Telser, A., Robinson, H.C., Dorfman, A.: The biosynthesis of chondroitin sulfate. Arch. Biochem. 116, 458 (1966).
Toole, B.P., Lowther, D.A.: Dermatan sulfate-protein: Isolation from and interaction with collagen. Arch. Biochem. 128, 567 (1968).
Tsiganos, C.P., Muir, H.: A hibrid protein-polysaccharide of keratan sulphate and chondroitin sulphate from pig laryngeal cartilage. Biochem. J. 104, 26C (1967).
Tsiganos, C.P., Muir, H.: Studies on protein-polysaccharides from pig laryngeal cartilage. Heterogeneity, fractionation and characterization. Biochem. J. 112, 885 (1969).
Varadi, D.P., Griffiths, C.: Oversulfated chondroitin sulfate in normal human urine. Biochim. biophys. Acta (Amst.) 230, 248 (1971).
Wilson, L.G., Bandurski, R.S.: Enzymatic reactions involving sulfate, sulfite, selenate, and molybdate. J. biol. Chem. 233, 975 (1958).

Isolation of Corneal Keratan-Chondroproteoglycans and Their Composition

H. W. Stuhlsatz, R. Kisters, A. Wollmer, and H. Greiling

Each different type of connective tissues in the human organism shows a specific distribution of proteoglycans. Further differences exist in the proteoglycan molecule itself. Besides variations in the polysaccharide chain, there are also differently composed protein cores containing the same type of polysaccharide chain. As found by several investigators proteochondroitin sulfates and proteokeratan sulfates of different organs vary in amino acid distribution of their protein or peptide matrix covalently bound (Anderson et al., 1967; Greiling and Stuhlsatz, 1966; Greiling et al., 1967). On the other hand, no proteoglycan could be separated with only one type of polysaccharide chain, even after numerous purification procedures. The heterogeneity of the polysaccharide portion in the same proteoglycan was pointed out by several authors in different types of connective tissues: cornea (Greiling and Stuhlsatz, 1966; Anseth and Laurent, 1961; Berman, 1970), trachea (Tsiganos and Muir, 1967), nasal septum (Gerber et al., 1960; Sajdera and Hascall, 1969; Franek and Dunstone, 1967), rib cartilage (Kleine and Hilz, 1968), knee joint cartilage (Greiling and Stuhlsatz, 1969), aorta (Kröz and Buddecke, 1967) and skin (Fransson, 1970).

In order to examine the heterogeneity of the polysaccharide chains of proteoglycans of bovine cornea, the latter were isolated as recently described (Stuhlsatz et al., 1971) including high speed homogenization and several consecutive chromatographic steps on ecteola cellulose, biogel P-150, P-30 and P-10 (Fig. 1 and 2). In extracting the fresh corneas and acetone-dried corneas with water under high speed homogenization there were distinct differences as shown in Table 1. The first extract of acetone-dried corneas contained nearly half of the total hexosamine in contrast to those which contained only one-third the extract of fresh corneas. The concentration of hexosamine in the first extract of acetone-dried corneas was even greater. The hexosamine not extractable with water amounts to 49% in acetone-dried corneas and to 57% in fresh corneas. This observation and the finding of more than 65% of the total hexosamine in peptidoglycosaminoglycans of papain digested corneal extracts led to the conclusion that in corneas there will also be proteoglycans not extractable with water, a statement being valid for several other types of connective tissue (Schubert, 1958).

Apart from the six fractions framed in Fig. 1 two more were isolated from acetone-dried corneas by a procedure consisting of column chromatography on Dowex 1 x 2 performed on the first aqueous extract after high speed homogenization. The column was eluted by a gradient of LiCl (Stuhlsatz et al., 1971) resulting in the two fractions PCSI and PCSII. In contrast to the fractions PGIR, PGIÜ, PGIIA and PGIIB, the proteoglycans PGIID2 and PCSI were homogeneous according to ultracentrifugal analyses, but not chromatographically with respect to their chondroitin sulfate and keratan sulfate contents.

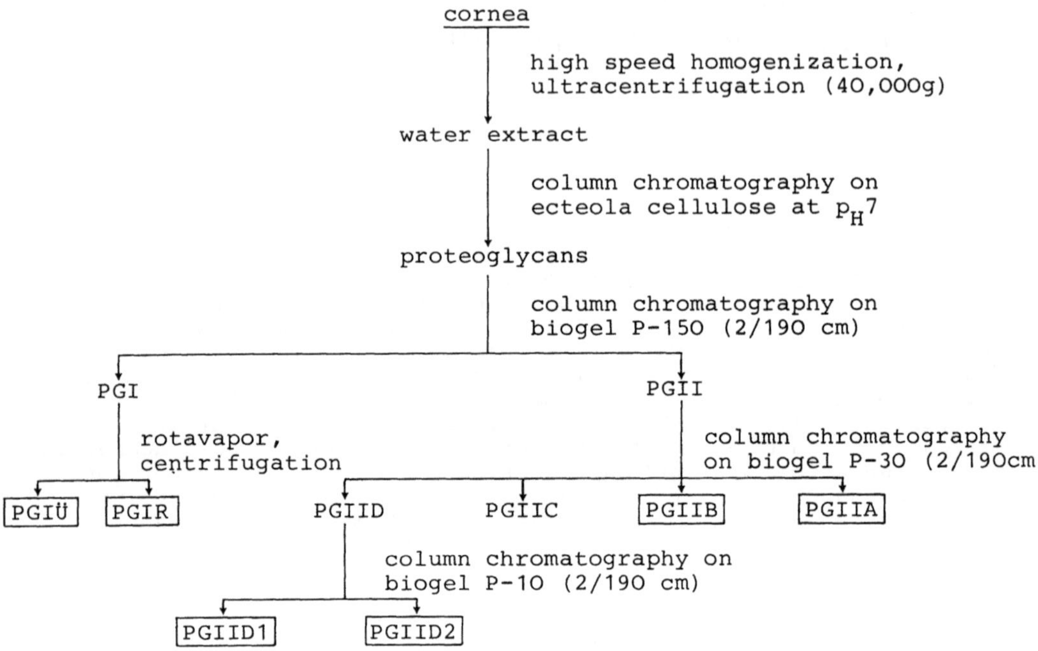

Fig. 1. Isolation of proteoglycans (PG) from bovine cornea. Ü = supernatant, R = residue

Table 2 shows the analytical data of these fractions. According to the glucosamine/galactosamine ratio the polysaccharide portion of the fractions PGIR, PGIIA, PGIIB and PGIID2 predominantly consists of keratan sulfate, whereas the fractions PGIID1, PCSI and PCSII predominantly contain chondroitin sulfate. Though both are keratan chondroproteoglycans, the fraction PGIID2 may be considered as "proteokeratan sulfate" and fraction PCSI as "proteochondroitin sulfate". These two fractions also differ in their amino acid content as shown in Table 3: the aspartic acid/serine ratio of fraction PGIID2 is distinctly higher than that of fraction PCSI (1.9 and 1.2, resp.).

In Table 2 it is seen that only the fractions PCSI and PCSII are fully sulfated in contrast to the other ones, especially PGIIB and PGIID1, the low sulfate content of which may be due to a high chondroitin content. The protein content of the fractions PGIR, PGIÜ, PGIIA and PGIIB reaches from 47 to 62%. Whereas the fractions PGIID1 and PGIID2 have nearly the same protein content (about 10%), the fractions PCSI and PCSII differ distinctly. As no hydroxyproline was detectable a contamination with collagen may be excluded.

The analyses of eight corneal fractions listed in Table 2 lead to the conclusion that besides proteochondroitinkeratansulfates, e.g. PGIID2 and PGIIA, and proteokeratanchondroitin sulfates, e.g. PCSI and PCSII, there exist other types of hybridization, e.g. in PGIIB, which may be built up by keratan sulfate and chondroitin chains, and in fraction PGIID1, which may contain still chondroitin sulfate chains besides the two foregoing.

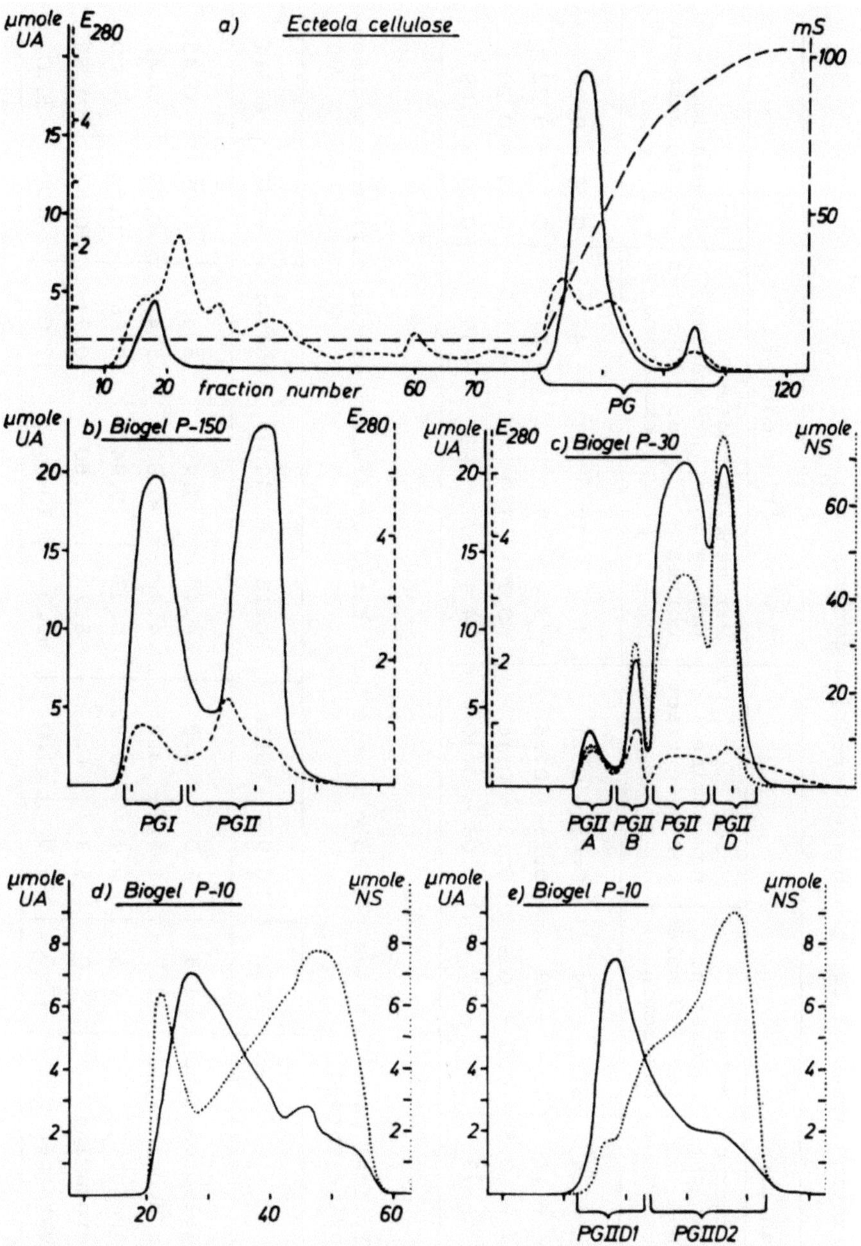

Fig. 2. Separation of proteoglycans from bovine cornea. a Chromatography on ecteola cellulose. b Chromatography of fraction PG on biogel P-150. c Chromatography of fraction PGII on biogel P-30. d Chromatography of fraction PGIIC on biogel P-10. e Chromatography of fraction PGIID on biogel P-10.
UA = uronic acid. NS = neutral sugar. mS = millisiemens (siemens = reciprocal ohm)

Table 1. Extraction steps of fresh and acetone-dried corneas

	Fresh cornea			Acetone-dried cornea		
	Per cent of dry weight cornea	Moles hexosamine per 100 moles hexosamine	Moles hexosamine per mg extract	Per cent of dry weight cornea	Moles hexosamine per 100 moles hexosamine	Moles hexosamine per mg extract
1. Extr.	21 - 24	31 - 34	0.13 - 0.21	16 - 19	41 - 47	0.35 - 0.36
2. Extr.	4 - 5	10 - 12	0.25 - 0.26	3 - 9	3 - 21	0.17 - 0.33
3. Extr.	0.4 - 0.5	0.5 - 0.9	0.13 - 0.16	0.7 - 1.0	0.7 - 1.5	0.14 - 0.20
Residue	71 - 73	55 - 57	0.08 - 0.10	74 - 77	37 - 49	0.07 - 0.09

Table 2. Analytical data in moles per mole (where not stated otherwise) of the different proteoglycans of bovine cornea

	PGIR	PGIÜ	PGIIA	PGIIB	PGIID1	PGIID2	PCSI	PCSII
Glucosamine/galactosamine	1.8	1.0	3.8	4.4	0.79	5.4	0.09	0.43
Neutral sugar/uronic acid	1.7	1.1	3.4	4.4	0.80	5.4	0.09	0.42
Sulfate/hexosamine	0.68	0.71	0.90	0.57	0.46	0.86	0.98	1.00
Amino Acid/hexosamine	5.0	5.3	6.0	8.7	0.60	0.71	0.52	1.55
Protein content (per cent)	47.0	48.0	51.0	62.0	10.0	11.0	8.0	20.0

It is of interest that none of these proteoglycans show as high a content of keratan sulfate as do the corneal peptidokeratan sulfates KS I to IV (Greiling and Stuhlsatz, 1966) isolated after digestion with papain, the glucosamine/galactosamine ratio of which is 160/1, whereas the "proteochondroitin sulfate" PCSI is rather similar to a corneal peptidochondroitin sulfate - called CS IV (Greiling et al., 1967) - also obtained by papain digestion, with respect to its high chondroitin sulfate content.

Circular dichroism (CD) spectra of two proteoglycans PGIID2 and PCSI have been recorded, tentatively called "proteokeratan sulfate" and "proteochondroitin sulfate", respectively (Fig. 3).

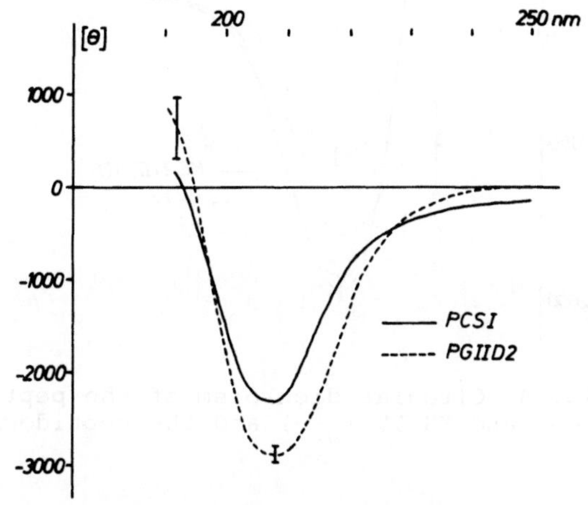

Fig. 3. Circular dicroism of the proteoglycans PGIID2 (---) and PCSI (——)

In Fig. 4 the CD spectra of the four peptidokeratan sulfates KS I to KS IV mentioned above and that of peptidochondroitin sulfate CS II (Greiling et al., 1967) are shown.

"Proteokeratan sulfate" corresponding to the peptido keratan sulfates KS I to II, fraction KS IV is more likely comparable to the peptidochondroitin sulfate CS II. This is also valid for the optical rotatory disperison (ORD) shown in Fig. 5.

The resemblance of peptidokeratan sulfate KS IV and peptidochondroitin sulfate CS II is also evident in their acridin orange (AO) complexes (Fig. 6) with respect to the visible absorption bands, whereas in the range of 350 to 220 nm the spectra qualitatively are the mirror image of each other.

The differences between the two proteoglycans are reflected even more in the CD spectra of their AO complexes than in the far Cotton effects (Fig. 7).

The proteochondroitin sulfate corresponds to the peptidokeratan sulfate KS IV, but the "proteokeratan sulfate" resembles the peptidokeratan sulfates KS I to III (Fig. 8).

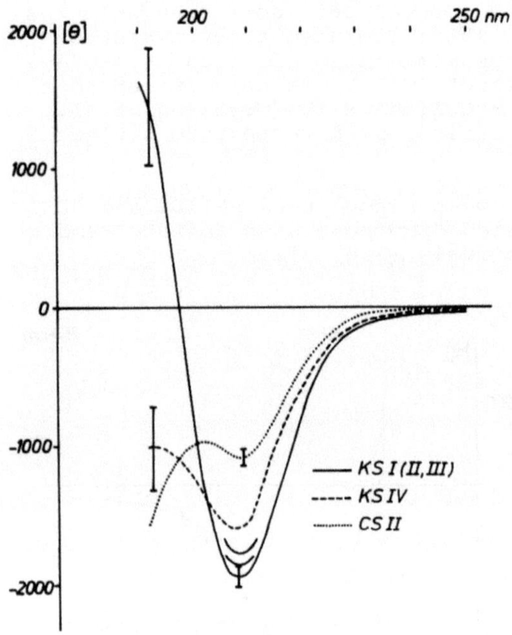

Fig. 4. Circular dichroism of the peptidokeratan sulfates KS I (II,III) (——) and KS IV (---) and the peptidochondroitin sulfate CS II (....)

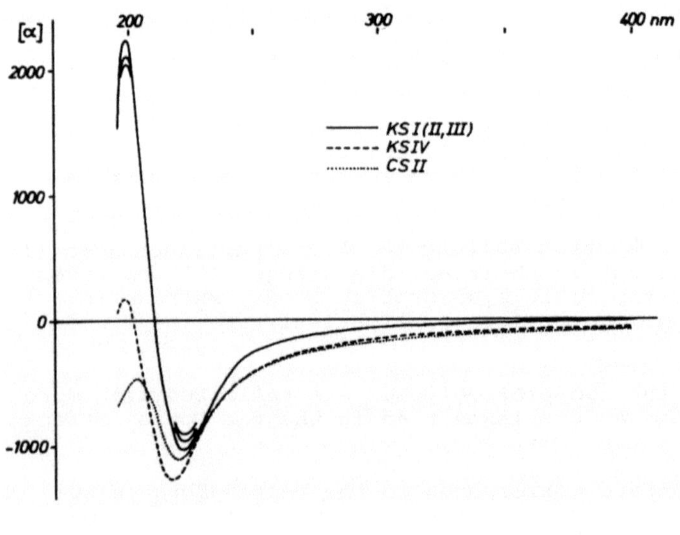

Fig. 5. Optical rotatory dispersion of the peptidokeratan sulfates KS I (II, III) (——) and KS IV (---) and the peptidochondroitin sulfate CS II (...)

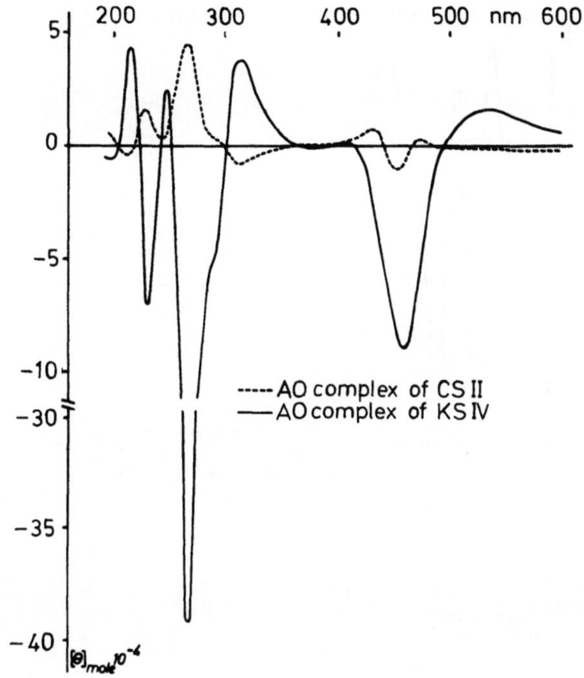

Fig. 6. CD spectra of the acridin orange complexes of the peptidokeratan sulfate KS IV (——) and the peptidochondroitin sulfate CS II(---)

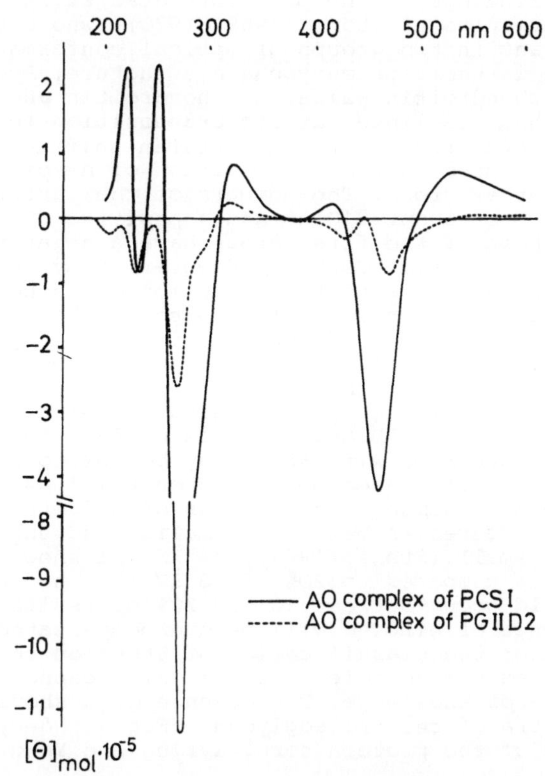

Fig. 7. CD spectra of the acridin orange complexes of the proteoglycans PGIID2 (---) and PCSI (——)

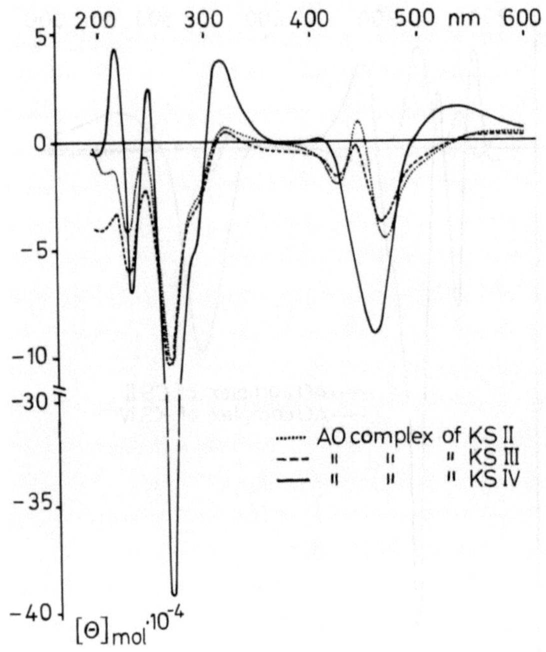

Fig. 8. CD spectra of the acridin orange complexes of the peptidokeratan sulfates KS II (...), KS III (---) and KS IV (——). For better lucidity the spectrum of KS I was not plotted

Our findings on the peptidokeratan sulfates are not in agreement with the results of Stone (1969, 1970), who classifies the glycosaminoglycans in two groups of optical conformation correlated to a difference in their primary chain structure. According to these criteria the chondroitin sulfates, chondroitin and hyaluronate, the hexosamine of which is linked at its C-3 position to a second monosaccharide, form one group. Heparin, heparan sulfate and keratan sulfate, being substituted in the C-4 position of hexosamine by a second sugar, form the other group. The members of the latter show a negative Cotton effect at about 210 nm and a positive CD band at about 190 nm, whereas those of the first group have a negative CD band at about 208 nm and no positive CD band. However, they show a second negative band below 190 nm, with the exception of chondroitin 6-sulfate of shark cartilage (Stone, 1970). The peptidokeratan sulfates KS I to IV are of the same type of substitution at C-4 position of hexosamine. Accordingly the alterations of their chemical and physicochemical data such as sulfate and amino acid content, optical rotation, and molecular weight (Greiling and Stuhlsatz, 1966) are nearly parallel. This is true for the ORD and CD bands of KS I to III only. The marked differences observed with KS IV may be due to the degree of oversulfatation, as the differences between chondroitin 4-sulfate and 6-sulfate must depend on number and position of sulfate groups. On the other hand the differences between mammalian and shark chondroitin sulfates are only small (Stone, 1969). As KS III shows an oversulfatation of about 10% as compared to 20% of KS IV the change to the other spectral type should occur between 10 and 20% oversulfatation. But whether the influence of binding will be overcompensated by other properties or whether the classification of Stone on the basis of substitution in the hexosamine moiety is not valid cannot be decided at the present state of knowledge. The absence of a shoulder at 222 nm in the CD spectra of the proteoglycans (Fig. 3) deny a helical secondary structure of the protein core. Eyring and Yang (1968) derived a randomly coiled structure from their results. It is interesting that the spectrum of the acridin orange complex of KS IV agrees with that of chondroitin 6-sulfate published by Eyring, Kraus and Yang (1968), though these substances differ in position of sugar substitution in the hexosamine moiety but agree in location of the sulfate group at C-6 position in the hexosamine moiety. The positive extrinsic band near 270 nm of the peptidochondroitin sulfate CS II (Fig. 6) is also found in the acridin orange complex of polyglutamate existing as a random coil at p_H 8 (Eyring et al., 1968). The "proteochondroitin sulfate" shows a negative band at this wavelength and thus agrees qualitatively with the peptidokeratan sulfate KS IV.

The high content of aspartic acid in all corneal proteoglycans listed in Table 3 leads to the question of a participation of this amino acid in the linkage region between keratan sulfate chain and protein core. Several investigators (Greiling and Stuhlsatz, 1966; Seno et al., 1965; Castellani, 1965; Mathews and Cifonelli, 1965) have shown aspartic acid as dominating amino acid in peptido- and proteokeratan sulfates. We found that under mild acidic conditions at p_H 2 after two

Table 3. Amino acid distribution in moles per 100 moles of the different proteoglycans of bovine cornea

	PGIR	PGIÜ	PGIIA	PGIIB	PGIID1	PGIID2	PCSI	PCSII
Asp	13.6	12.6	14.4	13.9	15.1	18.0	10.5	12.6
Thr	4.9	5.5	5.4	6.8	6.0	7.3	4.1	4.8
Ser	6.4	7.3	6.0	6.5	7.1	9.3	8.6	7.0
Glu	11.7	15.4	13.4	12.4	10.3	11.0	10.0	12.0
Pro	7.1	8.5	11.5	7.7	11.2	6.7	8.2	9.2
Gly	6.8	6.8	4.8	5.7	5.0	6.5	11.6	9.7
Ala	5.3	5.6	4.6	5.6	3.8	4.4	5.0	5.3
Val	4.1	5.2	3.6	6.0	6.1	3.6	4.9	4.8
Cys	0.7	0.9	1.1	1.3	1.1	0.8	0.5	0.5
Met	0.7	0.7	0.8	1.7	0.6	0.7	0.3	0.3
Ile	5.2	4.3	5.6	5.2	5.6	5.5	3.9	5.1
Leu	12.4	10.1	10.6	10.1	8.7	8.3	10.0	12.3
Tyr	3.7	2.9	3.7	3.9	3.1	2.2	2.0	2.3
Phe	4.3	3.5	3.8	4.1	4.8	5.8	3.9	3.9
Orn	0.4	0.2	0.2	0.3	2.3	1.7	5.6	0.5
Lys	7.0	5.5	6.6	4.9	4.7	4.0	5.2	4.8
His	2.1	1.9	1.6	1.8	2.0	1.6	1.9	1.8
Arg	3.6	3.0	2.2	2.2	2.6	2.6	3.8	3.1

hours nearly equimolar amounts of ammonia and aspartic acid were liberated. This and the simultaneous formation of free N-acetylglucosamine caused us to search for a type of linkage region between aspartic acid and polysaccharide chain described for certain glycoproteins by Neuberger et al. (1966). In hydrolysates of ovalbumine

Table 4. Analytical data of corneal peptidoglycosaminoglycans from sedimentation and diffusion analyses

Peptidoglycos-aminoglycan	Molecular weight	Optical rotation	per cent amino acid	Sulfate/disaccharide (molar ratio)	µmoles mannose per 100 mg
KS I	8600	$- 5.5^{\circ}$	6.1	0.9	20.0
KS II	12900	$- 4.5^{\circ}$	4.9	1.0	15.4
KS III	15500	$- 1.0^{\circ}$	3.8	1.1	13.5
KS IV	19800	0.0°	3.2	1.2	-
CS II	18500	$- 32.0^{\circ}$	1.6	0.8	-
CS IV	30200	$- 44.0^{\circ}$	1.7	1.1	-

these authors found 4'-L-aspartamido N-acetylglucosamin. After having synthesized this compound from 2-acetamido-1-amino-3,4,6,-tri-O-acetyl-1,2-dideoxyglucose and 2-benzyl-N-benzyloxycarbonyl-L-aspartate according to Marks et al. (1963) we could discover it in very small amounts in hydrolysates of corneal peptidokeratan sulfates by the following microanalytical method: The reference substance 2-acetamido-1-(4'-L-aspartamino)-1,2-dideoxy-β-D-glucose appears in our elution program (Greiling and Stuhlsatz, 1966) of the amino acid autoanalyzer (Technicon) after 88 minutes having a E_{440}/E_{570} ratio of 1.57. In hydrolysates ($p_H 2$, 3h., 100°C) of corneal peptidokeratan sulfates we found a substance of the same elution time and E_{440}/E_{570} ratio. Both substances were congruent when run simultaneously and showed the same behavior in high voltage electrophoresis. With a potential gradient of 62 V/cm they had a migration velocity of 15 cm/h. (formic acid: acetic acid buffer $p_H 1.85$, Whatman Nr. 3 MM). Both substances gave the same yellowbrown color with ninhydrin reagent. Both were hydrolysed after 3 h in 1 M hydrochloric acid to yield equimolar amounts of glucosamine, ammonia and aspartic acid. Recently Rodén (1970) was also able to isolate 4'-L-aspartamido-N-acetylglucosamine from corneal keratan sulfate.

With a gaschromatographic technique (Kisters and Greiling, 1968) we found in the peptidokeraten sulfates KS I to IV descending amounts of mannose. As shown by Fig 9 we could demonstrate fucose too.

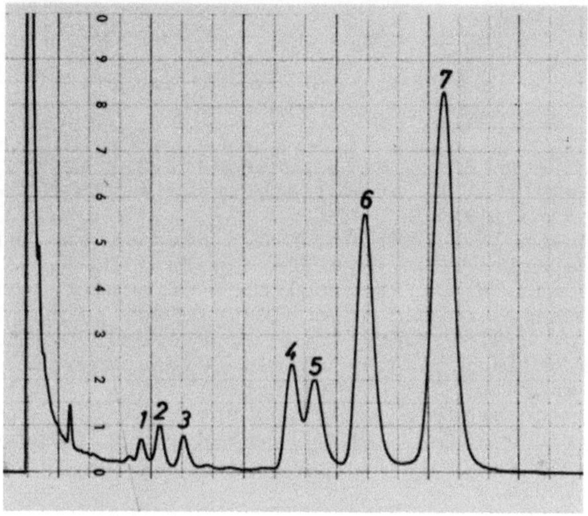

Fig. 9. Gaschromatographic separation of the trimethylsilylethers of neutral sugars from corneal peptidokeratan sulfates. 1, 2 and 3 = fucose, 4 = α-D-mannose, 5 = galactofuranose, 6 and 7 = α- and β-D-galactose, respectively

The mannose contents of KS I, KS II and KS III are in nearly the same relation to one another as their amino acid contents (Table 4).

Our results do not explain the structural significance of mannose in proteokeratan sulfates. But with respect to the "binding" aspartic acid - derived from the molecular weights experimentally found - we get aspartic/mannose ratios of 1/2 in KS I, KS II and KS III. Although we were not yet able to isolate a mannose containing oligosaccharide from hydrolysates of peptidokeratan sulfates, we tentatively postulate the following structure for the corneal keratanchondroproteoglycan (Fig. 10).

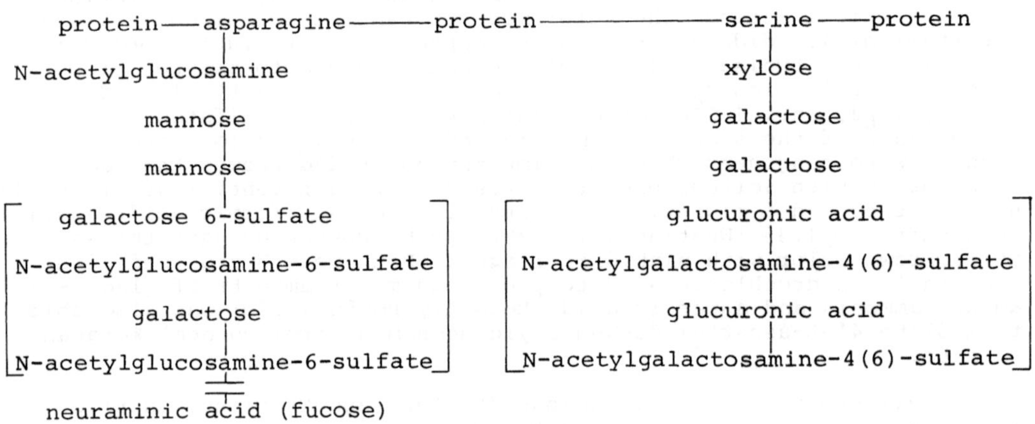

Fig. 10. Postulated structure of corneal keratanchondroproteoglycan

References

Anderson Bray, B., Liebermann, R., Meyer, K.: Structure of human skeletal keratosulfate - the linkage region. J. biol. Chem. 242, 3373 (1967).

Anseth, A., Laurent, T.C.: Studies on corneal polysaccharides. I. Separation. Exp. Eye Res. 1, 25 (1961).

Berman, E.R.: Proteoglycans of bovine corneal stroma. In: E.A. Balazs: Chemistry and Molecular Biology of the Intercellular Matrix. Vol. 2, p. 879. London: Academic Press 1970.

Castellani, A.A.: Protein-polysaccharide complexes containing keratansulfate; In: G. Quintarelli: The Chemical Physiology of Mucopolysaccharides, p. 107. Boston: Little Brown 1965.

Eyring, E.J., Kraus, H., Yang, J.T.: Induced optical activity of chondroitin sulfate C-acridine orange complexes. Biopolymers 6, 703 (1968).

Eyring, E.J., Yang, J.T.: Conformation of proteinpolysaccharide complex from bovine nasal septum. J. biol. Chem. 243, 1306 (1968).

Franek, M.D., Dunstone, J.R.: Connective tissue proteinpolysaccharides. Fractionation of the proteinpolysaccharides from bovine nasal cartilage. J. biol. Chem. 242, 3460 (1967).

Fransson, L.A.: Structure and metabolism of the proteoglycans of dermatan sulfate. In: E.A. Balazs: Chemistry and Molecular Biology of the Intercellular Matrix. Vol. 2, p. 823. London: Academic Press 1970.

Gerber, B.R., Franklin, E.C., Schubert, M.: Ultra centrifugal fractionation of bovine nasal chondromucoproteins. J. biol. Chem. 235, 2870 (1960).

Greiling, H., Stuhlsatz, H.W.: Struktur und Stoffwechsel von Glykosaminoglykan-Proteinen. I. Die Keratansulfat-Peptide der Rinder-Cornea. Hoppe-Seylers Z. physiol. Chem. 345, 236 (1966).

Greiling, H., Stuhlsatz, H.W.: Glykosaminoglykan-Peptide aus dem humanen Kniegelenkknorpel. Hoppe-Seylers Z. physiol. Chem. 350, 449 (1969).

Greiling, H., Stuhlsatz, H.W., Kisters, R.: Struktur und Stoffwechsel von Glykosaminoglykan-Proteinen. III. Die Chondroitin-Peptide der Rinder-Cornea. Hoppe-Seylers Z. physiol. Chem. 348, 970 (1967).

Greiling, H., Stuhlsatz, H.W., Plagemann, L.: Struktur und Stoffwechsel von Glykosaminoglykan-Peptiden. II. Die Chondroitin-4-sulfat-Peptide der Rinder-Cornea. Hoppe-Seylers Z. physiol. Chem. 348, 121 (1967).

Kisters, R., Greiling, H.: Die quantitative Gas-Chromatographie der Neutralzucker von Glykosaminoglykan-Peptiden. Z. analyt. Chem. 243, 359 (1968).

Kleine, T.O., Hilz, H.: Charakterisierung der Protein-Poly-saccharide aus Kälberrippenknorpel und ihre In-vitro-Markierung mit $^{35}SO_4$. Hoppe-Seylers Z. physiol. Chem. 349, 1027 (1968).

Kröz, W., Buddecke, E.: Chemische und makromolekulare Altersveränderungen von Polysaccharid-Proteinen aus menschlichem Rippenknorpel. Hoppe-Seylers Z. physiol. Chem. 348, 665 (1967).

Marks, G.S., Marshall, R.D., Neuberger, A.: Carbohydrate in protein. 6. Studies on the carbohydrate peptide bond in hen's egg albumine. Biochem. J. 87, 274 (1963).

Mathews, M.B., Cifonelli, J.A.: Comparative biochemistry of keratosulfates. J. biol. Chem. 240, 4140 (1965).

Neuberger, A., Marshall, R.D., Gottschalk, A.: Carbohydrate-peptide linkages in glycoproteins and methods for their elucidation. In: A. Gottschalk: Glycoproteins, their Composition, Structure and Function, p. 273. Amsterdam: Elsevier 1966.

Rodén, L.: Structure and metabolism of the proteoglycan of chondroitin sulfates and keratan sulfate. In: E.A. Balazs: Chemistry and Molecular Biology of the Intercellular Matrix, p. 805. London: Academic Press 1970.

Sajdera, Hascall, V.C.: Proteinpolysaccharide complex from bovine nasal cartilage. A comparison of low and high shear extraction procedures. J. biol. Chem. 244, 77 (1969).

Schubert, M.: Chondromucoprotein. Fed. Proc, 17, 1100 (1958).

Seno, N., Meyer, K., Anderson, B., Hoffman, P.: Variations in keratosulfates. J. biol. Chem. 240, 1005 (1965).

Stone, A.L.: Optical rotatory dispersion of mucopolysaccharides and mucopolysaccharide-dye-complexes. II. Ultraviolet cotton effects in the amide transition bands. Biopolymers 7, 173 (1969).

Stone, A.L.: Asymmmetric modes of metachromatic binding of dyes to glycosaminoglycans and the influence of polymer conformation on the induced cotton effects in vitro. In: E.A. Balazs: Chemistry and Molecular Biology of the Intercellular Matrix. Vol. 2, p. 1067. London: Academic Press 1970.

Stuhlsatz, H.W., Kisters, R., Wollmer, A., Greiling, H.: Zur Struktur von Proteoglykanen aus der Cornea. Hoppe-Seylers Z. physiol. Chem. 352, 289 (1971).

Tsiganos, C.P., Muir, H.: A hybrid protein-polysaccharide of keratan sulphate and chondroitin sulphate from pig laryngeal cartilage. Biochem. J. 104, 26 C (1967).

Glycopeptide Pattern of the Polymeric Macromolecular Complex of the Stroma Layers and of the Endothelial Basement Membrane (Descemet's Membrane) of the Embryonic and Postnatal Calf Cornea

M. Moczar and E. Moczar

We reported previously that the extracellular matrix of a connective tissue may be characterized by the ratio of the heterosaccharides (of the structural glycoproteins) to glycosaminoglycans and to hydroxylysineglycosides of collagen (Robert et al., 1965; Moczar and Moczar, 1970a; Moczar et al., 1970; Robert and Robert, 1970).

The corneal stroma possessing a homogenous population of specialized fibroblasts is a suitable material for the study of these parameters during the embryonic development of the connective tissue. The cell containing stromal layers can be separated from the adjacent cell free membrane (endothelial basement membrane) both deriving from mesenchymal cells.

The presence of collagen and carbohydrate in the Descemet's membrane was reported (Dohlman and Balazs, 1955) and its antigenic properties were attributed to peptides with an aminoacid and carbohydrate composition unlike that of collagen (Kefalides and Denduchis, 1969). Several studies were reported on the macromolecular composition of the whole corneal stroma. There is little information available, however, on the different types of protein linked sugar chains of the endothelial basement membrane and of the adjacent stromal layers during the maturation process.

We report here on our investigation of the biochemical parameters mentioned above, in the collagen-structural glycoprotein-proteoglycan complex of the embryonic and postnatal calf corneal Descemet's membrane and of the adjacent stromal layers.

The overall sugar composition and hydroxyproline content of the Descemet's membrane shown on Table 1 resembles those of the whole corneal stroma (Dohlman and Balazs, 1955). The stroma is slightly richer in glucose and mannose. These results indicate the presence of the same type of macromolecules in both stromal layers: collagen, glycoproteins and proteoglycans.

The extraction procedure with 1 M $CaCl_2$ solution buffered with tris and citrate to pH 7.6 according to Robert and Dische (1963), yields the same type of extracts (collagen -SGP-proteoglycans complexes), from Descemet's membrane and from separated stroma. The yield of the $CaCl_2$ insoluble fraction (polymeric stroma) increases steadily in the corneal stroma during the gestation. In the Descemet's membrane relatively constant values were found from the embryonic and postnatal material accounting to about 60% of the dry weight (Moczar and Moczar, 1970b).

The product of collagenase-pronase digestion of this insoluble fraction from the stroma or from Descemet's membrane when fractionated on Sephadex G 50 M/column using the conditions as reported earlier

Table 1. Composition of the Descemet's membrane and the stroma of the calf cornea. Results given as g/100 g dry material

	Descemet's membrane	Stroma
Hexoses	3.67	3.4
Hexosamines	2.3	2.3
Hydroxyproline	8.85	7.1
Hyl:HylGal:HylGalGlc	1 : 0.4 : 0.83	1 : 0.52 : 0.86
Galactose	1	1
Glucose	0.5	0.62
Mannose	0.24	0.39

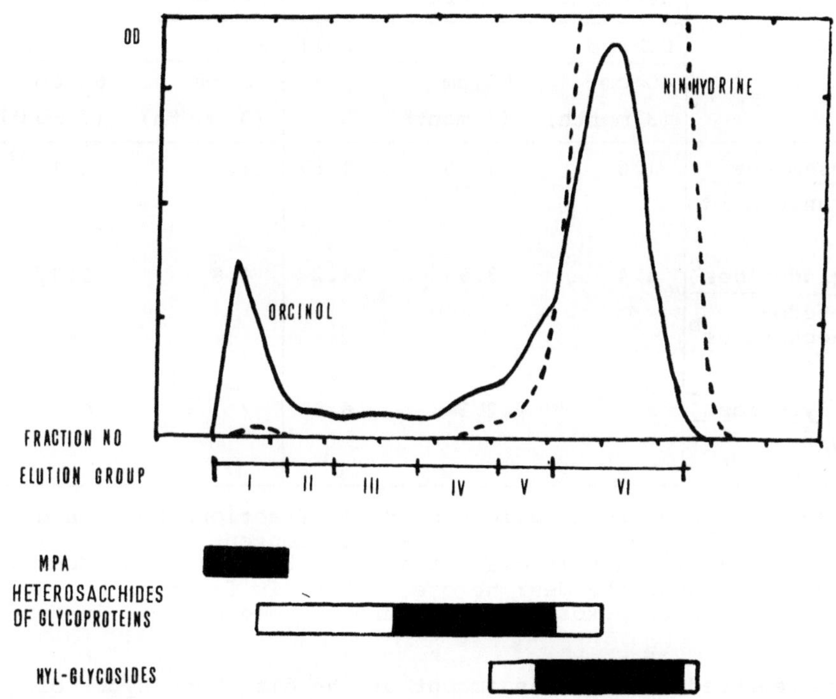

Fig. 1. Elution pattern on Sephadex G50 M column of the collagenase-pronase digest of the insoluble macromolecular complex of the endothelial basement membrane of calf cornea. ——— orcinol ----- trinitrobenzen sulfonic acid reaction. Column: 20 x 900 mm for 300 mg preparation. 3.5 ml fractions were collected. Eluant: acetic acid 0.1 M

(Moczar and Moczar, 1970c), gives a similar elution profile (Fig. 1).

The same type of elution curve was obtained from the embryonic and postnatal cornea (Moczar and Moczar, 1973). The results of the sugar analysis show that the orcinol positive fractions correspond to different types of carbohydrate chains linked to the protein of the corneal stroma. Table 2 shows that the first two peaks eluted near the void volume of the column contain the high molecular weight glycopeptides deriving mainly from the glycosaminoglycans (acid mucopolysaccharides). The fractions Nr. 4 and 5 correspond to the glycopeptides of structural glycoproteins. In the most retarded fraction (Nr. 6) the hydroxylysineglycosides can be identified (Moczar et al., 1970; Moczar and Moczar, 1970c; Moczar et al., 1968).

Table 2. Ratio of the amount of the different types of protein linked glycans in the polymeric collagen-proteoglycan-structural glycoprotein complex (insoluble in 1 M $CaCl_2$) of the developing Descemet's membrane and of stromal layers of the calf cornea (see Fig. 1)

	Descemet's membrane			Stroma		
	Embryo		Calf	Embryo		Calf
	40 cm (3 month)	65 cm (7 month)		40 cm (3 month)	65 cm (7 month)	
Hexosamines[a] / Uronic Acid	1.6	1.55	1.67	1.6	2.1	2.8
Hylglucosides[c] / Hetero-saccharides[b]	4.4	3.9	14.2	4.8	5.3	13.2
Hylglycosides[c] / MPA[d]	2	2.3	6.4	2	6	7.5

a Hexosamine and uronic acid content in fractions No. 1 and 2 (Fig. 1).
b Amount of heterosaccharides = mannose content in mg x 3 in the fraction No. 3, 4 and 5 (Moczar et al., 1968; Moczar and Moczar, 1973).
c Hydroxylysineglycosides: hexose content in fraction No. 6.
d MPA: galactose, hexosamine, and uronic acid content in fraction No. 1.

As we reported earlier the amount of the different types of glycans may be estimated from the sugar analysis and the yield of the individual glycopeptide fractions (Moczar and Moczar, 1970e) (see Table 2). Table 2 shows the ratio of the different glycans present in the insoluble macromolecular complex of embryonic and postnatal stroma and Descemet's membrane.

The increase of the hexosamine/uronic acid ratio in the high molecular weight peaks of the stroma with progressing age of the fetus is in contrast with the constancy of the corresponding values in the Descemet's membrane. This difference may reflect an increase of the keratan sulfate/chondroitinsulfate ratio in the polymeric stroma during the

Table 3. Spectrophotometric determination of the molar ratios of Hyl/HylGal/HylGalGlc (Hyl = 1) in the polymeric complex (insoluble in M $CaCl_2$) of the Descemet's membrane and the stromal layers of the embryonic and postnatal calf cornea (Moczar and Moczar, 1970f)

	Descemet's membrane				Stroma			
	Hyl	HylGal	HylGalGlc	Glycosylation[a]	Hyl	HylGal	HylGalGlc	Glycosylation[a]
Embryo 40 cm	1	0.22	0.83	51	1	0.36	0.71	51.5
65 cm	1	0.27	0.9	53.8	1	0.38	0.74	53
Calf	1	0.36	0.77	54.5	1	0.41	0.78	54.5

a $\frac{\text{glycosylated hydroxylysine}}{\text{total hydroxlysine}}$ residues × 100 embryonic development. The end of gestation is characterized by the relative accumulation of the hydroxylysineglycosides and by the decreasing heterosaccharide and glycosaminoglycan content of the basement membrane and of the stromal layers (Tables 2 and 3).

References

Dohlman, C., Balazs, E.A.: Chemical studies on Descemet's membrane of the bovine cornea. Arch. Biochem. 57, 445 - 457 (1955).

Kefalides, N.A., Denduchis, B.: Structural components of epithelial and endothelial basement membranes. Biochemistry 8, 4613 - 4621 (1969).

Moczar, E., Moczar, M.: Distribution of carbohydrates in the insoluble network of connective tissue. In: Chemistry and Molecular Biology of the Intercellular Matrix (Ed. E.A. Balazs), p. 243 - 250. New York, London: Academic Press 1970a.

Moczar, E., Moczar, M.: The glycopeptide pattern of the procine and rabbit corneal stroma. Comp. Biochem. Physiol. 36, 547 - 557 (1970c).

Moczar, E., Moczar, M.: Glycannes liés à la fraction insoluble de la cornee de roussette (Scyliorhinus canicula L). Europ. J. Biochem. 13, 28 - 36 (1970e).

Moczar, E., Moczar, M.: A micro-method for the determination of hydroxylysine and its glycosylated derivatives. J. Chromatog. 51, 277 - 282 (1970f).

Moczar, E., Moczar, M.: Protein linked carbohydrates in the developing corneal stroma of the cattle embryo. In: Protides of Biological Fluids (Ed. H. Peeters), Vol. 18, p. 121 - 125. Amsterdam: Elsevier 1971.

Moczar, M., Moczar, E.: Exp. Eye Res. 17, 5 - 17 (1973).

Moczar, M. Moczar, E., Robert, L.: Composition of glycopeptides obtained by proteolytic digestion of the media of porcine aorta. Atherosclerosis 12, 31 - 40 (1970).

Moczar, E., Robert, L., Moczar, M.: Etudes sur les fractions glucidiques de la trame fibreuse insoluble du stroma cornéen. Europ. J. Biochem. 6, 213 - 218 (1968).

Robert, L., Dische, Z.: Analysis of a sulfated sialofucoglucosaminogalactomannosidoglycane from corneal stroma. Biochem. biophys. Res. Commun. 10, 209 - 213 (1963).

Robert, L., Oudea, P., Zweibaum, A., Parlebas, J., Robert, B.: Immunochemistry of structural proteins and glycoproteins. In: Structure and Function of Connective and Skeletal Tissue, NATO Symposium, St. Andrews (Ed. R.G. Tristram), p. 406. London: Butterworth 1965.

Robert, L., Robert, B.: La biosynthèse des glycoprotéines du tissu conjonctif et sa régulation. In. Exp. Ann. Biochimie Méd., Vol. 30, p. 189 - 227. Paris: Masson 1970.

The Chemistry of Glycoproteins at the Surfaces of Tumor Cells

J. F. Codington and R. W. Jeanloz

Glycoproteins probably participate actively in most of the multiple and diverse functions attributed to the mammalian cell surface, such as active transport (Kraemer, 1971), contact inhibition (Abercrombie and Ambrose, 1962), histotypical recognition (Moscona, 1956), transplantation rejection (Sanderson et al., 1971), and antigen suppression (Apffel and Peters, 1970). Yet, there is little firm evidence implicating any particular glycoprotein of the cell surface in these activities, and, indeed, very little has yet been reported regarding the chemical structure or biological role of a cell surface glycoprotein (Kraemer, 1971). Since the cell surface is rich in carbohydrate material (Bennett, 1963), and since most of the glycoproteins of the cell appear in the surface region (Molnar, 1967), either as part of the external surface of the plasma or outer membrane or as macromolecules rooted in the membrane structure and extended outward from the cell, it would appear that progress in the isolation and characterization of these macromolecules would be rapid. Because of the extreme heterogeneity of surface glycoproteins, however, and because of the lack of technical means sufficiently sensitive to deal with the minute quantities of material generally available, this has not been so. Nevertheless, determinations of the chemical composition of glycoprotein material from the periphery of certain cells have been made, and some important information regarding chemical structures has been reported.

Three methods for the study of cell surface glycoproteins have been useful: (A) removal of glycoproteins or glycoprotein fragments from living cells by enzymes or other agents, and determination or fractionation of these materials from the supernatant solutions after spinning down the cells; (B) isolation of the outer cell membrane, followed by solubilization, fractionation, and determination of the materials; (C) use of immunological or immunologically-related methods, such as the use of antibodies or agglutinins to identify determinant structures on the cell surface. This short review will only consider results obtained by the use of Method (A). Although this approach is dependent upon the availability, in surface molecules, of linkages susceptible to cleavage by the agents used, the results obtained have been far more quantitative than those obtained to date by Method (B), since the isolation of membrane fragments generally results in poor yields of cell surface components (Wallach, 1967). Although the results obtained by Method (C) may be highly significant from a biological point of view (Inbar and Sachs, 1969), they have not yet yielded information of much value concerning chemical structures found in surface glycoproteins.

Neuraminidase removes essentially all surface sialic acid from most cells investigated (Kraemer, 1971). This carbohydrate component may be present as a terminal residue on either glycoprotein or glycolipid molecules at the cell surface. In the TA3 mammary adenocarcinoma ascites cell of the strain A mouse as much as 70% of the total sialic acid of the surface may be removed from living cells by proteolysis using TPCK-trypsin at 15 - 22°C (Codington et al., 1970). The proportion of this surface component removable by trypsin is considerably less for other cells studied. Shen and Ginsburg (1968) were able to remove 47 - 55% of surface sialic acid from HeLa cells with trypsin, but other investigators have been even less successful in removing surface sialic acid from tumor cells with this enzyme (Miller et al., 1963; Kraemer, 1966). Glycoproteins containing the highest proportion of sialic acid were cleaved initially by TPCK-trypsin from viable TA3 cells (Codington et al., 1970), as shown in Table 1. This result appears to contradict that of Kraemer (1966), who reported that the presence of sialic acid on cell surface glycoproteins of certain tumor cells inhibited cleavage by trypsin. These seemingly contradictory results may actually reflect a marked dissimilarity in the structure or conformation of the glycoproteins at the surfaces of different types of cells.

Table 1. Material isolated after successive incubations of TA3 cells with TPCK-trypsin (18 μg/ml in balanced salt solution) at 4°C

Incubation	No. of Cells	% Viable cells	Incub. period (min)	Sialic acid (μg)	Protein (mg)
I	1.8×10^9	100	20	150	2.44
II	1.6×10^9	99	20	69	2.34
III	1.7×10^9	99	20	37	2.34
IV	1.8×10^9	98	20	40	2.34
V	1.7×10^9	99	20	9	2.52

It is not yet clear whether any general relationship exists between malignancy and the amount, type, or linkages of the sialic acid at the surfaces of cells, although in certain studies the involvement of sialic acid in the malignant process has been demonstrated (Sanford and Codington, 1971; Simmons et al., 1971). In any case, it appears that no generalization can be made regarding the relative concentration of sialic acid at the surfaces of tumor or transformed cells, as compared to normal cells (Ohta et al., 1968). Although the amount of this component may vary several fold from one type of cell to another, the concentration found in certain tumor cells, namely, the hamster P-183 cell (Kraemer, 1966) and the strain A mouse TA3-Ha cell (Codington et al., 1970), 16×10^5 molecules per μ^2, exceeds that reported for normal cells.

Trypsin removed 28 - 64% of the total cellular carbohydrate components from the surface of the HeLa cell (Shen and Ginsburg, 1968), a result which suggests that a large proportion of the glycoprotein of this cell is found at the surface. Present were galactose, mannose, fucose, *N*-acetylgalactosamine, *N*-acetylneuraminic acid. Approximately the same relative amount of carbohydrate material (about 1% of the dry weight of the cells), but of different composition, was cleaved from viable TA3 ascites cells by TPCK-trypsin (Codington et al., 1970). It is noteworthy that no fucose, nor uronic acid, nor sulfate was detected in the material removed from TA3 cells, and the proportion of mannose was very small. In addition to aspartic acid, serine, and threonine, relatively large proportions of glutamic acid, glycine, alanine, and proline were removed from the TA3 cells.

Although several attempts to fractionate glycoproteins obtained from tumor cell surfaces have been reported (Langley and Ambrose, 1964; Walborg et al., 1969), only glycoproteins from Meth-A cells with H-2 histocompatibility activity have been isolated in relatively pure form (Muramatsu and Nathenson, 1970). It was surprising, therefore, that by limited proteolysis of living TA3 cells with TPCK-trypsin, followed by passage of the supernatant solution through columns of polyacrylamide gels, it was possible to obtain a fraction, Glycoprotein Fraction I, which was shown to be a mixture of only two structurally-related components, present in approximately equal amounts. After labelling Glycoprotein Fraction I with tritium (Lenten and Ashwell, 1971), these two components were partially separated on a column of Bio-Gel A-5M, using 6*M* guanidine hydrochloride as an eluent, as shown in Fig. 1. Sedimentation equilibrium studies on Glycoprotein Fraction I confirmed the presence of two components, or two families of closely related components, of molecular weights 138,000 and 308,000 daltons (Codington et al., 1972).

Glycoprotein Fraction I was found to be composed of approximately 70% carbohydrate and 30% protein. Its carbohydrate and amino acid composition is given in Table 2, as residues per molecule of the larger component of 308,000 daltons. Carbohydrate residues are present in the approximate molar proportions of galactose, 4, *N*-acetylgalactosamine, 2, *N*-acetylglucosamine, 1, and *N*-acetylneuraminic acid, 1 (Codington et al., 1971). Although Glycoprotein Fraction I is not homogeneous alkaline borohydride reduction established that all carbohydrate chains are attached by *N*-acetylgalactosamine to serine or threonine

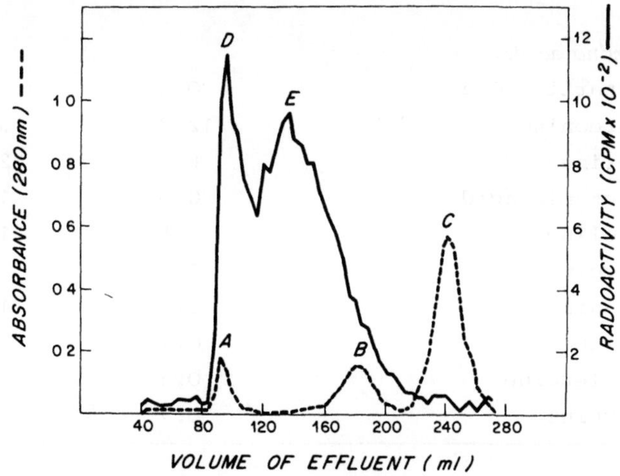

Fig. 1. Fractionation at 4°C of ^{3}H-labelled Glycoprotein Fraction I from a Bio-Gel P-100 column and proteins of known molecular weight on a Bio-Gel A-5M column (1.9 x 90 cm). Eluent, guanidine hydrochloride, 6M, and cysteine, 1m*M*. Broken line, absorbance at 280 nm; solid line, radioactivity in counts per minute. Proteins: A, myosin; B, bovine serum albumin; C, lysozyme; D and E., Glycoprotein Fraction I

residues of the protein core, suggesting the presence of at least two types of carbohydrate chains, since only half as many moles of each of two other components, *N*-acetylglucosamine and *N*-acetylneuraminic acid, are present. One terminal residue was found to be *N*-acetylneuraminic acid, which may be removed by neuraminidase, exposing a galactose residue, now made susceptible to periodate oxidation. A second terminal component is a β-linked galactose, which may be removed by the action of a β-galactosidase either before or after neuraminidase treatment.

The yield of Glycoprotein Fraction I, approximately 1.05 mg from 10^9 cells, or 0.5% of the weight of the lyophilized cells, establishes these two glycoproteins as major constitutents of the TA3 cell surface The size of the native molecules on the cell surface remains unknown. Based upon measurements from electron micrographs suggesting that the TA3 cell surface is approximately 2.5 times that of a perfect sphere (Codington et al., 1970), and upon the assumption that the isolated glycoprotein components of Glycoprotein Fraction I represent

Table 2. Composition of Glycoprotein Fraction I isolated from the TA3 mammary adenocarcinoma cell

Component carbohydrate	Percentage (weight)	Residues per molecule (308,000 d.)
Galactose	27	505
N-Acetylgalactosamine	19	295
N-Acetylglucosamine	9	133
N-Acetylneuraminic acid	13	145
Amino acids:		
Aspartic acid	0.3	9
Threonine	12.3	379
Serine	8.2	290
Glutamic acid	0.5	123
Proline	4.8	117
Glycine	1.6	86
Alanine	2.5	108
Valine	0.2	6
Isoleucine	0.1	3
Leucine	0.7	19

the approximate sizes of the native glycoproteins, each molecule on the cell surface would be separated from its neighbors by an average of only 220 Å. If, as expected, these molecules are in an extended configuration (Winzler, 1970), the larger one would extend outward from the cell membrane more than 3000 Å, and the smaller one more than 1300 Å. These lengths represent minimal figures, and the native molecules may be considerably larger in size, which would mean that their spacing on the cell surface would be correspondingly greater.

Acknowledgements: This is publication No. 559 of the Robert W. Lovett Memorial Group for the study of Diseases causing deformities, Harvard Medical School and Massachusetts General Hospital. This investigation was supported by Research Grant Ca-08418 from the National Cancer Institute, National Institutes of Health, U.S. Public Health Service.

References

Abercrombie, M., Ambrose, E.J.: The surface properties of cancer cells. A review. Cancer Res. 22, 525 (1962).

Apffel, C.A., Peters, J.H.: Regulation of antigenic expression. J. theor. Biol. 26, 47 (1970).

Bennett, H.S.: Morphological aspects of extracellular polysaccharides. J. Histochem. Cytochem. 11, 14 (1963).

Codington, J.F., Sanford, B.H., Jeanloz, R.W.: Glycoprotein coat of the TA3 cell. I. Removal of carbohydrate and protein material from viable cells. J. nat. Cancer Inst. 45, 637 (1970).

Codington, J.F., Sanford, B.H., Jeanloz, R.W.: Sialoglycopeptide fraction from the TA3 cell surface. Fed. Proc. 30, 1280 Abs (1971).

Codington, J.F., Sanford, B.H., Jeanloz, R.W.: Structural studies on a glycoprotein of the TA3 cell surface. Fed. Proc. 31, 465 Abs (1972).

Inbar, M., Sachs, L.: Structural difference in sites on the surface membrane of normal and transformed cells. Nature 223, 710 (1969).

Kraemer, P.M.: Sialic acid of mammalian cell lines. J. Cell Physiol. 67, 23 (1966).

Kraemer, P.: Complex carbohydrates of the cell surface. Biomembranes I (L.A. Manson, Ed.), p. 67. New York: Plenum Publishing 1971.

Langley, O.K., Ambrose, E.J.: Isolation of a mucopeptide from the surface of Ehrlich ascites tumour cells. Nature (Lond.) 204, 54 (1964).

Lenten, L.V., Ashwell, G.: Studies on the chemical and enzymatic modification of glycoproteins. J. biol. Chem. 246, 1889 (1971).

Miller, A., Sullivan, J.F., Katz, J.H.: Sialic acid content of the erythrocyte and of an ascites tumor cell of the mouse. Cancer Res. 23, 485 (1963).

Molnar, J.: Glycoproteins of Ehrlich ascites carcinoma cells. Incorporation of (^{14}C) glucosamine and (^{14}C) sialic acid into membrane proteins. Biochemistry 6, 3064 (1967).

Moscona, A.: Development of heterotypic combinations of dissociated chick cells. Proc. Soc. exp. Biol. Med. (N.Y.) 92, 10 (1956).

Muramatsu, T., Nathenson, S.G.: Comparison of glycopeptides from membrane located histocompatability-2 (H-2) alloantigens from different types of tumor cells. Fed. Proc. 29, 910 Abs (1970).

Ohta, N., Pardee, A.B., McAuslan, B.R., Burger, M.M.: Sialic acid contents and controls of normal and malignant cells. Biochim. biophys. Acta (Amst.) 158, 98 (1968).

Sanderson, A.R., Cresswell, P., Welsh, K.I.: Involvements of carbohydrates in the immunochemical determinant area of HL-A substances. Nature New Biology 230, 8 (1971).

Sanford, B.H., Codington, J.F.: Further studies on the effect of neuraminidase on tumor cell transplantability. Tissue Antigens 1, 153 (1971).

Shen, L., Ginsburg, V.: Release of sugars from HeLa cells by trypsin. In: Biological Properties of the Mammalian Surface Membrane. (L.A. Manson, Ed.), p. 67. Philadelphia: Wistar Inst. Press 1968.

Simmons, R.L., Rios, A., Lundgren, G., Ray, P.K., McKhann, C.F., Haywood, G.R.: Immunospecific regression of methylcholanthrene fibrosarcoma with the use of neuraminidase. Surgery 70, 38 (1971).
Walborg, E.F., jr., Lantz, R.S., Wray, V.P.: Isolation and chemical characterization of a cell-surface sialoglycopeptide fraction from Novikoff ascites cells. Cancer Res. 29, 2034 (1969).
Wallach, D.F.H.: Isolation of plasma membranes of animal cells. In: The Specificity of Cell Surfaces (B.D. Davis, L. Warren, Ed.), p. 129. Englewood Cliffs: Prentice-Hall 1967.
Winzler, R.J.: Glycoprotein antigens isolated from RBC. In: Blood and Tissue Antigens (D. Aminoff, Ed.), p. 117. New York: Academic Press 1970.

The Relationship between Chemical Structure and Fibrogenic Character of Glycosaminoglycans

M. Németh-Csóka

It has been established that chondroitin sulfate influences the growth of tissue culture and that Carrageenin type polysaccharides influence that of connective tissue (McCandless, 1965). In a working hypothesis McCandless postulated the structural characteristics of polysaccharides as polygalactans structure with no special configuration or degree of sulfation. On the other hand, it was proved by Meyer (1960) that the anionic groups of polysaccharides determine their influence on extracellular fibril formation. Therefore it seems to be important to deal with the significance of the chemical structure of polysaccharides of the polygalactic type.

As a method of investigation the in vitro gelation of neutral soluble collagen was chosen according to Bensusan and Hoyt (1958) as well as in our 1967 study. Investigated was the manner in which the rate of gelation (fibril formation) and the activation energy of collagen were influenced by the following polyanions: hyaluronic acid (HA), poly-galacturic acid (Po-Ga), pectic acid (Pe-Ac), chondroitin sulfate A (ChS-A) and -C (ChS-C), protein polysaccharide complex light (PPL), k-Carrageenin (k-Ca), l-Carrageenin (l-Ca), dextran sulfate (De-Su) and Heparin (He). On the basis of their effect polyanions could be divided into three subgroups: 1. ChS-A, ChS-C, PPL; 2. HA, k-Ca, Po-Ga, Pe-Ac; 3. l-Ca, He, De-S (Fig. 1). The first enhanced

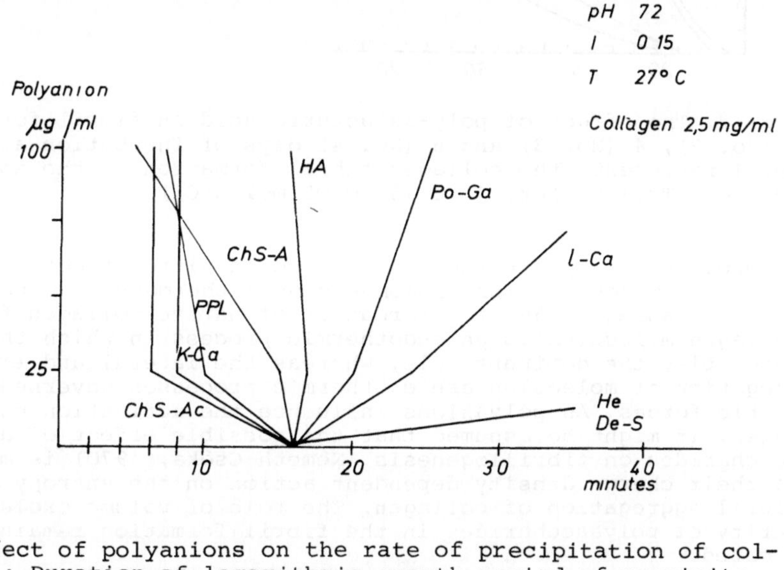

Fig. 1. The effect of polyanions on the rate of precipitation of collagen. Abscissa: Duration of logarithmic growth period of precipitation in min. Ordinate: Concentration of polyanions in µg/ml

the second did not influence, and the third delayed the rate of gelation. At the same time the activation energy (20-25 kCal)mol was decreased, remained unchanged, and increased, respectively. The chondroitin sulfates containing 4 - 5% sulfate-sulphur showed the greatest enhancing effect on fibril formation in vitro, whereas the heparin and dextran sulfate of 11 - 12 sulfate-sulphur content had a definite delaying effect. These results seemed to prove that the degree of sulfation determines the effect of polyanions.

Polygalactans free of sulfate like Po-Ga and Pe-Ac did not influence the in vitro fibril formation, but their character changed after preincubation at 0°C. After 5 - 6 days of incubation polygalactans delayed the fibril formation similarly to heparin (Fig. 2). As the three main subfractions of polyanions of 200,000, 85,000 and 15,000 molecular weight (fractions obtained on Sephadex G-200) seemed to change during incubation, namely the 80,000 mw,s subfraction increased, the cause of the changed character of polygalactans might be connected with the change of the composition of the subfractions. The fibrillogenic effect of polysaccharides free of sulfate can be explained by their volume exclusion effect (described by Laurent, 1963), which may influence the hydrophobic bond formation between tropocollagen molecules during fibril formation.

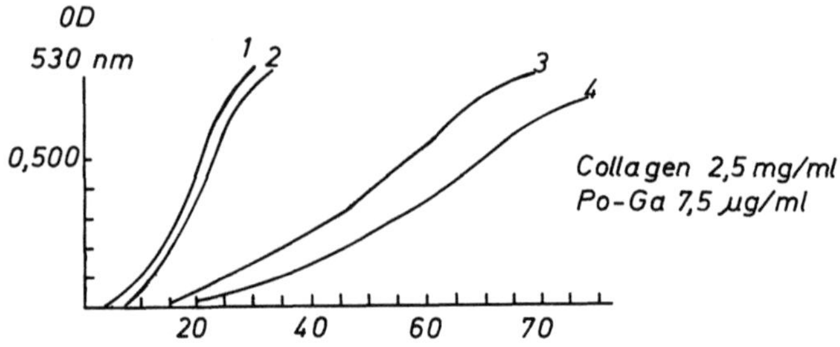

Fig. 2. The effect of poly-galacturic acid on fibril formation after 1 (No. 2), 4 (No. 3) and 6 (No. 4) days of incubation at 0°C. Curve No. 1 represents the collagen fibril formation in the solvent buffer (0.1 mol Tris buffer, 1 0.15, 0.05 mol NaCl)

We propose to explain the in vivo and in vitro effect of acid mucopolysaccharides on fibrillogenesis on a thermodynamic basis. According to Cassel (1966) the formation of native collagen from tropocollagen molecules is an endothermic process in which the hydrophobic bonds play the dominant role, whereas the lateral and end-to-end aggregation of molecules are exothermic processes governed by electrostatic forces. As polyanions influence the activation energy of collagen, it might be assumed that the possible effect of acid mucopolysaccharides on fibrillogenesis (Németh-Csóka, 1970) is mainly due to their charge density dependent action on the entropy change in the fibril aggregation of collagen. The role of volume exclusion peculiarity of polysaccharides in the fibril formation remains to be elucidated.

References

Bensusan, H.B., Hoyt, Barbara L.: The effect of various parameters on the rate of formation of fibers from collagen solutions. J. Amer. chem. Soc. 80, 719 - 724 (1958).
Cassel, J.M.: Collagen aggregation phenomena. Biopolymers 4, 989 - 997 (1966).
Laurent, T.C.: The interaction between polysaccharides and other macromolecules. The stability of protein in the presence of dextran. Biochem. J. 99, 253 (1963).
McCandless, E.L.: Chemical structural requirements for stimulation of connective tissue growth by polysaccharides. Ann. N.Y. Acad. Sci. 118, 867 - 882 (1965).
Meyer, K.: Nature and function of mucopolysaccharides of connective tissue. In: Molecular Biology (Nachmansohn, D., Ed.), p. 69. New York: Academic Press 1960.
Németh-Csóka, M.: The effect of acid mucopolysaccharides on the activation energy of collagen. Biochem. J. 104, 17 (1967).
Németh-Csóka, M.: Importance of sulphated acid mucopolysaccharides for fibrillogenesis of carrageenin granulomata of rats of different ages. Exp. Geront. 5, 67 - 75 (1970).

Periodate Sensitivity of Cross-Links of Mature Collagen

H. Hörmann

After the recent development concerning the structure of covalent cross-links in collagen, it appears desirable to mention former experiments on treatment of mature collagen with periodate or with hydroxylamine. From those, earlier conclusions on the chemistry of the cross-links had been drawn (Hörmann, 1960). The former results may now be discussed in the light of the new findings.

As early as 1955, Grassmann and Kühn had found that mature collagen from bovine hide slowly dissolves in a weakly alcaline solution of sodium periodate at 40°C. Reinvestigation by Hörmann and Klenk (1960) showed that the same collagen was dissolved at room temperature by a solution containing periodate, 10% acetic acid and 8 M urea. When urea was lacking, collagen was not dissolved at room temperature unless it was heat-shrunk.

At that time dissolution was already explained by the assumption that mature collagen may contain covalent cross-links which are sensitive to periodate (Hörmann, 1960).

This hypothesis primarily was founded on the following experimental facts: The dissolved material was non-dialysable. Amino acid analysis did not show considerable alterations besides lack of tyrosine, methionine and hydroxylysine which residues might be modified at their side chains. In addition, the hexoses were destroyed completely. Therefore, it was suggested that peptide bonds of collagen are not cleaved by periodate.

Additional experiments confirmed this conclusion. During dissolution by periodate a negligible increase of α-amino groups was observed (von Wilm, 1965). Furthermore, soluble collagen was treated during five days with periodate in 8 M urea containing 10% acetic acid (Hörmann, unpublished). After removal of the oxidating agent with bisulfite the material was dialysed against 0.075 M citrate buffer, pH 3.7 and analyzed in the ultracentrifuge. It showed two components with molecular weights of 2,000,000 and 474,000 (s_0 4.88 and 7.22 respectively). These weights represented multiples of the peptide chains of collagen e.g. dimers and polymers of α-chains. Neither free α-chains nor fragments of them were found which gave evidence that peptide chains were not degraded by periodate. Aggregation of α-chains was explained by the reaction of aldehyde structures formed during periodate oxidation.

As dissolution of collagen was explained by the presence of periodate sensitive cross-links, it was assumed originally that hexoses might be involved in cross-linking (Hörmann, 1960). Considerably later Bailey et al. (1969) showed that the acid resistant cross-links of collagen represented aldol-like structures. They originate from condensations of aldehydes which had been formed by modification of lysine and hydroxylysine side chains. The following structures are discussed:

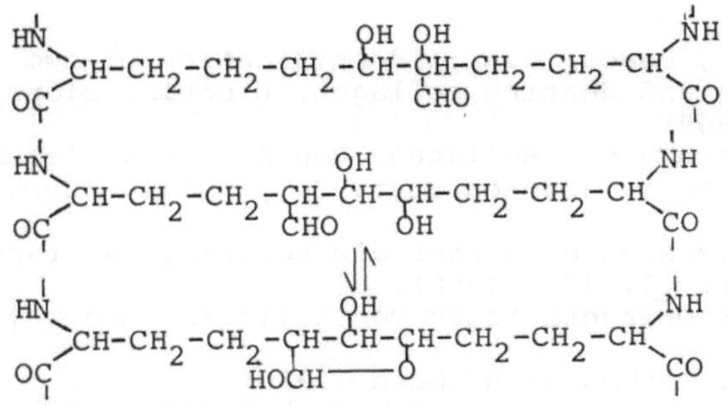

(I)

(IIa)

(IIb)

The formulas I and IIa contain adjacent hydroxy groups. Thus, the early hypothesis of periodate sensitive cross-links in mature collagen is confirmed.

Of special interest is structure IIa. It offers an explanation for the fact that mature collagen is dissolved very slowly by periodate/urea. Structure IIa may be able to form a five-membered hemiacetale ring (IIb). In this cyclic form a diol-structure is not present any more. Consequently, periodate sensitivity may be lost. The cross-link may only be cleaved, if the cyclic hemiacetale transforms to the open-chain aldehyde form. From analogy to carbohydrate chemistry we may expect that this reaction proceeds slowly.

Another reaction which was discussed concerning the structure of covalent cross-links was the dissolution of mature collagen by 0.72 M hydroxylamine at 20°C and pH 9.5 in presence of 2 M KCNS or 4 M LiCl (Hörmann, 1960, 1962). The collagen dissolved was non-dialysable. It showed an increased amount of protein-bound hydroxamic acids compared to soluble collagen treated by hydroxylamine under identical conditions. Therefore, it was concluded that mature collagen may contain additional hydroxylamine sensitive bonds and it was suggested that those bonds were located in the cross-links. At that time ester bonds between ω-carboxyl groups of aminodicarbonic acids and hydroxy groups of hexoses were discussed as involved in cross-linking.

However, recent comparison of the reaction products of soluble and of cross-linked collagen treated with hydroxylamine in presence of 4 M LiCl showed that under the same conditions the peptide chains of cross-linked collagen are attacked more severely and less specifically. The reaction product of soluble collagen, when analyzed by electrophoresis in acryl-amide gel at pH 4.8, resulted in several well-separated bands. Hydroxylamine dissolved mature collagen, on the other hand, revealed only diffuse bands with a considerable background of randomly cleaved material (Fig. 1). Consequently, the additional protein-bound hydroxamic acids present in hydroxylamine dissolved mature collagen may be due to a less specific cleavage of peptide chains in the cross-linked material swollen by 4 M LiCl.

Fig. 1. Acrylamide gel electrophoresis of the reaction products of soluble (a) and of mature collagen (b) obtained after 14 h. treatment with 0.72 M hydroxylamine in 4 M LiCl at pH 9.5 and at 37°C

a a

b b

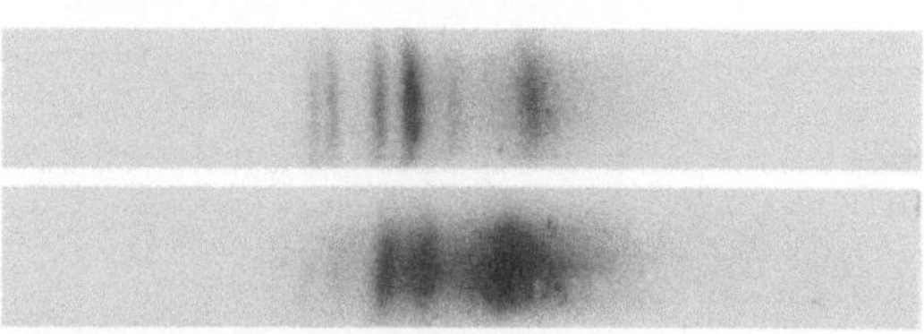

References

Bailey, A.J., Fowler, L.J., Peach, C.M.: Identification of two interchain crosslinks of bone and dentine collagen. Biochim. biophys. Res. Commun. 35, 663 (1969).

Grassmann, W., Kühn, K.: Abbau des Kollagens und des Prokollagens mit Natriumperjodat and Phenyljodosoacetat. Z. physiol. Chem. 301, 1 (1955).

Hörmann, H.: Chemische Untersuchungen über die Kohlenhydratgruppierung des Kollagens. Leder 11, 173 (1960).

Hörmann, H.: Zur Frage der Quervernetzung von Kollagen. Leder 13, 79 (1962).

Hörmann, H., Klenk, M.: Published by Hörmann (1960).

von Wilm, M.: Chemische Veränderungen an den funktionellen Gruppen des Kollagens. Doctor Thesis, Univ. Munich. 1965.

Variations in the Diameter of Collagen Fibrils, Bound Hexose and Associated Glycoproteins in the Intervertebral Disc.

F. Happey, A. Naylor, J. Palframan, C. H. Pearson, R. M. Render,

and R. L. Turner

In attempting to explain the different physical characteristics of collagen in various tissues, a number of possibilities have been proposed. These include variations in glycosaminoglycans, more firmly associated "structural glycoproteins" (Robert and Robert, 1967) and collagen hexose contents (Spiro, 1969; Grant et al., 1969). The location of the hexose on specific hydroxylysine residues may hinder the formation of intermolecular cross-links (Morgan et al.. 1970), possibly imposing a limit on fibril diameter. Analyses of different layers of bovine skin indicate that coarser fibres are characterised by an increased dermatan sulfate: hyaluronic acid ratio (Hoffman et al., 1957). This arises mainly from a deficiency in hyaluronic acid because, contrary to previous statements, the dermatan sulfate remains approximately constant (Pearson, 1970). Keratan sulfate may be implicated in the formation of very thin fibrils found in more transparent tissues like cornea and nucleus pulposus.

It is pertinent to ask whether any of the above factors are involved in the physical changes in collagen which occur on maturation and aging. The data presented here refer to two distinct regions of human lumbar intervertebral discs, the nucleus pulposus (N.P.) and annulus fibrosus (A.F.). Glycosaminoglycans were removed by repeated high speed homogenization in saline. Fibril diameter measurements were carried out on the insoluble residue, as part of a general electron microscopic investigation (Render, 1971). The bulk of the residue was digested with papain. Gel filtration separated the collagen glycopeptides (peak 2) from higher molecular weight material (peak 1), which was derived from non-collagenous substances apparently closely associated with the collagen. Both peaks were measured by neutral sugar determinations (Pearson et al., 1971).

A positively skewed distribution of diameters was found at all ages. Slight indications of bimodal distribution were observed in only two specimens, one of which was the 31 - 40 year-old pooled A.F. referred to below and the other was also A.F. (aged 7 years). Dahmen (1963) found two modes in mature N.P and A.F. In Fig. 1 it can be seen that the average diameter in the N.P. increased significantly after 20 years of age, although the modal values varied rather erratically.

The fibrils in the A.F. were consistently thicker than in the N.P., but surprisingly the average diameters decreased between 10 and 40 years of age. The major mode of the 31 to 40-year old A.F. had the lowest value (260 - 320 Å) found at any age in the A.F. and also the proportion of the thickest fibrils (700 Å) was the lowest.

A reduction in the strength of the A.F. in the mature tissue might therefore be expected. It should be mentioned that the present specimens of A.F. were drawn almost entirely from the anterior region. Measurements in the lateral posterior A.F., which is generally weaker

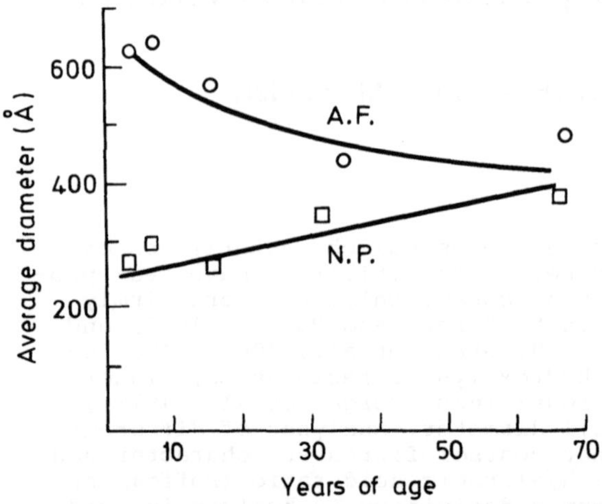

Fig. 1. Age variations in fibril diameter. 12 selected areas of each homogenate (phosphotungstic acid stained) were photographed at an indicated magnification of 15,000 or 20,000 times. 200 diameter measurements were made on enlarged prints. The difference between the average diameters of the youngest and oldest specimens was significant at the 0.1% level in both the N.P. and the A.F.

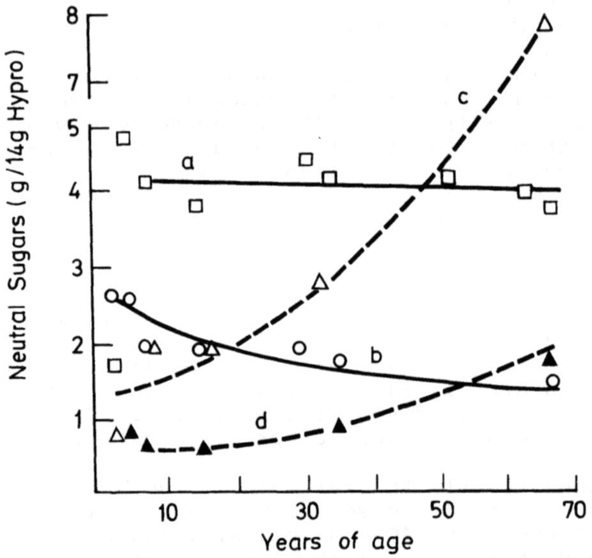

Fig. 2. Age variations in peak 1 and peak 2 (collagen hexose). a peak 2 N.P.; b peak 2 A.F.; c peak 1 N.P.; d peak 1 A.F.

and more liable to clefting (Ritchie and Fahrni, 1970) should be of interest.

The much higher hexose levels of the collagen in the N.P. than in the A.F., (Fig. 2), supports the idea of an inverse relationship with fibril diameter (Grant et al., 1969).

However, the increase in thickness in the N.P. on aging was not similarly explained, because the collagen hexose remained approximately constant, at least after the first decade. In the A.F. there was a noticeable decrease in collagen hexose, suggesting that a higher degree of cross-linking was permitted on aging, but nevertheless the fibrils tended to become thinner.

The associated non-collagenous material from which peak 1 is derived, contains a high proportion of glycoproteins as distinct from glycosaminoglycans, but keratan sulfate is also probably present (Pearson et al., 1971). We have so far failed to find evidence of dermatan sulfate in such fractions, including those from old normal specimens and prolapsed N.P., in which dermatan sulfate was expected from previous reports (Davidson and Woodhall, 1959; Lyons et al., 1964).

Peak 1 rises dramatically in the N.P. after the fourth decade and an increase also occurs in the A.F., although the general level is much lower (Fig. 2). The increase in fibril diameter in the oldest N.P. may be related to an accumulation of peak 1 material. Digestion with trypsin removes most of these non-collagenous components, whilst dissolving very little collagen, but preliminary results suggest that the average diameter of the fibrils is appreciably lowered (Render, 1971). This may indicate the removal of compounds very intimately associated with the fibrils, which were however, striated and generally normal in appearance. Alternatively, it is possible that digestion revealed a number of thin fibrils previously masked by amorphous non-collagenous material, as has been observed in skin (Sweeny et al., 1963). The larger average diameter recorded for the oldest N.P. would then be more apparent than real. Against this idea is the fact that a plot of the proportion of fibrils in the N.P. with the largest diameters (400 Å) corresponded closely with that obtained for the average diameter shown in Fig. 1.

It may yet emerge that age variations in specific hydroxylysine-linked hexose, rather than the total values considered here, are related to physical changes. More fundamental changes in the type of collagen are also feasible (Miller, 1971; Trelstad et al., 1970). However, at present a working hypothesis is that collagen hexose contents are important in determining gross physical modifications in different tissues and locations within a tissue, whereas age changes are more likely to arise from the association with non-collagenous components.

Acknowledgements: Financial assistance was received from the Medical Research Council, the Leeds Regional Hospital Board and the Laming Evans Fund of the Royal College of Surgeons of England.

References

Dahmen, G.: Submikroskopische Untersuchungen an Wirbelbandscheiben. Z. Rheumaforschung 22, 192 - 213 (1963).

Davidson, E.A., Woodhall, B.: Biochemical alterations in herniated intervertebral discs. J. biol. Chem. 234, 2951 - 2954 (1959).

Grant, M.E., Freeman, I.L., Schofield, J.D., Jackson, D.S.: Variations in the carbohydrate content of human and bovine polymeric collagens from various tissues. Biochem. biophys. Acta (Amst.) 177, 682 - 685 (1969).

Hoffman, P., Linker, A., Meyer, K.: The acid mucopolysaccharides of connective tissues II. Further experiments on chondroitin sulfate. Arch. Biochem. 69, 435 - 440 (1957).

Lyons, H., Jones, E., Quiin, F.E., Sprunt, D.H.: Protein-polysaccharide complexes of normal and herniated human intervertebral discs. Proc. Soc. exp. Biol. (N.Y.) 115, 610 - 614 (1964).

Miller, E.J.: Isolation and characterization of a collagen from chick cartilage containing three identical chains. Biochemistry 10, 1652 - 1659 (1971).

Morgan, P.H., Jacobs, H.G., Segrest, J.P., Cunningham, L.W.: A comparative study of glycopeptides drived from selected vertebrate collagens. A possible role of the carbohydrate in fibril formation. J. biol. Chem. 245, 5042 - 5048 (1970).

Pearson, C.H.: A study of the estimation and distribution of mucopolysaccharides in bovine skin. Thesis, University of Leeds 1970.

Pearson, C.H., Happey, F., Naylor, A., Turner, R.L., Palframan, J., Shentall, R.D.: Collagens and associated glycoproteins in the human intervertebral disc. Ann. rheum. Dis. In Press 1971.

Render, R.M.: An investigation using the electron microscope into the structure of the human intervertebral disc. Thesis, University of Bradford 1971.

Ritchie, J.H., Fahrni, W.H.: Age changes in lumbar intervertebral discs. Canad. J. Surg. 13, 65 - 71 (1970).

Robert, L., Robert, B.: Structural glycoproteins of membranes and connective tissue: biochemical and immunopathological properties. In: Protides of the Biol. Fluids, Vol. 15, p. 143 - 148. Amsterdam: Elsevier 1967.

Spiro, R.G.: Characterization and quantitative determination of the hydroxylysine-linked carbohydrate units of several collagens. J. biol. Chem. 244, 602 - 612 (1969).

Sweeny, P.R., Pearce, R.H., Vance, H.G.: The chemical anatomy of rat skin. Canad. J. Biochem. 41, 2307 - 2316 (1963).

Trelstad, R.L., Kang, A.H., Igarashi, S., Gross, J.: Isolation of two distinct collagens from chick cartilage. Biochemistry 9, 4993 - 4998 (1970).

II. Synthesis

Biosynthesis of Connective Tissue Proteoglycans

L. Rodén and N. B. Schwartz

Current investigations of the biosynthesis of connective tissue proteoglycans encompass a number of diverse aspects of this area, some of which are common to the study of biosynthetic processes in general, while others represent problems which are unique to the connective tissues. The scope of the field ranges from studies of individual enzyme reactions in cell-free systems to investigations of proteoglycan metabolism in the intact tissue or the whole animal under the influence of various metabolic regulators. In the limited space of this presentation, it is clearly impossible to cover all of these areas even in a superficial fashion. Emphasis will therefore be placed primarily on some recent developments in the study of individual enzymes participating in polysaccharide biosynthesis, especially the chain-initiating xylosyltransferase and other glycosyltransferases involved in the early stages of polysaccharide chain formation.

A vast fund of information has accumulated in the recent past concerning the biosynthesis of glycoproteins (for reviews, see Schachter and Rodén, 1972), and much of the work in this field has been paralleled by similar studies on proteoglycans (Rodén, 1970). It is not surprising, in view of the many similarities between glycoproteins and proteoglycans, that their biosynthetic pathways have evolved along similar patterns and that their metabolism is regulated by identical or analogous control mechanisms. The general mode of synthesis which is now recognized for both types of glycoconjugates involves formation of the protein core, followed by addition of the oligosaccharide or polysaccharide components. It has been assumed in the past that this process occurs by stepwise glycosyl transfer from nucleotide sugars, starting with the addition of a monosaccharide unit to a specific amino acid residue in the protein. Continued chain growth would then take place by a series of glycosyl transfer steps in which the sugars are transferred to the nonreducing terminal monosaccharide unit of the carbohydrate prosthetic group. Although it appears that the polysaccharide chains of the sulfated proteoglycans are formed in this manner, an alternative pathway involving lipid intermediates has recently been postulated for the formation of the many glycoproteins which contain an N-acetylglucosamine-asparagine linkage. If the first mode of chain initiation were operative, it should be possible to demonstrate transfer of N-acetylglucosamine from UDP-N-acetylglucosamine to asparagine residues of endogenous acceptors, in analogy with the finding that xylose is transferred from UDP-xylose to serine residues of the core protein of cartilage proteoglycan (see below). However, despite repeated attempts in several laboratories such a reaction has never been detected. Whereas it could be hypothesized that the pool of endogenous acceptor might be too small to permit significant transfer, the negative results could equally well be regarded as an indication that chain initiation occurs by an altogether different mechanism. Indeed, this possibility has received experimental support by the work of Behrens et al. (1972, 1973) and Heath et al. (1973), who have suggested that the synthesis of glycoproteins with a core region containing N-acetylglucosamine and mannose proceeds via lipid intermediates by the following reaction sequence:

$$\text{UDP-GlcNAc} + \text{DMP} \longrightarrow \text{DDP-GlcNAc} + \text{UMP}$$

$$\text{GDP-Man} + \text{DMP} \longrightarrow \text{DMP-Man} + \text{GDP}$$

$$\text{UDP-GlcNAc} + \text{DDP-GlcNAc} \longrightarrow \text{DDP-GlcNAc}_2 + \text{UDP}$$

$$\text{DDP-GlcNAc}_2 + \left\{\begin{array}{l}\text{DMP-Man}_x\\ \text{GDP-Man}_x\end{array}\right. \longrightarrow \text{DDP-GlcNAc}_2\text{Man}_x$$

$$\text{DDP-GlcNAc}_2\ \text{Man}_x + \text{protein} \longrightarrow \text{protein-GlcNAc}_2\ \text{Man}_x + \text{DDP}$$

(DMP, dolichol monophosphate; DDP, dolichol diphosphate)

It is evident from this reaction sequence that the carbohydrate-protein linkage would be formed by transfer of the entire core oligosaccharide to the protein moiety, and a separate transfer of N-acetylglucosamine to an asparagine residue would never occur. Whereas much evidence has already accumulated in favor of this mechanism of chain initiation, it should be emphasized that carbohydrate-protein linkages other than the N-acetylglucosamine-asparagine bond are not necessarily formed by the same route.

1. Subcellular Localization of Glycoprotein and Proteoglycan Biosynthesis

Most of our knowledge regarding the subcellular sites of glycoprotein biosynthesis comes from studies of liver cells which are amenable to fractionation of their different membrane structures to such an extent that the experimental data can be interpreted in a meaningful way. It is now generally accepted that the polypeptide backbones of liver glycoproteins are assembled on membrane-bound ribosomes, but there is still some controversy as to whether carbohydrate chain initiation occurs before the nascent peptide is released from the ribosomes. In any event, there is often a clear correlation between the position of a particular sugar residue in the finished molecule and the subcellular localization of the glycosyltransferase catalyzing its transfer to the growing chain. Thus, the internal core region, which in many glycoproteins consists of N-acetylglucosamine and mannose, is added at an early stage as the growing molecule traverses the channels of the endoplasmic reticulum from rough-surfaced to smooth-surfaced regions. In contrast, external segments, e.g., the trisaccharide units composed of sialic acid, galactose, and N-acetylglucosamine, are added in the Golgi zone by stepwise transfer (Schachter and Rodén, 1972). Studies of the subcellular localization of the biosynthesis of cartilage proteoglycan have yielded essentially similar results (Horwitz and Dorfman, 1968; Horwitz, 1972). The enzymes catalyzing transfer of the three sugars adjacent to the carbohydrate-protein linkage (xylose and two galactose residues) were found in higher concentration in the rough membranes than in the smooth fraction, whereas the polymerizing enzymes, involved in the formation of the repeating disaccharide units, were distributed approximately equally between the rough and the smooth membranes. It should be noted that a fraction of heavy membranes was observed in the cartilage homogenate which had no counterpart among the subcellular liver fractions and that the marked differences in fractionation patterns make a direct comparison between the two tissues quite difficult.

2. Substrate Specificity of Glycosyltransferases

Before proceeding to a discussion of the reactions involved in polysaccharide synthesis, some general features of glycosyl transfer reactions will be briefly mentioned. In most cases, the glycosyltransferases behave in accordance with the 'one enzyme-one linkage' concept (Hagopian and Eylar, 1968) which postulates that an enzyme is specific with regard to (a) the sugar transferred, (b) the acceptor, and (c) the position and anomeric configuration of the linkage formed. These rules are not without exception and, e.g., lactose synthetase may catalyze transfer to either glucose or N-acetylglucosamine. This situation is quite extraordinary, however, inasmuch as the enzyme normally catalyzes transfer to N-acetylglucosamine or glycoprotein-bound N-acetylglucosamine units and utilizes glucose as acceptor only in the presence of α-lactalbumin. The latter protein has therefore been described as a 'specifier protein' (Brew et al., 1968).

Whereas recognition of the proper monosaccharide unit at the nonreducing terminus of an oligo- or polysaccharide chain is a basic feature of the substrate specificity of all glycosyltransferases, an acceptor structure larger than a single monosaccharide unit may sometimes be required. In chondroitin sulfate synthesis, examples of both situations are observed. One of the two galactosyltransferases which participate in the formation of the carbohydrate-protein linkage region transfers galactose to xylosylated core protein but may also utilize the free monosaccharide, D-xylose, as acceptor. In contrast, the second galactosyltransferase is quite specific for the disaccharide, 4-O-β-D-galactosyl-D-xylose (free or protein-bound), and other disaccharides with nonreducing terminal galactose cannot serve as acceptors. Further details of the substrate specificities of these and other chondroitin sulfate glycosyltransferases are discussed elsewhere (Rodén, 1970).

3. Glycosyltransferases of Connective Tissue Polysaccharide Biosynthesis

From the 'one enzyme-one linkage' concept it may be inferred that each disaccharide unit that can be visualized as a component of a complex carbohydrate molecule corresponds to a specific glycosyltransferase capable of transferring the nonreducing terminal unit of this particular disaccharide. A survey of known carbohydrate structures in mammalian glycoproteins, glycolipids, and connective tissue polysaccharides shows that at least 57 different disaccharides can be distinguished as components of these substances, and, accordingly, the existence of 57 different glycosyltransferases can be postulated. Additionally, the entire group of glycosyltransferases includes the enzymes involved in formation of the carbohydrate-protein linkages, by direct transfer to specific amino acid residues of core proteins or via lipid intermediates, and those catalyzing transfer to the lipid moieties of gangliosides and other glycolipids. Approximately 10 glycosyltransferases have been recognized within these latter categories, but further clarification of the mechanism of oligosaccharide chain initiation is needed before the situation can be defined more precisely (e.g., if a carbohydrate-protein linkage is formed by direct glycosyl transfer to a protein acceptor, only one enzyme is required; however, if a lipid intermediate is involved, two transfer steps are necessary).

It must be emphasized at this point that the validity of the postulate that each disaccharide structure reflects the existence of a specific glycosyltransferase rests on the additional premise that the sugars do not undergo any major modifications following their incorporation into an oligo- or polysaccharide chain. This is not universally true, as it has recently been shown that, in a heparin-producing mastocytoma, iduronic acid is formed at the polymer level by epimerization of glucuronic acid residues (Lindahl et al., 1972). Only one glycosyltransferase, catalyzing transfer of glucuronic acid units, is therefore needed for the formation of both glucuronic acid and iduronic acid residues in heparin. Reasoning by analogy, it seems likely that the biosynthesis of the iduronic acid residues of dermatan sulfate occurs in a similar fashion, but definitive experimental evidence to this effect is still lacking. It should be noted that epimerization on the polymer level has previously been demonstrated in studies of alginic acid biosynthesis (Haug and Larsen, 1971), but the conversion of glucuronic acid to iduronic acid is evidently the only reaction of its kind in the formation of mammalian complex carbohydrates. As a consequence of the findings of Lindahl et al. (1972), there is now an apparent conflict with previous work by Jacobson and Davidson (1962) who have shown that an epimerase is present in skin which catalyzes the conversion of UDP-glucuronic acid to UDP-iduronic acid. This would seem to implicate UDP-iduronic acid as a precursor of the iduronic acid residues in polysaccharides, but such a role has never been demonstrated. It is presently difficult to visualize a function for the nucleotide sugar epimerase in view of the results of Lindahl et al., although it is possible that two alternative pathways of iduronic acid biosynthesis may exist or that the iduronic acid residues of dermatan sulfate are formed by a different route than those of heparin.

It is seen from Table 1 that 15 disaccharides can be derived from the structures of the connective tissue polysaccharides. Assumptions regarding the number of glycosyltransferases involved in the formation of these disaccharides obviously must take into account the fact that the iduronic acid residues of heparin and most likely also those of dermatan sulfate are formed after polymerization. Thus, the epimerase (or epimerases) catalyzing these reactions serve the function of the hypothetical iduronosyltransferases which would have been postulated from the 'one disaccharide-one glycosyltransferase' concept (e.g., the disaccharide unit 8B with a terminal glucuronic acid residue is presumably the primary biosynthetic product and is subsequently modified to yield disaccharide 7A and its O-sulfated derivatives).

Another possible deviation from the anticipated pattern concerns the acceptor specificity of the N-acetylgalactosaminyl transfer reactions in dermatan sulfate biosynthesis. Unless epimerization of glucuronic acid residues occurs only after addition of N-acetylgalactosamine has taken place, the existence of two N-acetylgalactosaminyltransferases catalyzing transfer to either glucuronic or iduronic acid units would have to be postulated in view of the normal specificity requirements. It has already been pointed out that the acceptor specificity of the glycosyltransferases is not always absolute and it is therefore possible that a single enzyme could catalyze transfer to both uronic acids. Indeed, N-acetylgalactosaminyl transfer to a dicasaccharide with a nonreducing terminal iduronic acid unit has been demonstrated by Malmström and Fransson (1971), and it appears from preliminary evidence that the enzyme catalyzing this reaction can also utilize glucuronic acid residues as acceptors (Malmström and Fransson, personal communication). Further detailed studies are obviously required concerning the acceptor specificity of the N-acetylhexosaminyltransferases and the time relationship between epimerization and N-acetyl-

Table 1. Disaccharide components of connective tissues polysaccharides[a]

Disaccharide component	Parent polysaccharide
1. β-GlcUA-(1→3)-GlcNAc	Hyaluronic acid
2. β-GlcNAc-(1→4)-GlcUA	
3. β-GlcUA-(1→3)-GalNAc	Chondroitin 4- and 6-sulfate
4. β-GalNAc-(1→4)-GlcUA	Dermatan sulfate
5. α-IdUA-(1→3)-GalNAc	Dermatan sulfate
6. β-GalNAc-(1→4)-IdUA	
7A. α-IdUA-(1→4)-$GlcNSO_4$	Heparin
7B. α-IdUA-(1→4)-GlcNAc	Heparan sulfate
8A. β-GlcUA-(1→4)-$GlcNSO_4$	
8B. β-GlcUA-(1→4)-GlcNAc	
9A. α-$GlcNSO_4$-(1→4)-IdUA	
9B. α-GlcNAc-(1→4)-IdUA	
10A. α-$GlcNSO_4$-(1→4)-GlcUA	
10B. α-GlcNAc-(1→4)-GlcUA	
11. β-Gal-(1→4)-GlcNAc	Keratan sulfate
12. β-GlcNAc-(1→3)-Gal	
13. β-Gal-(1→4)-Xyl	Carbohydrate-protein linkage regions of chondroitin 4- and 6-sulfate, dermatan sulfate, heparin, and heparan sulfate
14. β-Gal-(1→3)-Gal	
15. β-GlcUA-(1→3)-Gal	

[a] Sulfate groups have been omitted with the exception of N-sulfate groups which replace most of the N-acetyl groups in the course of biosynthesis of heparin and are also present to a varying extent in heparan sulfate.

hexosaminyl transfer before a conclusion can be reached regarding the formation of the structures under consideration, i.e., disaccharides 4, 6, 9B, and 10B.

Finally, the polysaccharide chain initiating reactions must be briefly considered. So far, three types of carbohydrate-protein linkages have been recognized in the connective tissue proteoglycans: (1) an O-glycosidic linkage between xylose and serine hydroxyl groups; (2) an O-glycosidic linkage between N-acetylgalactosamine and the hydroxyl groups of serine or threonine; and (3) an N-glycosylamine linkage between N-acetylglucosamine and the amide group of asparagine. The first linkage type is found in the chondroitin sulfates, dermatan sulfate, heparin and heparan sulfate; the second is involved in the linkage of skeletal keratan sulfate (keratan sulfate II) to protein, and the third is present in corneal keratan sulfate (keratan sulfate I).

The detailed structures of the carbohydrate-protein linkage regions of the two types of keratan sulfate have not yet been established.

However, the xylose-linked polysaccharides all have a specific linkage region which is identical from one polysaccharide to another and has the structure shown in Fig. 1.

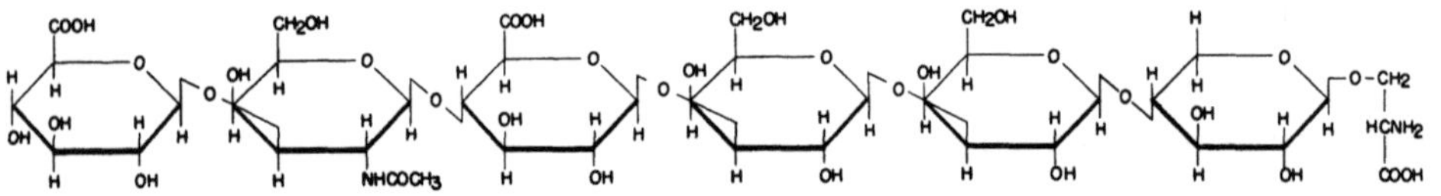

Fig. 1. Structure of chondroitin sulfate. Only one repeating disaccharide unit is shown. The sulfate group may be located on C-4 or C-6 of the N-acetylgalactosamine residue

Nothing is known about the keratan sulfate chain-initiating reactions, but it seems likely that the N-acetylglucosamine-asparagine linkage in corneal keratan sulfate proteoglycan is formed via lipid intermediates by a series of reactions analogous to those discussed previously.

Again reasoning by analogy, it may be suggested that the N-acetylgalactosamine-threonine/serine linkages in keratan sulfate II are formed by direct transfer of single N-acetylgalactosamine residues to the protein moiety of the proteoglycan,since it has been previously shown that the core protein of submaxillary mucin and a basic protein from bovine myelin serve as substrates for N-acetylgalactosamine transfer to threonine residues (reviewed in Schachter and Rodén, 1972). Participation of lipid intermediates in these reactions cannot be ruled out on the basis of available evidence, but it seems significant that transfer of single N-acetylgalactosamine residues could readily be demonstrated in the studies mentioned, whereas similar experiments aimed at detecting N-acetylglucosamine transfer to asparagine residues have never been successful.

In contrast to the present lack of knowledge regarding keratan sulfate synthesis, a considerable amount of information has now accumulated concerning the formation of the xylose-serine linkage which is most prevalent among the connective tissue proteoglycans. This work will be discussed in detail below.

The picture of connective tissue polysaccharide biosynthesis that emerges from the preceding discussion is in essential agreement with the basic notion that the polysaccharide chains are synthesized by a series of glycosyltransferases which catalyze the stepwise transfer of monosaccharide units from the appropriate nucleotide sugars. It is still somewhat uncertain whether the transfer occurs exclusively to the growing carbohydrate chain of the proteoglycan or whether part of the process might take place by assembly of an oligosaccharide chain on a lipid intermediate prior to transfer to the proteoglycan molecule. Also, the mechanism of iduronic acid formation represents a unique feature of proteoglycan biosynthesis which has no counterpart in other areas of mammalian complex carbohydrate metabolism. As a consequence, the simple assumption that the 15 disaccharides listed in Table 1 would signify the existence of a number of corresponding glycosyltransferases, has to be modified, and we may now suggest that the following enzymes participate in the formation of the individual polysaccharides:

hyaluronic acid transferases: one N-acetylglucosaminyltransferase and one glucuronosyltransferase;

chondroitin sulfate transferases: (i) a chain-initiating xylosyltransferase catalyzing transfer of xylose to the core protein of the proteoglycan; (ii and iii) two galactosyltransferases which consecutively transfer the two galactosyl residues of the linkage region; (iv) a glucuronosyltransferase which completes the formation of the specific carbohydrate-protein linkage region; (v) an N-acetylgalactosaminyltransferase and (vi) a second glucuronosyltransferase which jointly synthesize the bulk of the polysaccharide chain composed of 30 - 50 repeating disaccharide units;

dermatan sulfate transferases: one glucuronosyltransferase catalyzing transfer to N-acetylgalactosamine units, one epimerase, and one or two N-acetylgalactosaminyltransferases catalyzing transfer to iduronic acid and glucuronic acid residues;

heparin and heparan sulfate transferases: one glucuronosyltransferase catalyzing transfer to N-acetylglucosamine residues, one epimerase, and one or two N-acetylglucosaminyltransferases catalyzing transfer to glucuronic acid and iduronic acid;

keratan sulfate transferases: one galactosyltransferase and one N-acetylglucosaminyltransferase; since keratan sulfate contains several sugars other than galactose and N-acetylglucosamine, i.e., sialic acid, fucose, mannose and galactosamine (keratan sulfate II) as well as galactose in excess over the amount present in the repeating disaccharide units, glycosyltransferases catalyzing the transfer of these sugars must also be postulated.

The existence of most of the enzymes mentioned above has already been well documented. In general, the initial experimental approach has been to study glycosyl transfer from the various nucleotide sugars to endogenous acceptors which are usually present in the crude tissue homogenates or subcellular fractions. Subsequently, exogenous acceptors - mostly oligosaccharides of well-defined structure - have been used to investigate the characteristics of a particular reaction and the properties of the individual enzymes in more detail. Much of the information obtained by such studies has been reviewed in detail elsewhere (Rodén, 1970), and only a brief comment will be made here concerning the formation of hyaluronic acid which remains one of the unresolved problems in this area.

Synthesis of high molecular weight hyaluronic acid from the two nucleotide sugars, UDP-N-acetylglucosamine and UDP-glucuronic acid, has been demonstrated with both mammalian and bacterial enzyme preparations, which contain endogenous acceptors (Rodén, 1970). However, transfer of single monosaccharide units to exogenous hyaluronic acid oligosaccharides cannot be obtained under the same conditions that have been used successfully for analogous studies of chondroitin sulfate synthesis. This is quite unexpected, and it must therefore be considered possible that the biosynthesis of hyaluronic acid proceeds by a different pathway than that of chondroitin sulfate. Especially, the participation of lipid intermediates in this process cannot be ruled out, although the available evidence speaks against this possibility (Stoolmiller and Dorfman, 1969).

4. Some Properties of Glycosyltransferases

As a group, the glycosyltransferases possess certain characteristic properties which create considerable experimental difficulties, including (a) an often firm association with cellular membranes which makes solubilization an obligatory first step in any purification procedure, (b) a tendency of solubilized and partially purified preparations to aggregate, sometimes irreversibly and with loss of activity, if the ionic strenth is lowered beyond a certain level or if the enzyme concentration is raised above approximately 1 mg per ml, and (c) dependence of enzymatic activity on the presence of lipids or detergents, sometimes to the point that complete inactivation occurs on removal of lipids (Rothfield and Romeo, 1971).

The glycosyltransferases involved in connective tissue polysaccharide biosynthesis appear to be especially recalcitrant to various solubilization procedures, and only recently have methods for effective solubilization of these enzymes been developed. It was shown by Helting (1971) that treatment with detergent (Tween 20) in the presence of ammonium hydroxide resulted in extensive solubilization of the galactosyltransferases involved in heparin synthesis in a mouse mastocytoma. The same method has been applied successfully to the solubilization of the analogous chondroitin sulfate galactosyltransferases from embryonic chick cartilage (Rodén et al., 1972). A modified procedure has also been developed in which the nonionic detergent, Nonidet P-40, and 0.5 M potassium chloride are used instead of alkali and Tween 20 (Rodén et al., 1972; Schwartz and Rodén, 1974).

Dependence of glycosyltransferase activity on the presence of lipids has been shown most clearly for certain glycosyltransferases (a glucosyl- and a galactosyltransferase) participating in the formation of bacterial lipopolysaccharides (Rothfield and Romeo, 1971). In this case, the enzymes were completely inactive in the absence of lipids, and the activity was restored on addition of phosphatidylethanolamine or other lipids with a diacylglycerophosphate backbone. Interestingly, the substrate was the primary target for the action of the phospholipids rather than the enzyme, and the sequence of events may be depicted as follows:

$$\text{PE} + \text{LPS} \rightarrow \text{LPS}\cdot\text{PE}$$

$$\text{LPS}\cdot\text{PE} + \text{Enzyme} \rightarrow \text{Enzyme}\cdot\text{LPS}\cdot\text{PE}$$

$$\text{Enzyme}\cdot\text{LPS}\cdot\text{PE} + \text{UDP-Gal} \rightarrow \text{Gal-LPS}\cdot\text{PE} + \text{UDP} + \text{Enzyme}$$

(LPS, lipopolysaccharide; PE, phosphatidylethanolamine)

After formation of a multimolecular binary complex between the lipopolysaccharide substrate and phosphatidylethanolamine under carefully controlled conditions, interaction with the enzyme yielded a ternary complex, which was enzymatically active and catalyzed incorporation of galactose into its substrate portion.

Recent work in our laboratory has shown that the chondroitin sulfate glycosyltransferases are also affected by phospholipids, particularly UDP-galactose:D-xylose galactosyltransferase. This enzyme is extensively inactivated by treatment with phospholipase C and can be reactivated by addition of phospholipids, the most efficient being lysolecithin and lysophosphatidylethanolamine (Schwartz and Rodén, 1972). Investigations of the role of phospholipids in connective tissue polysaccharide synthesis are only now beginning and have not reached the degree of sophistication attained in the studies of

Rothfield and Romeo (1971), partly because highly purified enzymes have not been previously available.

5. Polysaccharide Chain Initiation by Xylosyltransferase

The xylosyl transfer reaction, which initiates the formation of most of the connective tissue polysaccharides, was first demonstrated by Grebner et al. (1966a) who showed that hen's oviduct contains an enzyme which transfers xylose from UDP-xylose to endogenous acceptors. This enzyme has subsequently also been found in a mouse mastocytome (Grebner et al., 1966b), embryonic chick cartilage and brain (Robinson et al., 1966; Stoolmiller et al., 1972) and, in lower concentration, in some other tissues. The nature of the endogenous acceptor has not been studied in any detail, but it would be expected that it consists of the core protein of the proteoglycans, possibly with polysaccharide chains in various stages of completion.

A particular difficulty in the study of xylosyltransferase has been the problem of finding a suitable exogenous substrate for use in purification of the enzyme. This was resolved by the work of Baker et al. (1972) who showed that the core protein could be prepared from cartilage proteoglycan by Smith degradation and was an excellent acceptor for xylosyl transfer. Smaller acceptors were also found, e.g., the tripeptide, serylglycylglycine, but the K_m value of this substrate was much higher than that of the core protein - 20.0 mM as compared to 0.064 mM, expressed in terms of concentration of serine residues.

Xylosyltransferase is unique among the polysaccharide glycosyltransferases in being more loosely bound to the membranes of the endoplasmic reticulum than the other enzymes of this group and, on thorough homogenization, 80 - 90% of the total activity of a homogenate of embryonic chick cartilage is found in the supernatant fraction after high speed centrifugation (Stoolmiller et al., 1972). More than the apparent proportion of the activity is left in the particulate fraction, however, and repeated digestion of the pellet with ribonuclease (Baker et al., 1971) or treatment with Nonidet P-40 and potassium chloride yields a considerable amount of soluble enzyme approaching that present in the initial high speed supernatant fraction.

Once an exogenous substrate for xylosyltransferase was available, the enzyme could be reliably quantitated and was purified 40- to 50-fold by a procedure involving ammonium sulfate fractionation and gel chromatography on Sephadex G-200 or centrifugation in a sucrose density gradient (Stoolmiller et al., 1972). Further purification has been obtained by chromatography on an affinity matrix consisting of Sepharose-bound core protein (Schwartz and Rodén, 1973, 1974). After removal of endogenous acceptors by the purification steps used by Stoolmiller et al. (1972), the partially purified enzyme was adsorbed to the affinity matrix and was subsequently eluted by a solution of core protein, yielding quantitative recovery of activity. However, since the degree of purification could not readily be estimated in the presence of the protein substrate, alternative methods of elution were sought. It was found that the enzyme was also eluted, in lower yield (30 - 40%), by buffer containing 0.25 M KCl. Interestingly, it was also observed that once a column had been saturated with enzyme and eluted with salt, further batches could be adsorbed and quantitatively eluted with salt alone. The binding capacity of such a preconditioned column was correspondingly lower - approximately 1/3 of the initial capacity - indicating that 2/3 of the binding sites were still occupied by enzyme which could only be desorbed by the substrate itself. The purification obtained in the affinity step was in the order of 160-fold, resulting in an overall purification over the crude homogenate of approximately 4000-fold. Rechromatography of some preparations yielded enzyme of even higher specific activity, respresenting close to 7000-fold purification.

The affinity-purified enzyme was homogeneous on analysis by polyacrylamide gel electrophoresis, immunodiffusion and immunoelectrophoresis. The molecular weight was 95,000 - 100,000 daltons, as determined by gel chromatography, and the enzyme consisted of two pairs of non-identical subunits, as indicated by the finding that gel electrophoresis after pretreatment with SDS and mercaptoethanol yielded two bands with migration rates corresponding to molecular weights of 23,000 and 27,000 daltons, respectively.

The behavior of xylosyltransferase on affinity chromatography suggested that two types of interaction are involved in its binding to the matrix: 1) the enzyme-substrate interaction which can be dissociated only by the substrate itself; and 2) an interaction between individual enzyme molecules which can be dissociated by salt. The latter type of interaction is in keeping with the observation that the enzyme readily aggregates at low ionic strength even at relatively low protein concentration to form cloudy suspensions which may again be clarified by an increase of the ionic strength. It is of interest to note that the xylosyltransferase may also interact with UDP-galactose: D-xylose galactosyltransferase which catalyzes the second step in chondroitin sulfate synthesis. This interaction will be discussed in more detail below.

6. Other Linkage Region Glycosyltransferases

The second reaction in chondroitin sulfate synthesis, i.e., the transfer of galactose to the xylosyl-protein formed by xylosyltransferase, was first demonstrated by Robinson et al. (1966), who showed that galactose was incorporated from UDP-galactose into endogenous acceptors present in a particulate enzyme preparation from embryonic chick cartilage. The product was identified by isolation of radioactively labeled galactosylxylitol after cleavage of the xylose-serine linkage with alkaline borohydride. Evidence for the second galactosyl transfer step was obtained by Helting and Rodén (1969), who isolated 3-O-β-D-galactosyl-D-galactose from a partial acid hydrolysate of a similar reaction mixture.

The study of the two galactosyltransferases catalyzing these reactions has been greatly facilitated by the solubilization procedures discussed previously (Helting, 1971; Rodén et al., 1972; Schwartz and Rodén, 1974) and by the finding that D-xylose and 4-O-β-D-galactosyl-D-xylose can be used as acceptors (Helting and Rodén, 1969). The galactosyltransferases from chick cartilage have now been separated by gel chromatography on Sephadex G-200, and UDP-galactose:D-xylose galactosyltransferase has been further purified by chromatography on an affinity matrix containing enzymatically xylosylated core protein. This procedure has yielded approximately 600-fold purification over the crude homogenate (Schwartz and Rodén, 1974).

Formation of the specific carbohydrate-protein linkage region is completed by addition of a glucuronic acid residue, and the enzyme catalyzing this reaction has been purified 330-fold by Helting (1972) from a homogenate of a mouse mastocytoma. This glucuronosyltransferase is presumably involved in the synthesis of heparin, and it should be pointed out that it is not yet known whether analogous reactions in the formation of the linkage regions of different polysaccharides are catalyzed by the same or different enzymes.

A considerable body of information is available concerning some of the properties of the "polymerizing" glycosyltransferases, particularly those responsible for the biosynthesis of the repeating disaccharide units of chondroitin sulfate (Rodén, 1970; Stoolmiller and Dorfman, 1969), but these enzymes have not yet been purified to any significant extent. However, the recent advances in methodology make it likely that rapid progress will be made in this area in the near future.

7. Interactions between Glycosyltransferases

It has previously been suggested that the chondroitin sulfate glycosyltransferases may be arranged on the membranes of the endoplasmic reticulum in such a fashion that enzymes catalyzing consecutive transfer steps are located in adjacent positions (Horwitz and Dorfman, 1968; Stoolmiller et al., 1972). An arrangement of this type, which would be expected to increase the efficiency of the biosynthetic process, could conceivably result from specific interactions between the various glycosyltransferases. Evidence in support of this idea has recently been obtained by Schwartz et al. (1974), who observed that the core protein matrix designed to adsorb xylosyltransferase was also capable of binding UDP-galactose:D-xylose galactosyltransferase, after having been equilibrated with xylosyltransferase. This interaction was apparently at least partially specific, since the bulk of the protein was not adsorbed and the galactosyltransferase obtained on subsequent elution had been purified approximately 10-fold over the starting material. The specific activity of the purified enzyme was similar to that of the preparation isolated by affinity chromatography on a xylosylated core protein matrix as described earlier. It should be noted that the enzyme could be eluted by Nonidet P-40 in the presence of salt but not by salt alone, suggesting that hydrophobic forces were involved in its binding to xylosyltransferase.

As highly purified preparations of other chondroitin sulfate glycosyltransferases become available, it will obviously be of great interest to determine whether specific interactions occur between these as well, and it may be envisioned that the affinity systems described here will be of great value not only as tools for purification but, perhaps more importantly, as simplified model systems for membrane assembly. The use of such systems will make it possible to establish the existence and nature of specific interactions more conclusively, and the simplicity afforded by model systems with a limited number of variables will obviate many ambiguities and difficulties in the interpretation of the behavior of the more complex intact membranes.

References

Baker, J.R., Rodén, L., Stoolmiller, A.C.: J. Biol. Chem. 247, 3838 (1972).
Baker, J.R., Rodén, L., Yamagata, S.: Biochem. J. 125, 93P (1971).
Baynes, J.W., Hsu, A.-F., Heath, E.C.: J. Biol. Chem. 248, 5693 (1973).
Behrens, N.H.: Abstracts of the Sixth Annual Miami Winter Symposia, p. 24. Department of Biochemistry, University of Miami 1973.
Behrens, N.H., Carminatti, H., Staneloni, R.J., Leloir, L.F., Cantarella, A.I.: Proc. Nat. Acad. Sci. U.S.A., in press.
Brew, K., Vanaman, T.C., Hill, R.L.: Proc. Nat. Acad. Sci. U.S.A. 50, 491 (1968).
Grebner, E.E., Hall, C.W., Neufeld, E.F.: Biochem. Biophys. Res. Commun. 22, 672 (1966a)
Grebner, E.E., Hall, C.W., Neufeld, E.F.: Arch. Biochem. Biophys. 116, 391 (1966b).
Hagopian, A., Eylar, E.H.: Arch. Biochem. Biophys. 128, 422 (1968).
Haug, A., Larsen, B.: Carbohydrate Res. 17, 297 (1971).
Heath, E.C., Baynes, J.W., Hsu, A.-F.: Abstracts of the Sixth Annual Miami Winter Symposia, p. 27, Department of Biochemistry, University of Miami 1973.
Helting, T.: J. Biol. Chem. 246, 815 (1971).
Helting, T.: J. Biol. Chem. 247, 4327 (1972).
Helting, T., Rodén, L.: J. Biol. Chem. 244, 2790 (1969).
Horwitz, A.L.: Ph. D. Thesis, University of Chicago, 1972.
Horwitz, A.L., Dorfman, A.: J. Cell Biol. 38, 358 (1968).
Jacobson, B., Davidson, E.A.: J. Biol. Chem. 237, 638 (1962).
Leloir, L.F., Staneloni, R.J., Carminatti, H., Behrens, N.H.: Biochem. Biophys. Res. Commun. 52, 1285 (1973).
Lindahl, U., Bäckström, G., Malmström, A., Fransson, L.-Å.: Biochem. Biophys. Res. Commun. 46, 985 (1972).
Malmström, A., Fransson, L.-Å.: FEBS Letters 16, 105 (1971).
Parodi, A.J., Behrens, N.H., Leloir, L.F., Carminatti, H.: Proc. Nat. Acad. Sci. U.S.A. 69, 3268 (1972).
Robinson, H.C., Telser, A., Dorfman, A.: Proc. Nat. Acad. Sci. U.S.A. 56, 1859 (1966).
Rodén, L.: In: Metabolic Conjugation and Metabolic Hydrolysis (Fishman, W.H., Ed.) Vol. II, p. 345 - 442. New York: Academic Press 1970.
Rodén, L., Baker, J.R., Helting, T., Schwartz, N.B., Stoolmiller, A.C., Yamagata, S., Yamagata, T.: Methods Enzymol. 28, 638 (1972).
Rothfield, L., Romeo, D.: In: Structure and Function of Biological Membranes (Rothfield, L.I., Ed.) p. 251 - 284. New York: Academic Press 1971.
Schachter, H., Rodén, L.: In: Metabolic Conjugation and Metabolic Hydrolysis (Fishman, W.H., Ed.) Vol. III, p. 1 - 149. New York: Academic Press 1972.
Schwartz, N.B., Rodén, L.: Fed. Proc. 31, 434Abs (1972).
Schwartz, N.B., Rodén, L.: Fed. Proc. 32, 560Abs (1973).
Schwartz, N.B., Rodén, L.: Carbohydrate Res., in press.
Schwartz, N.B., Rodén, L.: Submitted for publication (1974).
Schwartz, N.B., Rodén, L., Dorfman, A.: Biochem, Biophys. Res. Commun. 56, 717 (1974).
Stoolmiller, A.C., Dorfman, A.: In: Comprehensive Biochemistry (Florkin, M., Stotz, E.H., Eds.) Vol. 17, p. 241 - 275. Amsterdam: Elsevier 1969.
Stoolmiller, A.C., Dorfman, A.: J. Biol. Chem. 244, 236 (1969).
Stoolmiller, A.C., Horwitz, A.L., Dorfman, A.: J. Biol. Chem. 247, 3525 (1972).

Synthesis of Sulfated Glycosaminoglycans by Cultured Cells, with Particular Reference to the Mouse Fibroblast Line 3T6

C. J. Bates and C. I. Levene

During the past decade, a number of mammalian fibroblast lines have been isolated, stabilized and maintained for several years in tissue culture. They have provided a new tool for the study of connective tissue formation, starting with a homogeneous and well-defined cell type which will grow in a chemically defined and controllable environment.

One of these lines, the 3T6 mouse fibroblast, which was originally developed in Green's laboratory (Todaro and Green, 1963; Goldberg and Green, 1964) has proved particularly useful in studies on the control of collagen biosynthesis (Goldberg et al., 1963; Green and Goldberg, 1964; Goldberg and Green, 1967, 1968, 1969; Priest and Bublitz, 1967; Davies et al., 1968; Priest and Davies, 1969; Aleo, 1969; Levene and Bates, 1970; Bachra and van der Eb, 1970; Manner, 1971; Ramaley and Rosenbloom, 1971). 3T6 cultures form a multi-layered tissue on Falcon plastic petri dishes, and the collagen which they produce can be quantitated by microchemical methods.

Like many cell lines in culture, the 3T6 fibroblast produces, and secretes into growth medium, substantial amounts of hyaluronic acid (Green and Hamerman, 1964). In considering the formation of connective tissue matrix, however, we felt that a study of sulfated glycosaminoglycan production might be more informative, and an observation by Dr. J. Steinberg (personal communication), suggested that 3T6 cultures could incorporate ^{35}S-sulfate into macromolecules with some of the properties of sulfated glycosaminoglycans.

We therefore devised methods of labelling and analysis of the compounds, and were able to isolate a labelled macromolecular, protease-resistant fraction from both the cell layer and the growth medium of the cultures (Bates and Levene, 1971). Of the labelled material in the cell layer compartment, about half had the properties of haparan sulfate; the other half had the properties of a chondroitin sulfate mixture. In the growth medium compartment the same species were found, but the chondroitin sulfate component predominated, and the chains were shorter. If proteolytic digestion were omitted, most of the sulfated glycosaminoglycans were insoluble in tricheoroacetic acid and were therefore probably present as proteoglycans in the native state.

The production of heparan sulfate is in accord with its recent identification (Kraemer, 1971; Dietrich and Montes de Oca, 1970) in cultures of a wide variety of cell lines of very diverse origins. In view of its consistent occurrence and apparent relationship to the cells' mitotic cycle, Kraemer has suggested a role for heparan sulfate in cell division. At present there seems to be no definite evidence that this particular sulfated glycosaminoglycan is closely associated with connective tissue production, either *in vivo* or in tissue culture.

The formation of chondroitin sulfate by cultured cells has been reported widely: for newly isolated strains of fibroblastic type (Grossfeld et al., 1957; Berenson et al., 1958; Morris, 1960; Schafer et al., 1967; Matalon and Dorfman, 1966; Dorfman and Ho, 1970; Conrad, 1970; Dietrich and Montes de Oca, 1970) for chondrocytes (Glick and Stockdale, 1964; Nameroff and Holtzer, 1967; Shulman and Meyer, 1968) for leukocytes (Olsson et al., 1968) and for a variety of established cell lines of diverse origin (Dietrich and Montes de Oca, 1970; Suzuki et al., 1970; Kraemer, 1971; Saito and Uzmann, 1971). However, none of these studies have clarified the involvement of chondroitin sulfate, or its proteoglycan, in connective tissue formation. Many of the cell lines, which produce chondroitin sulfates, are also known to synthesize collagen, and a few of the lines studied, such as the lymphocytic tumor cell L5178Y seem not to produce either collagen (Green, 1968) or chondroitin sulfate (Kraemer, 1971). Collagen and chondroitin sulfate synthesis by chick embryo chondrocytes can be switched on or off together by altering the culture conditions (Prockop et al., 1964), suggesting a possible relationship between the synthesis of these two connective tissue components. Clearly these problems are far from being understood, but since recent investigations have shown that a wide variety of cells in culture produce several different types of sulfated glycosaminoglycan in addition to hyaluronic acid, the way should now be open to a more detailed study of their biological role.

References

Aleo, J.J.: Collagen synthesis in cultured cells: the influence of beta-aminoproprionitrile. Proc. Soc. exp. Biol. (N.Y.) 130, 451 - 454 (1969).

Bachra, B.N., van der Eb, A.J.: Site of proline hydroxylation during collagen synthesis in mouse fibroblasts. Biochemistry 9, 3001 - 3010 (1970).

Bates, C.J., Levene, C.I.: The synthesis of sulphated glycosaminoglycans by the mouse fibroblast line, 3T6. Biochim. biophys. Acta (Amst.) 237, 214 - 226 (1971).

Berenson, G.S., Lumpkin, W.M., Shipp, V.G.: Study of the time-course production of acid mucopolysaccharides by fibroblasts in a synthetic Medium. Anat. Rec. 132, 585 - 596 (1958).

Conrad, G.W.: Collagen and mucopolysaccharide biosynthesis in mass cultures and clones of chick corneal fibroblasts *in vitro*. Develop. Biol. 21, 611 - 635 (1970).

Davies, L.M., Priest, J.H., Priest, R.E.: Collagen synthesis by cells synchroneously replicating DNA. Science 159, 91 - 93 (1968).

Dietrich, C.P., Montes de Oca, H.: Production of heparin related mucopolysaccharides by mammalian cells in culture. Proc. Soc. exp. Biol. (N.Y.) 134, 955 - 962 (1970).

Dorfman, A., Ho., P.-L.: Synthesis of acid mucopolysaccharides by glial tumour cells in tissue culture. Proc. nat. Acad. Sci. (Wash.) 66, 495 - 499 (1970).

Glick, M.C., Stockdale,F.E.: Differences in glycosaminoglycans derived from chick embryo chondrocytes grown in vitro and in vivo. Biochim. biophys. Acta (Amst.) 83, 61 - 68 (1964).

Goldberg, B., Green, H.: An analysis of collagen secretion by established mouse fibroblast lines. J. Cell Biol. 22, 227 - 258 (1964).

Goldberg, B., Green, H.: Collagen synthesis on polyribosomes of cultured mammalian fibroblasts. J. molec. Biol. 26, 1 - 18 (1967).

Goldberg, B., Green, H.: The synthesis of collagen and protocollagen hydroxylase by fibroblastic and non-fibroblastic cell lines. Proc. nat. Acad. Sci. (Wash.) 59, 1110 - 1115 (1968).
Goldberg, B., Green, H.: Relation between collagen synthesis and collagen proline hydroxylase activity in mammalian cells. Nature (Lond.) 221, 267 - 268 (1969).
Goldberg, B., Green, H., Todaro, G.J.: Collagen formation *in vitro* by established mammalian cell lines. Exp. Cell Res. 31, 444 - 447 (1963).
Green, H., Goldberg, B.: Collagen and cell protein synthesis by an established mammalian fibroblast line. Nature (Lond.) 204, 347 - 349 (1964).
Green, H., Hamerman, D.: Production of hyaluronate and collagen by fibroblast clones in culture. Nature (Lond.) 201, 710 (1964).
Green, H., Warren, K.B.: In: Symposium of the International Society for Cell Biology, Vol. 7, p. 123 - 134. New York, London: Academic Press 1968.
Grossfeld, H., Meyer, K., Godman, G.C., Linker, A.: Mucopolysaccharides produced in tissue culture. J. biophys. biochem. Cytol. 3, 391 - 396 (1957).
Kraemer, P.: Heparin sulphates of cultured cells II. Acid-soluble and -precipitable species of different cell lines. Biochemistry 10, 1445 - 1450 (1971).
Levene, C.I., Bates, C.J.: Growth and macromolecular synthesis in the 3T6 mouse fibroblast. I. General description and the role of ascorbic acid. J. Cell. Sci. 7,671 - 682 (1970).
Manner, G.: Cell division and collagen synthesis in cultured fibroblasts. Exp. Cell Res. 65, 49 - 60 (1971).
Matalon, R., Dorfman, A.: Hurler's syndrome: Biosynthesis of acid mucopolysaccharides in tissue culture. Proc. nat. Acad. Sci. (Wash.) 56, 1310 - 1316 (1966).
Morris, C.C.: Quantitative studies on the production of acid mucopolysaccharides by replicate cell cultures of rat fibroblasts. Ann. N.Y. Acad. Sci. 86, 878 - 914 (1960).
Nameroff, M., Holtzer, H.: The loss of phenotypic traits by differentiated C. IV. Changes in polysaccharides produced by dividing chondrocytes. Develop. Biol. 16, 250 - 281 (1967).
Olsson, I., Gardell, S., Thunell, S.: Biosynthesis of glucosaminoglycans (mucopolysaccharides) in human leukocytes. Biochim. biophys. Acta (Amst.) 165, 309 - 323 (1968).
Priest, R.E., Bublitz, C.: The influence of ascorbic acid and tetrahydropteridine on the synthesis of hydroxyproline by cultured cells. Lab. Invest. 17, 371 - 379 (1967).
Priest, R.E., Davies, L.M.: Cellular proliferation and synthesis of collagen. Lab. Invest. 21, 138 - 142 (1969).
Prockop, D.J., Pettengill, O., Holtzer, H.: Incorporation of sulphate and the synthesis of collagen by cultures of embryonic chondrocytes. Biochim. biophys. Acta (Amst.) 83, 189 - 196 (1964).
Ramaley, P.B., Rosenbloom, J.: Inhibition of proline and lysine hydroxylation prevents normal extrusion of collagen by 3T6 fibroblasts in culture. FEBS Letters 15, 59 - 64 (1971).
Saito, H., Uzman, B.G.: Production and secretion of chondroitin sulphates by permanent mammalian cell lines. Fed. Proc. 30, 457 (1971).
Schafer, I.A., Silverman, L., Sullivan, J.C., Robertson, W. van B.: Ascorbic acid deficiency in cultured human fibroblasts. J. Cell Biol. 34, 83 - 95 (1967).
Shulman, H.J., Meyer, K.: Cellular differentiation and the aging process in cartilaginous tissues. Mucopolysaccharide synthesis in Cell cultures of chondrocytes. J. exp. Med. 128, 1353 - 1362 (1968).
Suzuki, S., Kojima, K., Utsumi, K.R.: Production of sulphated mucopolysaccharides by established cell lines of fibroblastic origin. Biochim. biophys. Acta (Amst.) 222, 240 - 243 (1970).

Todaro, G.J., Green, H.: Quantitative studies of the growth of mouse embryo cells in culture and their development into established lines. J. Cell Biol. 17, 299 - 313 (1963).

The Subcellular Site of the Biosynthesis of N-Glycoloyl- and N-Acetyl-O-Acetylneuraminic Acid Components of Porcine, Bovine and Equine Submaxillary Gland Glycoproteins

R. Schauer

1. Chemistry and Occurrence of N-Acetyl-, N-Glycoloyl- and N-Acetyl-O-Acetylneuraminic Acids

Acylneuraminic acids bound to glycoproteins and to gangliosides are widely distributed in cells and fluids of animals. As their biological functions are manifold (Gottschalk, 1972), they have gained the interest of many chemists and physiologists in the last years. Acylneuraminic acids, which have been called "sialic acids" by Blix et al. (1957), contain acyl residues bound to the amino-group and in some cases also to the hydroxyl groups. The best known and most widely distributed sialic acid is N-acetylneuraminic acid, systematically named β-D(-)-5-acetamido-3,5-dideoxy-D-glycero-D-galacto-nonulosaminic acid (Fig. 1).

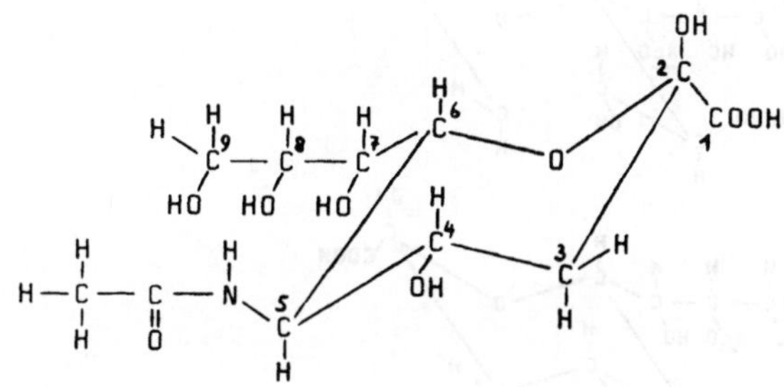

Fig. 1. N-acetylneuraminic acid

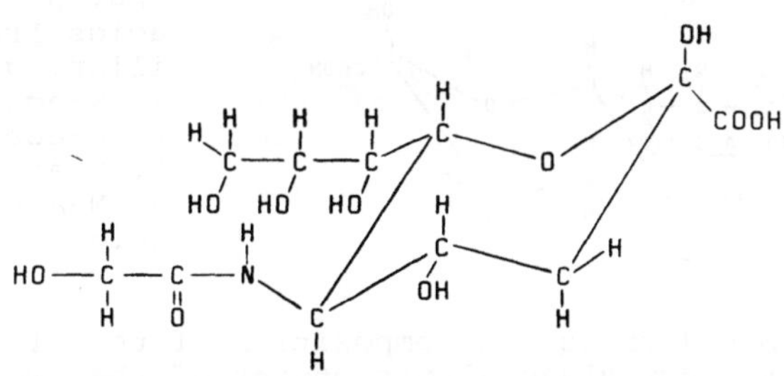

Fig. 2. N-glycoloylneuraminic acid

Glycoproteins of many animals contain N-glycoloylneuraminic acid (Fig. 2) in varying amounts, which have been found to be species- and tissue-specific (Gottschalk, 1960; Schoop et al., 1969).

The sialic acid fraction of porcine submaxillary gland glycoproteins consists, for instance, of about 90% N-glycoloyl- and 10% N-acetylneuraminic acid.

N-acetylneuraminic acid containing in addition one to three acetyl residues, bound as ester to different hydroxyl groups of neuraminic acid, have been found in varying amounts in some glycoproteins. 60 - 70% of the sialic acids of glycoproteins of submaxillary glands of cow and horse are O-acetylated. While in equine glycoprotein the O-acetyl groups are located predominantly in position 4 of the 9-C-chain of neuraminic acid, those are positioned at C-7 and/or C-8 in the corresponding bovine glycoprotein (Blix and Lindberg, 1960; Schauer and Faillard, 1968), (Fig. 3).

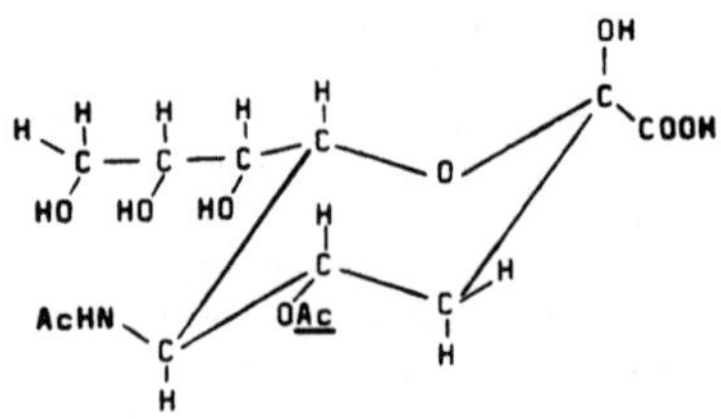

1

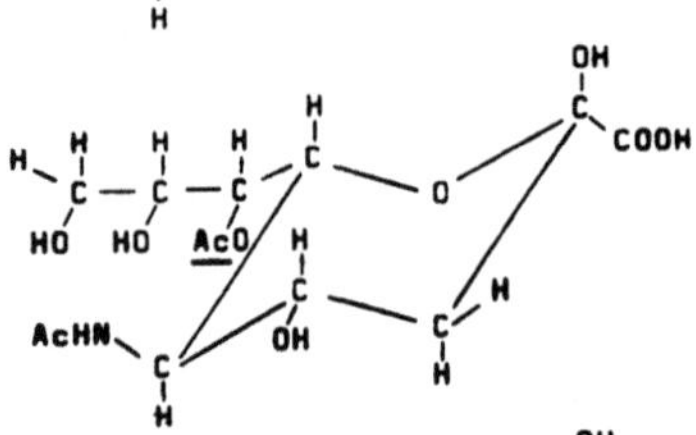

2

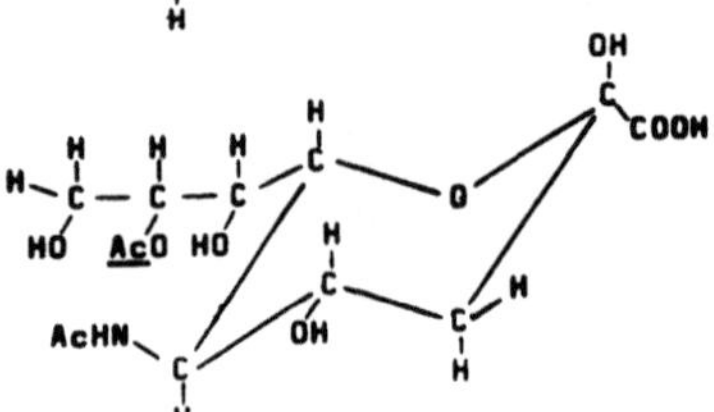

3

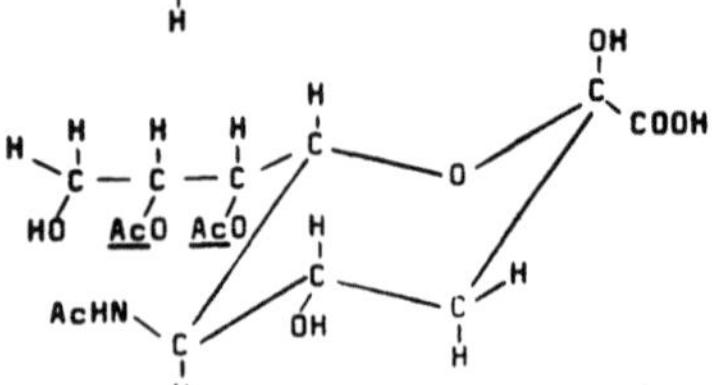

4

Fig. 3. N-acetyl-O-acetylneuraminic acids from bovine and equine submaxillary gland glycoproteins
1. N-acetyl-4-O-acetylneuraminic acid
2. N-acetyl-7-O-acetylneuraminic acid
3. N-acetyl-8-O-acetylneuraminic acid
4. N-acetyl-7,8-di-O-acetylneuraminic acid

Table 1 shows the composition of the sialic acid fractions of submaxillary gland glycoproteins of the two animals (Blix and Lindberg, 1960; Schauer and Faillard, 1968; Schauer, 1970a).

Table 2 shows the isolation procedure for sialic acids. Great care must be taken as to acid concentration and temperature, especially with the rather labile O-acetylated sialic acids.

Table 1. Composition (%) of sialic acid fractions of bovine and equine submaxillary gland glycoproteins

Sialic acid	Bovine	Equine
N-glycoloyl-neuraminic acid	15	1 - 2
N-acetyl-neuraminic acid	20 - 25	15 - 20
N-acetyl-4-O-acetyl-neuraminic acid	-	70 - 75
N-acetyl-7-O-acetyl-neuraminic acid	∿ 25	-
N-acetyl-8-O-acetyl-neuraminic acid	∿ 25	-
N-acetyl-7,8-di-O-acetyl-neuraminic acid	5 - 10	-
N-acetyl-4,?-di-O-acetyl-neuraminic acid	-	2 - 5
N-acetyl-?-tri-O-acetyl-neuraminic acid	∿ 3	-

Table 2. Isolation, purification and analysis of sialic acids of glycoproteins (for further details see Blix and Lindberg, 1960; Schauer and Faillard, 1968; Schauer et al., 1968; Schauer, 1970a)

1. Hydrolysis of the glycoside bonds
   by neuraminidase or
   by dilute HCl:
   0.05 n HCl, 80°C, 1 hour, for release of N-acetyl-and N-glycoloylneuraminic acids; 0.01 n HCl, 70°C, 1 hour, for N-acetyl-(pluri)-O-acetylneuraminic acids (for nearly quantitative release of sialic acids, hydrolysis under these conditions should be repeated after dialysis);
2. Dialysis against water (3 times, 2°C, 24 hours)
3. Anion-exchange chromatography on Dowex 2 × 8, $HCOO^-$-form; elution of sialic acids by a gradient of 0 - 0.8 n HCOOH, 2°C;
4. Concentration of the eluate at 30°C under reduced pressure, and freeze-drying;
5. Fractionation of sialic acids by chromatography on a column of cellulose powder in n-butanol/n-propanol/water (1:2:1), 2°C. (Small amounts (500 µg) of sialic acid mixtures are fractionated by preparative thin-layer chromatography.);
6. Analysis of the sialic acids by
   a) Orcinol method
   b) Thiobarbituric acid method
   c) Thin-layer radio-chromatography
   d) Determination of the amount (Hestrin's method, using hydroxylamin) and localization (using perjodic acid, polarographic method: Schauer and Faillard, 1968) of the O-acetyl groups;
   e) Estimation of the specific radioactivity and intramolecular localization of the radioactive label in biosynthesis experiments;
7. Crystallization from ethanol/water or acetic acid/water

2. Biosynthesis of Acylneuraminic Acids

In mammalian tissues, N-acetylneuraminate is synthesized from N-acetylmannosamine-6-phosphate and phosphoenolpyruvate (PEP) by action of N-acetylneuraminate-9-phosphate-synthetase (Warren and Felsenfeld,

1961; Roseman et al., 1961). A specific phosphatase then removes the phosphate group (Jourdian et al., 1964). These reactions, shown in Fig. 4, supposedly occur in the cytosole; they have been demonstrated in particle-free homogenates of liver (Warren and Felsenfeld, 1961a and b), in sheep colonic mucosa (Kent and Draper, 1968), and in submaxillary glands (Schauer, 1970b).

N-acetyl-D-glucosamine
↓
N-acetyl-D-mannosamine
↓ ATP → ADP
N-acetyl-D-mannosamine-6-phosphate
↓ PEP → P
N-acetyl-neuraminate-9-phosphate
↓ H_2O → P
N-acetyl-neuraminate
↓ CTP → PP
N-acetyl-neuraminate-CMP-glycoside
↓ → CMP
N-acetyl-neuraminate-heterosaccharide
(glycoproteins
glycolipids, etc.)

Fig. 4. Biosynthesis of N-acetylneuraminic acid and its transfer into glycosidic linkage in mammalian tissues

However, the enzymes responsible for activation and transfer of N-acetylneuraminic acid to glycoprotein or glycolipid acceptors probably are attached to subcellular particles. Thus, N-acetylneuraminate-CMP-synthetase has been found in nuclei of porcine retina tissue (Kean, 1969). The occurrence of sialyltransferases in subcellular membranes has been described by Roseman (1968).

For a long time the biosynthetic pathways of N-glycoloylneuraminic acid and also of N-acetyl-O-acetylneuraminic acids were unknown. *In vitro*, the isolated enzymes catalyzing the epimerization of N-acetylglucosamine to N-acetylmannosamine, the phosphorylation and condensation steps leading to N-acetylneuraminate (Fig. 4), allow the same reactions to occur with N-glycoloylhexosamine, resulting in the synthesis of N-glycoloylneuraminate (Warren and Felsenfeld, 1961a and b; Roseman et al., 1961). *In vivo*, however, this pathway does not exist. Experiments made in our laboratory establish the fact that N-acetylneuraminic acid is an intermediate for both N-glycoloyl- and N-acetyl-O-acetylneuraminic acids (Schauer et al., 1968; Schoop et al., 1969; Schauer, 1970a, b and c).

The glycoloyl group of N-glycoloylneuraminate is derived directly from the acetyl group by a hydroxylating enzyme. The synthesis of N-glycoloylneuraminic acid ^{14}C-labelled only in the glycoloyl group, was observed by incubating surviving slices of porcine submaxillary glands with several precursors of N-acylneuraminic acids containing N-(1-^{14}C)-acetyl groups such as N-(1-^{14}C)acetyl-mannosamine, N-(1-^{14}C)-acetyl-glucosamine or (1-^{14}C)acetate itself (Schauer et al., 1968; Schoop et al., 1969). The incorporation of ^{18}O from molecular oxygen specifically into the methyl group of the N-acetyl substituent of N-acetylneuraminic acid was also demonstrated (Schoop and Schauer, 1969). In particle-free homogenates of porcine submaxillary gland tissue free N-(1-^{14}C)acetyl-neuraminate is hydroxylated in the presence of ascorbate or NADPH and O_2. The enzyme was therefore called *"N-acetylneuraminate, ascorbate or NADPH:oxygen-oxidoreductase (N-acetyl-hydroxylating)"* (EC 1.14.1 or 2) Schauer, 1970c).

Investigating the biosynthesis of N-acetyl-O-acetylneuraminic acids in surviving slices of bovine and equine submaxillary glands, a rapid and uniform incorporation of (1-^{14}C)acetate into the N- and the 4-, 7-, or 8-O-acetyl groups of the various N-acetyl-O-acetylneuraminic acids (Table 1) was observed (Schauer, 1970a). Acetyl groups from acetyl coenzyme A are incorporated specifically into the positions 7 and/or 8 of N-acetylneuraminate in particle-free homogenates of bovine submaxillary glands (Schauer, 1970b). In the respective equine system N-acetyl-4-O-acetylneuraminate is synthesized. These experiments demonstrate the existence of specific O-acetyltransferases in the two animals. They have been called *"acetyl-CoA:N-acetylneuraminate-4-O-acetyltransferase"* (horse)and *"acetyl-CoA:N-acetylneuraminate 7 and/or 8-O-acetyltransferase(s)"* (cow); EC 2.3.1.? (Schauer, 1970a, b). Whether the bovine O-acetyltransferase system consists of one enzyme able to incorporate acetate into the positions 7 *and* 8, or of two different enzymes, is not known.

3. Subcellular Localization and Substrates of Enzymes Hydroxylating and O-Acetylating N-Acetylneuraminate

In Fig. 5 a scheme is given for the fractionation of submaxillary gland tissues used for preparation of the following subcellular compartments at a good yield: 1. subcellular membranes, 2. the 120,000 × g supernatant consisting predominantly of the soluble glycoproteins of the cytosole and 3. the pool of free sialic acids, liberated from the cytosole compartment by dialysis.

On the basis of preliminary results, both subcellular site and substrate specificity seem to be identical for the N-acetylneuraminate-hydroxylase and for the N-acetylneuraminate-O-acetyltransferases. From surviving slices of porcine or bovine submaxillary glands incubated with (1-^{14}C)acetate, free N-acetyl-, N-glycoloyl- and N-acetyl-O-acetylneuraminic acids not yet glycoprotein-bound, can be isolated in small amounts (about 1% of total sialic acids; Schoop et al., 1969; Schauer, 1970b). The highest specific radioactivity was found in this pool of free sialic acids, followed by the corresponding acylneuraminates of glycoproteins of subcellular membranes. The lowest specific radioactivity was shown to occur in the sialic acids of glycoproteins from the cytosole compartment (Fig. 6 and 7; Schauer, 1970b; Schauer and Wember, 1970). From Fig. 7 it can be seen that after 5 minutes of incubation the specific radioactivity of N-glycoloylneuraminic acid of membrane glycoproteins already amounted to about 5 dpm/μg sialic acid, whereas at this time radioactivity was not yet detectable in N-glycoloylneuraminic acid of cytosole glycoproteins. Similar specific radioactivities were observed for N-acetylneuraminic acid. The specific radioactivities of the small amounts of free N-acetyl- and N-glycoloylneuraminic acids isolated - not delineated in Fig. 7 - in several experiments were up to 150 times higher than that of the N-acetyl- and N-glycoloylneuraminic acids of cytosole glycoproteins after 2 or 4 hours of incubation (Schoop et al., 1969).

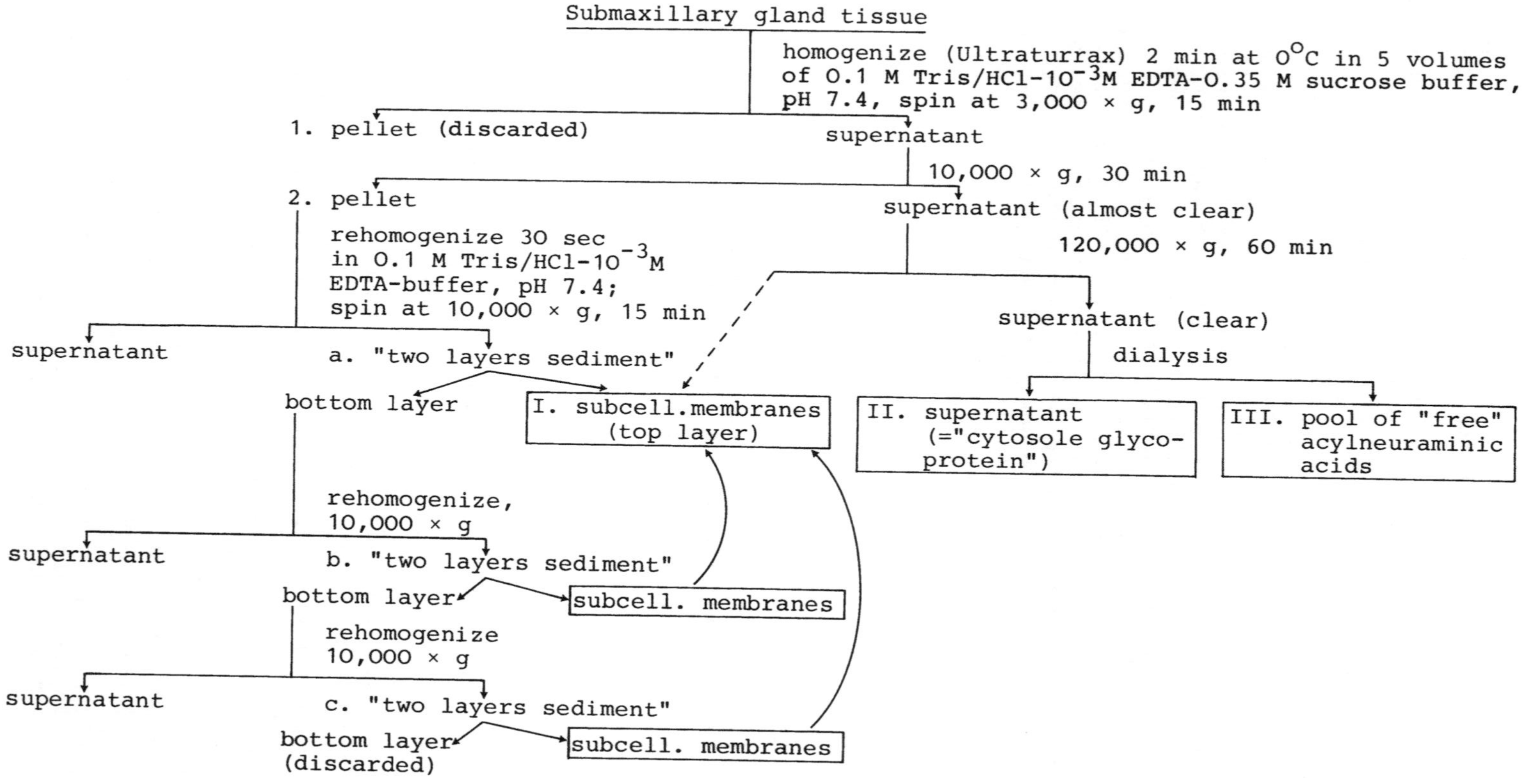

Fig. 5. Scheme for fractionation of bovine and porcine submaxillary gland tissues by differential centrifugation. Preparation of I. subcellular membranes ("microsomes"; 10,000 × g membrane fraction), II. the 120,000 × g supernatant, corresponding to the cytosole compartment and III. the pool of "free" sialic acids. - The 10,000 × g-sediment is composed of a yellowish top layer consisting predominantly of membranes, and a bottom layer from which additional membrane-material can be obtained after rehomogenization. The purity of the membranes was proved as described by Schauer (1970b), Schauer and Wember (1970), and by electron microscopy

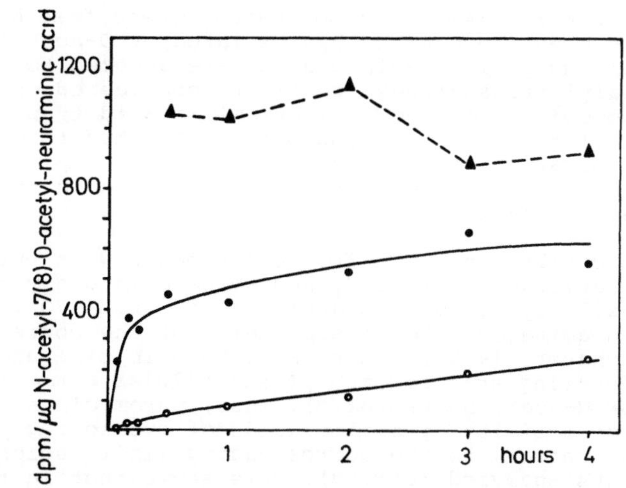

Fig. 6. Course of the specific radioactivity (dpm/μg) of N-acetyl-7 (8)-mono-O-acetylneuraminic acid in subcellular compartments of 5 g bovine submaxillary gland slices incubated with 50 μCi (1-^{14}C)acetate for several periods (Schauer, 1970b). N-acetyl-mono-O-acetylneuraminic acid, ▲---▲ in the pool of free sialic acids; •——• bound to glycoproteins of subcellular membranes; and o——o as component of cytosole glycoproteins. For preparation of the subcellular compartments, and for isolation and identification of the sialic acid see Table 2 and Fig. 5

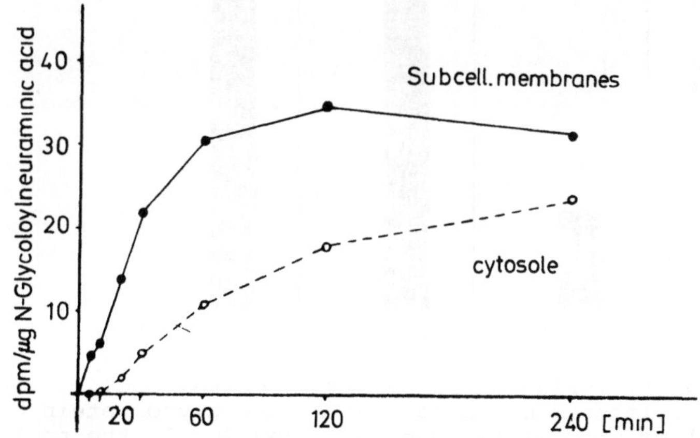

Fig. 7. Course of the specific radioactivity of N-glycoloylneuraminic acid of glycoproteins of subcellular membranes (•——•) and of cytosole compartment (o----o). Incubation of 5 g slices of porcine submaxillary gland with 25 μCi (1-^{14}C)acetate for several periods (Schauer and Wember, 1970)

These experiments indicate that free N-acetylneuraminate, and probably also its CMP-glycoside (we are testing this substrate at present), is converted in the living cell to free N-glycoloyl- and to free N-acetyl-O-acetylneuraminates. This is in accordance with the experiments made with particle-free homogenates, where free N-(1-^{14}C)acetylneuraminate has been shown to be hydroxylated or O-acetylated analogously. The various acylneuraminic acids are assumed to be then transferred (by sialyl-transferases, which are considered specific with regard to the composition of the acylneuraminic acid transferred; Schoop et al., 1969) to growing glycoproteins attached to microsomes. After the incorporation of the sialic acids, the complete glycoproteins may be released from the membranes and stored in the cytosole (Schauer and Wember, 1970).

It can be concluded from experiments with puromycin that the incorporation of N-glycoloylneuraminate into glycoproteins of porcine submaxillary glands (Schauer and Wember, 1970), and of N-acetyl-O-acetylneuraminates into glycoproteins of the corresponding bovine tissue (Schauer, 1970b), is related both in time and site to protein synthesis occurring at ribosomes of subcellular membranes. (The same was observed for N-acetylneuraminate). Thus, puromycin suppresses the specific radioactivity of sialic acids of cytosole glycoproteins at about 85% in contrary to the corresponding sialic acids in the other compartments analyzed (Fig. 8). This shows that in the intact cell N-acetylneuraminate bound to soluble glycoproteins present in high amounts in the cytosole (about 95% of total sialic acids of the tissue), seems not to serve as acceptor of oxygen or of acetyl groups.

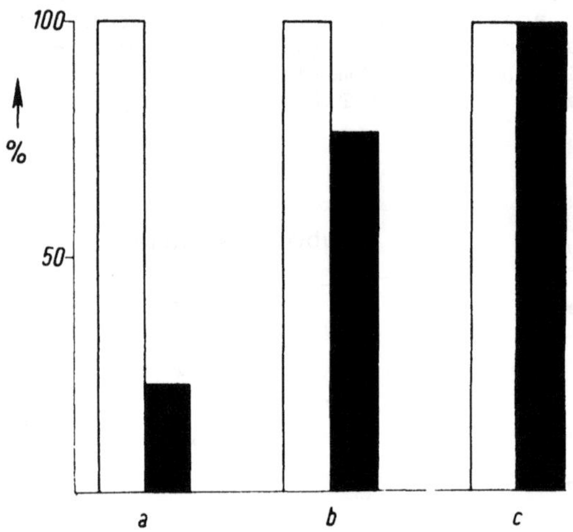

Fig. 8. Influence of puromycin on the specific radioactivity of N-glycoloylneuraminic acid: (a) in the glycoprotein of the cytosole, or (b) of subcellular membranes and (c) in the pool of free sialic acids. Porcine submaxillary gland, incubated with (1-^{14}C)acetate for 60 minutes. Black columns: 5×10^{-4}M puromycin; white columns: control (= 100%; (a) 6 dpm, (b) 22 dpm, and (c) 146 dpm/μg N-glycoloylneuraminic acid) (Schauer and Wember, 1970). - Puromycin has an equivalent influence on the specific radioactivities of N-acetyl-O-acetylneuraminic acids from the corresponding compartments (a), (b) and (c) of bovine submaxillary gland (Schauer, 1970b)

In addition to the chemical modifications of free N-acetylneuraminate occurring before the transfer to growing glycoproteins attached to subcellular membranes, as is indicated by the experiments presented, an incorporation of N-acetylneuraminate into subcellular membranes prior to the hydroxylation or O-acetylation reactions should be taken into consideration. A prerequisite for such a mechanism would be a group of enzymes arranged in subcellular membranes in such a way that it could hydroxylate or O-acetylate both free N-acetylneuraminate (or its 9-phosphate or CMP-glycoside) and also N-acetylneuraminate already incorporated into microsomal glycoproteins. Indeed, a complex composed of cooperating enzymes seems to exist in subcellular membranes. Washed bovine and porcine submaxillary gland microsomes have been found to incorporate (radioactive) N-acetyl-neuraminate (from its CMP-glycoside) into their glycoproteins and, thereafter, to O-acetylate (bovine microsomes) or hydroxylate (porcine and bovine microsomes) it there. The same membrane fractions are also able to O-acetylate or hydroxylate free N-acetylneuraminate.

Two recent experiments demonstrate the presence of the following enzymes in submaxillary gland microsomes: N-acetylneuraminate-transferase; N-acetylneuraminate, ascorbate or NADPH:oxygen-oxidoreductase; and acetyl-CoA:N-acetylneuraminate-O-acetyltransferases. Subcellular membranes from porcine and bovine submaxillary glands, in which only N-acetylneuraminate was radioactively labelled, were incubated in the presence of O_2 and ascorbate. In course of 6 hours, N-(1-^{14}C)-acetyl-neuraminate was converted to N-(1-^{14}C)glycoloyl-neuraminate (Fig. 9).

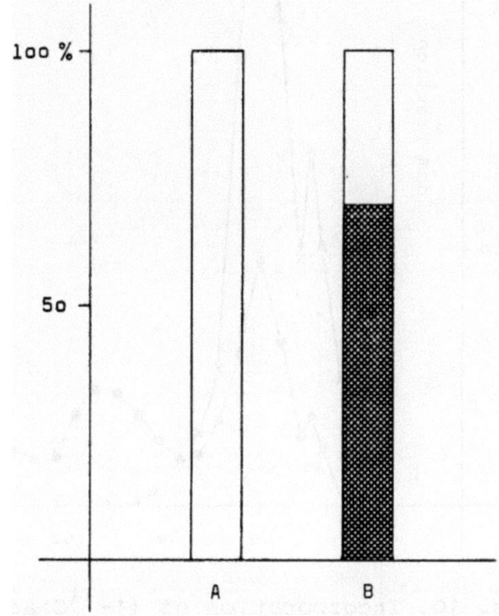

Fig. 9. Hydroxylation of N-(1-^{14}C)acetyl-neuraminate of subcellular membrane glycoproteins of porcine submaxillary gland. The incubation mixture consisted of 0.5 ml of 10,000 × g sediment of subcellular membranes labelled with N-(1-^{14}C)acetylneuraminate (= 100% of radioactive sialic acids), and 1.5 ml of 0.2 M Tris/ HCl - 10^{-3}M EDTA-buffer, pH 7.4, 10 µ moles of ascorbate added after every 2 hours; atmosphere: oxygen; time of incubation: 6 hours. For isolation of the sialic acids see Table 2. A: Control experiment: incubation mixture without ascorbate. B: Enzyme assay: complete incubation mixture. White column: N-(1-^{14}C)acetyl-neuraminate. Black column: N-(1-^{14}C)glycoloyl-neuraminate

The yields of glycoloylneuraminate were high (at an average of 70% in 10 experiments; ranging from 10 to 95%) in studies with microsomes of porcine submaxillary glands. They were low (5 to 10%) with bovine subcellular membranes. (N-(1-^{14}C)acetyl-neuraminate-labelled microsomes had been prepared by two methods: 1. Transfer of CMP-N-(1-^{14}C)acetyl-neuraminate to isolated subcellular membranes; this demonstrates the existence of N-acetylneuraminate-transferase in these particles. 2. Isolation of microsomes from tissue slices incubated anaerobically with (1-^{14}C)acetate (atmosphere: N_2) which yielded exclusively (1-^{14}C)-acetylated neuraminic acids.)

By incubation of subcellular membranes of bovine submaxillary gland with (1-^{14}C)acetyl-CoA, N-acetylneuraminate bound to membrane glycoproteins was O-acetylated. The formation of N-acetyl-O-(1-^{14}C)-acetyl- and of N-acetyl-di-O-(1-^{14}C)acetyl-neuraminic acids in these experiments is shown in Fig. 10.

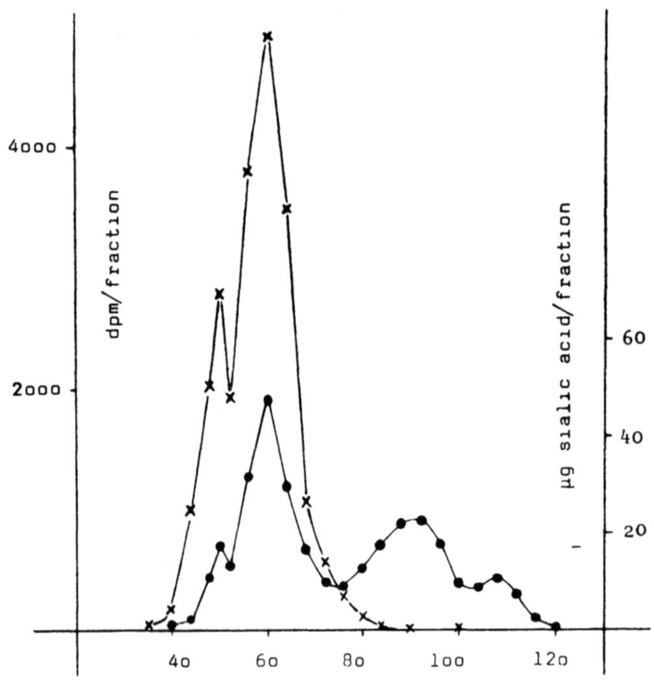

Fig. 10. Incorporation of (1-^{14}C)acetate into N-acetylneuraminate bound to subcellular membrane glycoproteins of bovine submaxillary glands. The incubation mixture consisted of 20 ml of 10,000 × g sediment of subcellular membranes, 20 ml of 0.2 M Tris/HCl-10^{-3}M EDTA-buffer, pH 7.4, 80 µ moles of $MgCl_2$, 100 µ moles of glutathione and 10 µ moles of (1-^{14}C)acetyl-CoA (20 µCi). The latter two substances were added in two portions at the beginning and after 2 hours of incubation. Time of incubation: 4 hours. The isolation of the sialic acids was performed according to Table 2. Analyses of the eluate were made for sialic acid content (Orcinol reagent, •—•), and for radioactivity (×—×). Fractions 40 - 52: N-acetyl-7,8-di-O-acetylneuraminic acid, 53 - 74 N-acetyl-7- and N-acetyl-8-O-acetylneuraminic acids, 75 - 102 N-acetylneuraminic acid, 103 - 120 N-glycoloylneuraminic acid

The results of these experiments with bovine and porcine submaxillary glands, which demonstrate chemical modifications of both free and membrane-bound N-acetyl-neuraminic acid by enzyme action, are summarized in a model presented in Fig. 11.

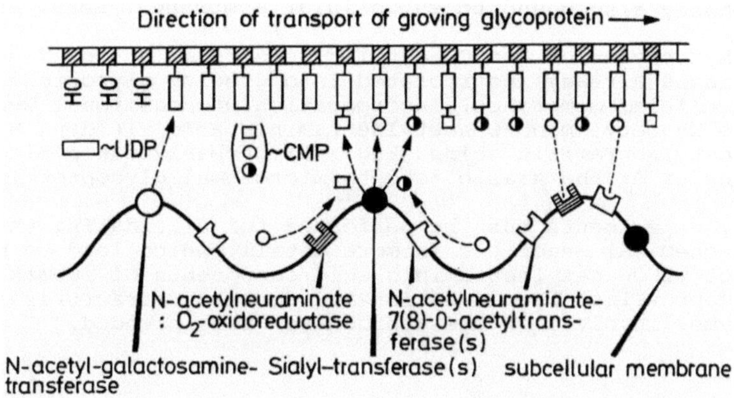

Fig. 11. "Multienzyme Complex" modifying and incorporating sialic acids into subcellular membranes of bovine submaxillary gland. For details see text. (O N-acetyl-, □N-glycoloyl-, ◑ N-acetyl-7(8)-mono and di-O-acetylneuraminic acids; ▭ N-acetyl-D-galactosamine; ▨ serine or threonine

A sequence of reactions is assumed to occur in subcellular membranes of bovine submaxillary gland leading to the carbohydrate moiety of the glycoprotein of this tissue. Bovine submaxillary glands synthesize both N-glycoloylneuraminic acid and several N-acetyl-(mono- and pluri-) O-acetylneuraminic acids (Schauer, 1970a) (Table 1), which are linked predominantly to N-acetylgalactosamine residues (Hashimoto and Pigman, 1962) bound to serine and threonine of the peptide chain (Downs and Pigman, 1969). In contrast to the disaccharide units of bovine glycoprotein, the composition of the carbohydrate moieties of porcine submaxillary gland glycoproteins was shown to be more complicated (Carlson, 1968). However, our experiments indicate that mechanism and subcellular localization of the synthesis and of the transfer of N-acetyl- and of N-glycoloylneuraminic acids to these oligosaccharide chains may correspond to those occurring in bovine tissue.

In summary, the membrane-bound "multienzyme complex" of bovine submaxillary gland is believed to catalyze the following reactions (Fig. 11):

1. Transfer of N-acetylgalactosamine to the protein core being transported through the "channels" of the endoplasmatic reticulum. Such an incorporation of N-acetylgalactosamine into glycoproteins of subcellular membranes was shown to occur by N-acetylgalactosamine transferases localized in subcellular membranes of bovine (Hagopian and Eylar, 1969) and of ovine (Rossignol et al., 1969) submaxillary glands.

2. Hydroxylation and O-acetylation of free, not yet glycoprotein-bound N-acetylneuraminate molecules or their C-9-phosphate or CMP-derivatives at the surface of subcellular membranes. (N-acetylneuraminate is assumed to be synthesized in cytosole compartment, as was already discussed above, possibly in the vicinity of subcellular membranes.)

3. Transfer of N-acetyl-, N-glycoloyl- and N-acetyl-O-acetyl-neuraminic acids to N-acetylgalactosamine residues of the glycoprotein by acylneuraminate-specific (?; Schoop et al., 1969) transferases. (Acylneuraminate-CMP-synthetase(s,?) responsible for the activation of the various sialic acids before their transfer to the glycoprotein, are probably also bound to subcellular membranes; Kean, 1969.)

4. Hydroxylation and O-acetylation of some of the N-acetylneuraminate residues already incorporated into growing glycoproteins attached to subcellular membranes. Incorporation of additional O-acetyl groups into N-acetyl-mono-O-acetylneuraminic acid yielding N-acetyl-pluri-O-acetylneuraminic acids (Fig. 10) probably takes place after the transfer of the sialic acid to microsomal glycoproteins.

Many experiments must be performed for elucidating the mechanism of the chemical reactions in more detail, which lead to N-glycoloyl- and N-acetyl-O-acetylneuraminic acid components of submaxillary gland glycoproteins. The isolation and further characterization of the enzymes involved in these processes are intended.

References

Blix, G., Gottschalk, A., Klenk, E.: Proposed nomenclature in the field of neuraminic and sialic acids. Nature 179, 1088 (1957).

Blix, G., Lindberg, E.: The sialic acids of bovine and equine submaxillary mucins. Acta chim. scand. 14, 1809 (1960).

Carlson, D.M.: Structures and immunochemical properties of oligosaccharides isolated from pig submaxillary mucins. J. biol. Chem. 243, 616 (1968).

Downs, F., Pigman, W.: Preparation of glycopeptides from bovine submaxillary mucin by chemical degradation. Biochemistry 8, 1760 (1969).

Gottschalk, A.: The chemistry and biology of sialic acids. Cambridge University Press 1960.

Gottschalk, A.: Glycoproteins; their Composition, Structure and Function, 2nd Ed. Amsterdam-London-New York: Elsevier 1972.

Hagopian, A., Eylar, E.H.: Glycoprotein biosynthesis: the purification and characterization of a polypeptide: N-acetylgalactosamine transferase from bovine submaxillary glands. Arch. Biochem. 129, 515 (1969).

Hashimoto, Y., Pigman, W.: A comparison of the composition of mucins and blood-group substances. Ann. N.Y. Acad. Sci. 93, 541 (1962).

Jourdian, G.W., Swanson, A.L., Watson, D., Roseman, S.: Isolation of sialic-acid-9-phosphatase from human erythrocytes. J. biol. Chem. 239, 8 (1964).

Kean, E.L.: Sialic acid activating enzyme in ocular tissue. Exp. Eye Res. 8, 44 (1969).

Kent, P.W., Draper, P.: Biosynthesis of intestinal mucins. Sialic acids of sheep colonic epithelial mucin. Biochem. J. 106, 293 (1968).

Roseman, S., Jourdian, G.W., Watson, D., Rood, R.: Enzymatic synthesis of sialic acid 9-phosphates. Proc. nat. Acad. Sci. (Wash.) 47, 958 (1961).

Roseman, S.: Biochemistry of glycoproteins and related substances. In: Cystic Fibrosis, Part II (Ed. E. Rossi, E. Stoll), p. 244. Basel: S. Karger 1968.

Rossignol, B., Herman, G., Clauser, H.: Tentative identification of N-acetylgalactosamine incorporating membranes from ovine submaxillary glands. Biochem. biophys. Res. Commun. 34, 111 (1969).

Schauer, R.: Biosynthese von N-Acetyl-O-acetylneuraminsäuren, I. Inkorporation von (^{14}C)Acetat in Schnitte der Unterkieferspeicheldrüse von Rind und Pferd. Z. physiol. Chem. 351, 595 (1970a).
Schauer, R.: Biosynthese von N-Acetyl-O-acetylneuraminsäuren, II. Untersuchungen über Substrat und intracelluläre Lokalisation der Acetyl-Coenzym A:N-Acetylneuraminat-7- und 8-O-Acetyltransferase vom Rind. Z. physiol. Chem. 351, 749 (1970b).
Schauer, R.: Biosynthese der N-Glykoloylneuraminsäure durch eine von Ascorbinsäure bzw. NADPH abhängige, N-acetyl-hydroxylierende "N-Acetylneuraminat:O_2-Oxidoreduktase" in Homogenaten der Unterkieferspeicheldrüse vom Schwein. Z. physiol. Chem. 351, 783 (1970c).
Schauer, R., Faillard, H.: Das Verhalten isomerer N,O-Diacetylneuraminsäureglykoside im Submaxillarismucin von Pferd und Rind bei Einwirkung bakterieller Neuraminidase. Z. physiol. Chem. 349, 961 (1968).
Schauer, R., Schoop, H.J., Faillard, H.: Zur Biosynthese der Glykoloylgruppe der N-Glykoloylneuraminsäure. Die oxidative Umwandlung der N-Acetyl-Gruppe zur Glykoloyl-Gruppe. Z. physiol. Chem. 349, 645 (1968).
Schauer, R., Wember, M.: Studies on the subcellular site of the biosynthesis of N-glycoloylneuraminic acid in porcine submaxillary gland. Z. physiol. Chem. 351, 1353 (1970).
Schoop, H.J., Schauer, R.: Über den Einbau von molekularem Sauerstoff bei der Biosynthese der N-Glycoloylneuraminsäure durch eine Neuraminsäure-N-Acetyl-Hydroxylase. Z. physiol. Chem. 350, 13 (1969).
Schoop, H.J., Schauer, R., Faillard, H.: Die oxidative Entstehung von N-Glykoloylneuraminsäure aus N-Acetylneuraminsäure. Z. physiol. Chem. 350, 155 (1969).
Warren, L., Felsenfeld, H.: The biosynthesis of N-acetylneuraminic acid. Biochem. biophys. Res. Commun. 4, 232 (1961a).
Warren, L., Felsenfeld, H.: N-acetylmannosamine-6-phosphate and N-acetylneuraminic-acid-9-phosphate as intermediates in sialic acid biosynthesis. Biochem. biophys. Res. Commun. 5, 185 (1961b).

N-Acetyl-Hydroxylation in the Biosynthesis of Glycoproteins

H. J. Schoop, Eva H. Stockhorst, and Mathilde Merkel

N-acetyl hydroxylases are involved in the formation of N-glycoloyl groups in N-acetylated substrates. Glycoloylated products known, so far, to occur in nature are N-glycoloylneuraminic acid and N-glycoloylmuramic acid. N-glycoloylneuraminic acid is the terminal sugar in glycoproteins and glycolipids of animals (Blix et al., 1956; Klenk and Uhlenbruck, 1957; Klenk and Padberg, 1962; Tettamanti et al., 1965). It is found in animals from echinodermata to mammals (Warren, 1963), with the exception of man, where N-acetylneuraminic acid is found exclusively (Schoop, 1970). N-glycoloylmuramic acid has recently been detected in the cell wall of mycobacteria (Adam et al., 1969; Azuma et al., 1970; Petit et al., 1970).

There is much evidence for N-acetyl hydroxylation in the biosynthesis of N-glycoloylneuraminic acid. 1. The radioactivity of (^{14}C) labelled N-acetyl groups of precursors in the biosynthesis of acylneuraminic acids has been recovered in the (^{14}C) labelled N-glycoloyl group of the N-glycoloylneuraminic acid (Schauer et al., 1968; Schoop et al., 1969). 2. The source of oxygen in the -hydroxyl function of the N-glycoloyl group of N-glycoloylneuraminic acid is molecular oxygen, and not oxygen of water (Schoop and Schauer, 1969; Schoop, 1971). This could be demonstrated in experiments using gaseous (^{18}O) oxygen and ($H^{18}OH$) water. An analogous N-acetyl hydroxylation will therefore be discussed in the biosynthesis of the N-glycoloylmuramic acid (Adam et al., 1969; Petit et al., 1970).

N-acetyl hydroxylation is known to occur in the detoxification processes of liver; N-acetylated 4-substituted anilides are in that case converted to the corresponding N-glycoloyl derivatives (v. Jagow et al., 1968; Kiese and Lenk, 1969, 1970). The N-acetyl hydroxylation reaction is also proved by recovering N-(^{14}C) glycoloyl derivatives after application of N-(^{14}C) acetyl anilides (Kiese and Lenk, 1970).

In spite of the similarities - formation of N-glycoloyl out of N-acetyl groups - it is assumed that there are different enzymes involved in these reactions. The acetanilide N-acetyl hydroxylase seems to be a typical *catabolic* xenobiotics hydroxylating mixed-function oxidase. In contrast the acetylneuraminate N-acetyl hydroxylase is involved in the *anabolic* biosynthesis of glycoproteins (Schoop and Schauer, 1969; Schoop, 1971). The acetylneuraminate hydroxylase is specific for N-acetylneuraminic acid (or the CMP or 9-phosphate derivatives) as substrate(s) (Schoop et al., 1969). The N-acylhexosamines are found to be non-glycoloylated in spite of being precursors of N-glycoloylneuraminic acid or being components of such glycoproteins, which contain N-glycoloylneuraminic acid (Schoop and Faillard, 1967).

The typical xenobiotics hydroxylating mixed-function oxidases using cytochrom P-450 are found in microsomal preparations (Estabrook et al., 1968). They are not to be solubilized without decrease of activity, and react with substrates which possess, as a rule, highly lipophilic properties, as e.g. in the case of acetanilide and derivatives (v. Jagow et al., 1968; Kiese and Lenk, 1969). N-acetylneuraminic acid, however, is a very hydrophilic substrate, and the acetylneuraminate N-acetyl hydroxylating activity may be found in the soluble supernatant of cell fragments (Schauer, 1970).

We may contribute another difference between the cited enzyme systems that are involved in the hydroxylation of N-acetyl groups. We could not find any evidence for the characteristic properties of cytochrom P-450 in acetyl-neuraminate hydroxylation. It was not possible to inhibit this reaction by carbon monoxide (vol/vol CO: O_2 = 1:1 or 9:1), in either the presence or absence of light (wavelength 450 nm). But it can be inhibited by agents that form complexes with ionizable iron.

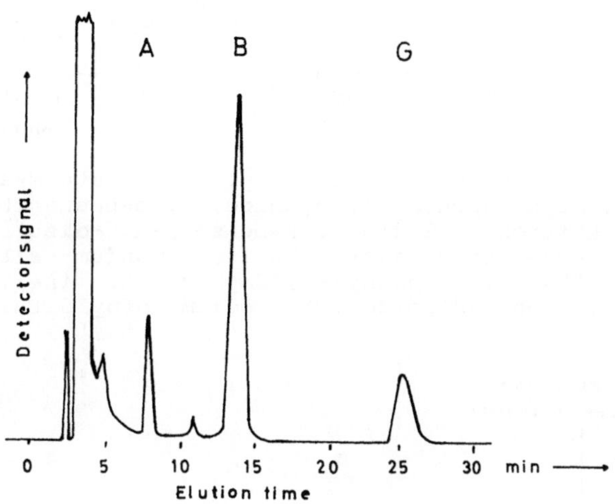

Fig. 1. Estimation of the N-acetyl (A) and the N-glycoloyl (G) groups of N-acylneuraminic acids by gas chromatography. Benzene (B) as internal standard. Gas chromatograph Hewlett a. Packard 5750, coiled glass colums 1/4" o.d., 6 ft., packed with polyaromatic resin (PAR 1) 80 - 120 mesh. 155°C isotherm, carrier gas: Nitrogen 50 ml/min., split 1:7, FID

For measuring the enzymic activity of the acetylneuraminate N-acetyl hydroxylase radio gas chromatography is used. The living slices of submaxillary glands of pigs were incubated for 8 hrs. in the presence of ($1-^{14}C$) acetate and pure oxygen. The labelled (^{14}C) acetylated and (^{14}C) glycoloylated neuraminic acids were isolated as described earlier (Schauer et al., 1968) and then ethanolyzed for gas chromatography of the acyl esters. The specific radioactivity of the (^{14}C) glycolate in relation to the (^{14}C) acetate is equivalent to the hydroxylating activity.

Fig. 2 summarizes the effects of several inhibitors. Cyanide is very effective, but the inhibition of N-acetyl hydroxylation is accompanied by a proportional decrease of the cell respiration. α,α-dipyridyl and o-phenanthroline, which are known to form complexes preferably with ferrous ions, are strong inhibitors of N-acetyl hydroxylation without effect on the cell respiration. The effects of the other chelating

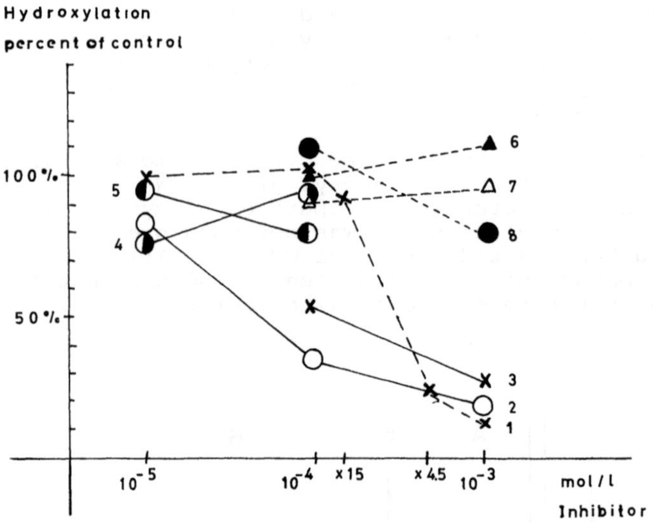

Fig. 2. Inhibition of acetylneuraminate N-acetyl hydroxylation by chelating agents: 1. cyanide, 2. o-phenanthroline, 3. α,α-dipyridyl, 4. tiferon (1,2-dihydroxybenzene-3,5-disulfonic acid), 5. kupferron (phenyl-nitroso-hydroxylamine-ammonium-salt), 6. bathocoproin (2,9-dimethyl-4,7-diphenyl-1,10-phenanthroline), 7. oxalic acid-bis-(cyclohexylidenehydrazide), 8. sodiumdiethyldithiocarbaminate

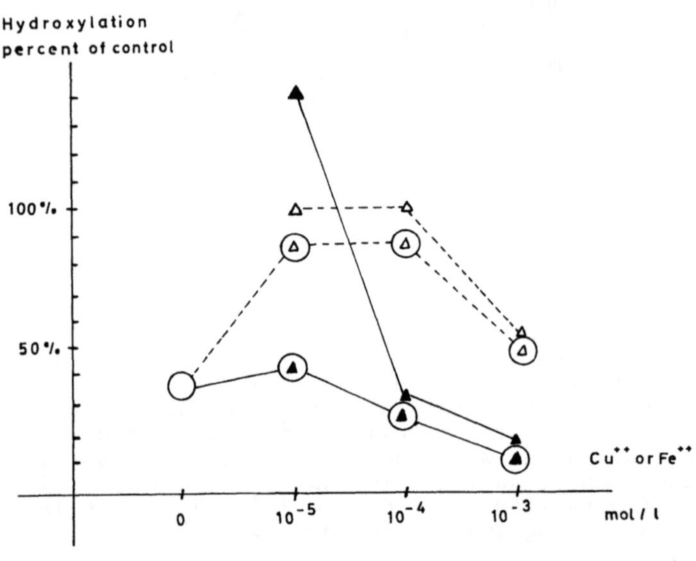

Fig. 3. The effects of Cu^{++} (——) and Fe^{++} (----) on the N-acetyl hydroxylation, and on the inhibition of the hydroxylation by o-phenanthroline (10^{-4}M) (⊙ = o-phenanthroline, (▲) = o-phenanthroline plus Cu^{++}, (△) = o-phenanthroline plus Fe^{++}, and ▲ = Cu^{++} resp. △ = Fe^{++} without o-phenanthroline

agents are not significant. There is no inhibition especially by those agents that have a greater affinity to form chelates with copper.

In order to get further insight into the role of iron we added Fe^{++} and Cu^{++} to inhibiting amounts of o-phenanthroline (Fig. 3). Fe^{++} in 10^{-5} and 10^{-4} molar concentrations prohibits the inhibition by o-phenanthroline, whereas Cu^{++} does not reactivate the N-acetyl hydroxylation. In the absence of o-phenanthroline 10^{-5}M Cu^{++} shows a remarkable activation, but 10^{-4} to 10^{-3}M Cu^{++} is toxic both for N-acetyl hydroxylation and cell respiration, whether in presence of o-phenanthroline or not.

These results give evidence that acetylneuraminate N-acetyl hydroxylase needs ferrous ions for activity. The missing properties of cytochrom P-450 as well as the easy exchange of the iron indicate that these ferrous ions are not heme bound.

References

Adam, A., Petit, J.F., Wietzerbin-Falszpan, J.: L'acide N-glycolylmuramique, constituant des parois de mycobacterium smegmatis: identification par spectrometrie de masse. FEBS Letters 4, 87 (1969).

Azuma, I., Thomas, D.W., Adam, A., Ghuysen, J.-M., Bonaly, R., Petit, J.-F., Lederer, E.: Occurrence of N-glycolylmuramic acid in bacterial cell walls. Biochim. biophys. Acta (Amst.) 208, 444 (1970).

Blix, G., Lindberg, E., Odin, L., Werner, J.: Studies on sialic acids. Acta Soc. Med. Upsalien. 61, 1 (1956).

Estabrook, R.W., Hildebrandt, A., Remmer, H., Schenkman, J.B., Rosenthal, O., Cooper, D.Y.: The role of cytochrome P-450 in microsomal mixed function oxidation reactions. In: Biochemie des Sauerstoffs (B. Hess, Hj. Staudinger, Ed.), p. 142 - 177. 19. Colloquium der Gesellschaft für Biologische Chemie. Berlin-Heidelberg-New York: Springer 1968.

Jagow, R. von, Kiese, M., Lenk, W.: Hydroxylation of acetic to glycolic acid in rabbits. Biochim. biophys. Acta (Amst.) 158, 45 (1968).

Kiese, M., Lenk, W.: Oxydation of acetanilides to glycolanilides and ocanilic acid in rabbits. Biochem. Pharmacol. 18, 1325 (1969).

Kiese, M., Lenk, W.: Hydroxylation of N-(4-chlorophenyl) ($^{14}C_1$) acetamide to N-(4-chlorophenyl)-($^{14}C_1$) glycolamide in rabbits. Biochim. biophys. Acta (Amst.) 222, 549 (1970).

Klenk, E., Padberg, G.: Über die Ganglioside von Pferdeerythrocyten. Hoppe-Seylers Z. physiol. Chem. 327, 249 (1962).

Klenk, E., Uhlenbruck, G.: Über die Abspaltung von N-Glykolyl-neuraminsäure (P-Sialinsäure) aus dem Schweinesubmaxillarismucin durch das "Receptor-Destroying Enzyme". Hoppe-Seylers Z. physiol. Chem. 307, 266 (1957).

Petit, J.F., Adam, A., Wietzerbin-Falszpan, J.: Isolation of UDP-N-glycolylmuramyl- (Ala, Glu, DAP) from mycobacterium phlei. FEBS Letters 6, 55 (1970).

Schauer, R.: Biosynthese der N-Glykoloylneuraminsäure durch eine von Ascorbinsäure bzw. NADPH abhängige N-Acetyl-hydroxylierende "N-Acetylneuraminat: O_2-Oxidoreduktase" in Homogenaten der Unterkieferspeicheldrüse vom Schwein. Hoppe-Seylers Z. physiol. Chem. 351, 783 (1970).

Schauer, R., Schoop, H.J., Faillard, H.: Zur Biosynthese der Glykolyl-Gruppe der N-Glykolyl-neuraminsäure. Hoppe-Seylers Z. physiol. Chem. 349, 645 (1968).

Schoop, H.J.: Die Acetylneuraminat-N-Acetyl-Hydroxylase, ein "sauerstoffübertragendes Ferment". Untersuchungen zu Vorkommen und Biosynthese der N-Glykolyl-neuraminsäure. Habilitationsschrift, Bochum 1970.

Schoop, H.J.: The incorporation of molecular oxygen in the biosynthesis of N-glycolylneuraminic acid. Hoppe-Seylers Z. physiol. Chem. in preparation (1971).

Schoop, H.J., Faillard, H.: Beitrag zur Biosynthese der Glykolyl-Gruppe der N-Glykolyl-neuraminsäure. Hoppe-Seylers Z. physiol. Chem. 348, 1518 (1967).

Schoop, H.J., Schauer, R.: Über den Einbau von molekularem Sauerstoff bei der Biosynthese der N-Glykolyl-neuraminsäure durch eine Neuraminsäure-N-Acetyl-Hydroxylase. Hoppe-Seylers Z. physiol. Chem. 350, 13 (1969).

Schoop, H.J., Schauer, R., Faillard, H.: Die oxydative Entstehung von N-Glykolyl-neuraminsäure aus N-Acetyl-Neuraminsäure. Hoppe-Seylers Z. physiol. Chem. 350, 155 (1969).

Tettamanti, G., Bertona, L., Berra, B., Zambotti, V.: Glycolyl-neuraminic acid in ox brain gangliosids. Nature (Lond.) 206, 192 (1965).

Warren, L.: The distribution of sialic acids in nature. Comp. Biochem. Physiol. 10, 153 (1963).

Biosynthesis of Collagen

K. I. Kivirikko

The collagen fibres in connective tissue are composed of tropocollagen molecules, which are rod-like macromolecules with a diameter of about 1.4 nm and a length of about 280 nm. These molecules are built up from three polypeptide chains called α-chains, which are coiled around one another in a superhelical conformation. The molecular weight of the individual α-chain is about 95,000, and thus the molecular weight of the tropocollagen molecule is about 300,000. In most of the body collagen there are two chains of one type, designated α1, and one of another type designated α2, each having a similar but distinct amino acid sequence (for reviews on collagen structure, see Ramachandran, 1967; Grant and Prockop, 1972).

The tropocollagen molecules are synthesized within the fibroblasts or their variants, such as osteoblasts and chondroblasts, and collagen biosynthesis follows the basic principles of protein biosynthesis in general. However, in the case of this protein there are a number of unusual reactions, and one of the special features is that the initial polypeptide chains are modified by several steps after all the information available in messenger RNA is translated. The general steps in the biosynthesis of collagen are: (a) assambly on ribosomal complexes of a polypeptide precursor of collagen, called protocollagen; (b) hydroxylation of appropriate proline and lysine residues in protocollagen to hydroxyproline and hydroxylysine residues; (c) substitution of some of the hydroxylysine residues with galactose or glucosylgalactose in O-glycosidic linkage; (d) secretion of a transport form of collagen into the extracellular matrix, and (e) specific aggregation and cross-linking of collagen to form the extracellular collagen fibres. The biosynthesis of collagen has been summarized in several recent reviews (e.g. Chvapil and Hurych, 1968; Gould, 1968; Prockop, 1970; Grant and Prockop, 1972), and, accordingly, the present review will deal mainly with the most recent studies on the biosynthesis of collagen. For more detailed references to the older work, the reader is referred to the above-mentioned reviews. The cross-linking and fibre formation of collagen are discussed in other articles in this book and, therefore, will not be reviewed here.

1. Ribosomal Assembly of Protocollagen

As with other proteins, the polypeptide chains of collagen are synthesized on ribosomal complexes containing messenger RNA. Early isotopic experiments of Stetten and Schoenheimer (1944) and of Stetten (1949) demonstrated that, unlike most of the amino acids found in proteins, free hydroxyproline is not incorporated into collagen. Instead, the hydroxyproline of collagen is derived from proline which is hydroxylated in bound form. These observations have since been confirmed by a number of investigators. Some studies suggested that small amounts of hydroxyproline can be incorporated directly into collagen, but more recent work has indicated that these findings were probably due to the contamination of the labelled hydroxyproline by small amounts of *cis*-hydroxyproline which was incorporated (Rosenbloom, 1971). Moreover, it has been found that the hydroxylysine of collagen originates from lysine, and not from the free hydroxylysine of the tissues. It has further been shown that hydroxyprolyl- and hydroxylysyl-tRNAs are not intermediates in the biosynthesis of collagen, and that the hydroxylation reactions do not occur until after the synthesis of a proline-rich and lysine-rich polypeptide precursor of collagen called protocollagen (for references, see above-mentioned reviews).

Since the size of the α-chains of collagen is large compared to that of the peptide chains of most other proteins, it was suggested in early studies that the α-chains might be composed of subunits which are synthesized independently and then joined by non-peptidic covalent bonds after they have been released from the ribosomes (see Gallop et al., 1967; Hodge, 1967). Recent studies have indicated, however, that the α-chains are synthesized as continuous single chains, and not as separate subunits (Vuust and Piez, 1970). Accordingly, the mRNA required for coding the α-chains containing about 1,000 amino acids must be of large size (Grant and Prockop, 1972). Lazarides and Lukens (1971) have recently reported data suggesting that the size of the mRNA of chick embryo polysomes synthesizing collagen corresponds to that expected for a monocistronic mRNA molecule coding for a single α-chain.

At the present time, it is not known whether the two α1-chains and one α2-chain come together to form the triple helical structure found in the tropocollagen molecule while the chains are being assembled on the ribosomes or at some later stage (see Grant and Prockop, 1972). Also, the question of whether the hydroxylations of proline and lysine take place while polypeptide chains are being assembled on ribosomes or after the release of complete α-chains is unsettled (see below).

2. Hydroxylations of Proline and Lysine

Hydroxylation of Proline. The synthesis of the hydroxyproline in collagen by the hydroxylation of proline in protocollagen is catalyzed by a soluble (Hutton and Udenfriend, 1966; Kivirikko and Prockop, 1967b) enzyme called protocollagen proline hydroxylase, and the hydroxylation reaction has been shown to require molecular oxygen, α-ketoglutarate, ferrous iron and a reducing substance, which can be ascorbate (e.g. Fujimoto and Tamiya, 1962; Prockop et al., 1963; Hurych and Chvapil, 1965; Prockop and Juva, 1965; Hutton et al., 1967; Kivirikko and Prockop, 1967b). The requirement of oxygen, initially reported in experiments with intact tissues, has been confirmed with purified preparations of the enzyme, and the K_m has been found to correspond to an atmosphere of about 3 to 6 per cent oxygen (Hutton et al., 1967; Kivirikko et al., 1968). Ferrous iron has been shown to be a specific requirement with a K_m of 1 to 5 × 10^{-6}M (Kivirikko and Prockop, 1967c, 1967d; Rhoads and Udenfriend, 1970). α-Ketoglutarate is likewise a specific requirement with a K_m of 5 to 10 × 10^{-6}M (Kivirikko and Prockop, 1967c, 1967d; Rhoads and Udenfriend, 1970), and a stoichiometric conversion of α-ketoglutarate to succinate and carbon dioxide has been shown to be coupled to the hydroxylation of proline residues (Rhoads and Udenfriend, 1968). During this reaction one atom of the oxygen molecule is incorporated into the hydroxyproline, while the other atom is incorporated into succinate (Cardinale et al., 1971).

The requirement for ascorbate is not specific, and in experiments with purified enzyme ascorbate can be replaced partly by several reducing compounds (Hutton et al., 1967; Rhoads and Udenfriend, 1970). However, it is not known whether these compounds can replace ascorbate *in vivo*.

Protocollagen proline hydroxylase has recently been isolated in highly purified form from chick embryo extracts (Halme et al., 1970a) and new-born rat skin (Rhoads and Udenfriend, 1970). The enzyme is an acidic protein with its isoelectric point at pH 4.4 (Pänkäläinen et al., 1970) and containing large amounts of acidic amino acids (Halme et al., 1970a; Rhoads and Udenfriend, 1970). The absorption spectrum shows no absorption maxima between 280 and 600 nm, which suggests the absence of certain prosthetic groups such as flavin nucleotides (Pänkäläinen and Kivirikko, 1971). Assays for thiamine pyrophosphate have demonstrated the absence of this coenzyme, indicating that the decarboxylation of α-ketoglutarate by protocollagen proline hydroxylase is not related to the decarboxylation of α-ketoglutarate which occurs in the Krebs cycle in mitochondria (Rhoads and Udenfriend, 1970; Pänkäläinen and Kivirikko, 1971). Highly purified preparations of protocollagen proline hydroxylase have been found to contain far less than one mole of iron per mole of enzyme, indicating that the enzyme does not contain stoichiometric amounts of firmly bound iron (Pänkäläinen and Kivirikko, 1971). Thiol reagents such as p-mercuribenzoate inhibit protocollagen proline hydroxylase, suggesting that free sulphydryl groups are required for the activity of the enzyme (Popenoe et al., 1969; Halme et al., 1970a). It is possible that iron is bound to the enzyme by sulphydryl linkages in a non-heme form (Prockop, 1971). The enzyme purified from chick embryos has a molecular weight of about 250,000 (Pänkäläinen et al., 1970), and recently it has been found to be a tetramer consisting of two kinds of monomers with molecular weights of about 60,000 and 64,000 (see Olsen et al., 1973). Electron microscopic studies have shown that the monomers are rod-shaped with a diameter of about 33 Å and a length of about 70 Å, and the tetramers are squares about 80 by 80 Å. The size and the different shapes of the structures seen by electron microscopy could be accounted for a model of the tetramer in which two V-shaped dimers were interlocked (Olsen et al., 1973).

Studies on substrate requirements of protocollagen proline hydroxylase have indicated that the enzyme does not hydroxylate free proline, the tripeptide Gly-Pro-Pro, or poly-L-proline (Kivirikko and Prockop, 1967b), but it catalyzes the hydroxylation of proline in the position preceding glycine in synthetic peptides with the structure of (X-Pro-Gly)$_n$ in which "X" is proline, alanine or a variety of other amino acids but not glycine (Prockop et al., 1967; Kivirikko and Prockop, 1967d; Nordwig et al, 1967; Hutton et al., 1968; Kivirikko et al., 1969; Suzuki and Koyama, 1969; Kikuchi et al., 1969). The minimum sequence requirement for the hydroxylation of proline is an intact -X-Pro-Gly-triplet (Rhoads and Udenfriend, 1969; McGee et al., 1971b). Data obtained in studies with a number of different peptide substrates and inhibitors indicate that the amino acid sequence, charge, chain length and helix of the peptide affect the interaction with protocollagen proline hydroxylase (see McGee et al., 1971b; Kivirikko et al., 1972b).

The effect of chain length is seen in that longer peptide chains of random-coil forms of (Ala-Pro-Gly)$_n$ and (Pro-Pro-Gly)$_n$ are hydroxylated more readily than shorter one is (Kivirikko and Prockop, 1967d; Hutton et al., 1968; Kivirikko et al., 1969, 1972b). A similar effect of chain length is also seen in the competitive inhibition of the enzyme by poly-L-proline in that the K_i decreases progressively when the chain length of poly-L-proline is increased from 12 to 150 prolyl residues (Prockop and Kivirikko, 1969; Kivirikko et al., 1969). The effect of chain length may be explained either by an unusually large active site of the enzyme or lateral movement of the enzyme along the polypeptide substrate (Prockop and Kivirikko, 1969; Prockop, 1970). The effect of helix on the interaction with the enzyme is seen in that certain polypeptides are not hydroxylated by the enzyme before their helix has been destroyed by thermal denaturation (Fujimoto and Prockop, 1968; Rhoads and Udenfriend, 1968; Rhoads et al., 1971), and certain other peptides are hydroxylated with a lower rate in the triple helical conformation than in a random-coil form (Kikuchi et al., 1969). Experiments on the inhibition of the enzyme by poly-L-proline have indicated that only form II of poly-L-proline, which has an extended single-stranded helix resembling the triple-stranded helix of collagen is a competitive inhibitor of the enzyme, whereas form I of poly-L-proline is not inhibitory (Kivirikko et al., 1967). On the other hand, studies with protocollagen, the natural substrate for the enzyme, have not shown an inhibitory effect of helix on hydroxyproline formation (Hutton et al., 1967; Kivirikko et al., 1968; Nordwig and Pfab, 1968). It seems evident that further studies are required before we know all the factors which affect the interaction of polypeptide substrates and inhibitors with protocollagen proline hydroxylase.

The activity of protocollagen proline hydroxylase has recently been measured in a number of animal and human tissues in various experimental and clinical conditions. The activity of the enzyme is highest in embryonic tissues (Mussini et al., 1967), and the activity in the tissues of young animals (Juva and Heikkinen, 1968) and children (Uitto et al., 1969) is higher than that in mature tissues. The enzyme activity increases with the rate of collagen synthesis in developing granulomas (Juva, 1968) and chick embryos (Halme, 1969). A similar parallelism between the enzyme activity and collagen formation is also found in the uterus during pregnancy and post-partum involution (Kao et al., 1969; Halme and Jääskeläinen, 1970). In experimental hepatic (Takeuchi and Prockop, 1969) and pulmonary (Halme et al., 1970b) fibrosis marked elevations are observed in the protocollagen proline hydroxylase activity, and the enzyme activity increases prior to an increase in the collagen content. Increased protocollagen proline hydroxylase activity has further been reported in the skin of some patients with scleroderma or other connective tissue disorders (Uitto et al., 1969, 1970a), in synovial tissue from patients with rheumatoid arthritis (Uitto et al., 1970b), and in rabbit aorta in experimental arteriosclerosis (Fuller and Langner, 1970). Addition of lactate to the medium of early log-phase fibroblasts increases the activity of protocollagen proline hydroxylase several-fold (Comstock and Udenfriend, 1970), and this effect may be due to an activation of an inactive precursor of the enzyme (McGee et al., 1971a). Thus, it is not known whether the changes in the enzyme activity in various conditions summarized above are due to the synthesis of increased amounts of the enzyme protein or to an activation of the inactive precursor molecule.

Hydroxylation of Lysine. The synthesis of hydroxylysine in collagen by the hydroxylation of lysine in protocollagen requires, like the synthesis of hydroxyproline, molecular oxygen, α-ketoglutarate, ferrous iron and a reducing substance which can be ascorbate (Prockop et al., 1966; Kivirikko and Prockop, 1967b, 1967c; Hausmann, 1967; Hurych and Nordwig, 1967). The K_m values for these cofactors or cosubstrates are only slightly different from those observed in the hydroxylation of proline (Kivirikko and Prockop, 1972), and α-ketoglutarate has been shown to be stoichiometrically converted to succinate and carbon dioxide during the hydroxylation of lysine (Kivirikko et al., 1972a) as during the hydroxylation of proline (see above).

Initial studies on the hydroxylation of lysine indicated that partially purified preparations of protocollagen proline hydroxylase also hydroxylated lysine residues in protocollagen, and it was suggested that both hydroxylations might be performed by a single enzyme (Kivirikko and Prockop, 1967c). Subsequently, however, highly purified preparations of protocollagen proline hydroxylase were shown not to contain any lysine hydroxylase activity (Halme et al., 1970a), and more recently , partially purified preparations of protocollagen lysine hydroxylase were prepared which were free of any proline hydroxylase activity (Miller, 1971; Kivirikko and Prockop, 1972). Experiments with competing polypeptides likewise indicated that the enzymic site for the hydroxylation of lysine is separate from that for the hydroxylation of proline (Weinstein et al., 1969). Thus it can be concluded that protocollagen proline hydroxylase and protocollagen lysine hydroxylase are two separate enzymes.

Protocollagen lysine hydroxylase has not yet been obtain in pure form, but recently the enzyme was purified up to 300-fold from chick embryo extracts, and was found to have several properties similar to those of protocollagen proline hydroxylase (Kivirikko and Prockop, 1972). Some initial information is available as to the determinants for the hydroxylation of lysine in peptide linkage (Kivirikko et al., 1972a). Data obtained so far demonstrate that both the amino acid sequence around lysine and the length of the peptide chain are critical determinants in the hydroxylation of lysine by protocollagen lysine hydroxylase.

Intracellular Site of the Hydroxylation Reactions. The question of whether the hydroxylations of proline and lysine take place while protocollagen polypeptide chains are being assembled on the ribosomes or whether they occur after the complete polypeptides have been released from the ribosomes has been subject to a number of investigations. Experiments using anaerobic conditions or chelators for ferrous iron indicated that protocollagen synthesis continued at 40 to 90 per cent of the control rate for several hours when the hydroxylation reactions were inhibited (Juva et al., 1966; Bhatnagar and Prockop, 1966), and the size of the protocollagen peptide chains was comparable to the size of complete α-chains of collagen (Lukens, 1966; Kivirikko and Prockop, 1967a; Lukens, 1970). When the inhibition of the hydroxylation reactions was reversed by exposing the tissues to oxygen or iron, protocollagen which had accumulated during the period of inhibition was hydroxylated to the same extent as the protocollagen in control tissues (Juva et al., 1966). If protocollagen had stayed on the ribosomes during the whole period of inhibition, the molecules would have quickly filled all the active ribosomes, and protocollagen synthesis would not have continued at 40 to 90 per cent of the control rate (see Prockop, 1970). Thus these experiments clearly indicate that during the reversal period protocollagen was hydroxylated in the cells of intact tissues after complete polypeptide chains were released from the ribosomes.

These experiments did not indicate, however, whether the hydroxylations occurred after the release of complete polypeptide chains from ribosomes under conditions where the hydroxylations had not been inhibited. Data from several experiments suggested that even under such conditions most or all of the hydroxylation of proline residues takes place after complete polypeptide chains have been released from ribosomes (Rosenbloom et al., 1967; Bhatnagar et al., 1967a, 1967b, 1967c; Bachra and van der Eb, 1970). Other experiments suggest, however, that about half of the hydroxylation of proline residues occurs before the release of the polypeptide chains from the ribosomes (Fernández-Madrid, 1970; Miller and Udenfriend, 1970; Lazarides and Lukens, 1971), and some experiments suggest that the major part of the hydroxylation of proline occurs on nascent peptides (Lazarides et al., 1971). This question is thus unsettled (for more detailed discussion, see Prockop, 1970). An explanation for the controversial results may be that normally part of the hydroxylation of proline takes place before the release of the peptide chains from ribosomes, and part after the release of the peptides, and in some cells or conditions a larger part of the hydroxylation of proline takes place before the release of the peptide chains than in others. Most of the studies mentioned above indicate that the hydroxylation of proline does not begin before the peptide chain has reached a certain minimum length, and that the hydroxylation of proline residues lags clearly behind the incorporation of proline.

The intracellular site of the hydroxylation of lysine residues has not been studied as extensively as the site of the hydroxylation of proline, but the hydroxylation of lysine has been shown to parallel the hydroxylation of proline (Blumenkrantz et al., 1969). It seems likely, therefore, that the above conclusions about the intracellular site of the hydroxylation of proline are also applicable to the hydroxylation of lysine.

3. Glycosylation of Hydroxylysine

All collagens contain small but variable amounts of carbohydrate in the form of galactose or glucosylgalactose in O-glycosidic linkage to hydroxylysine (e.g. Butler and Cunningham, 1966; Spiro, 1967, 1969). Since hydroxylysine is synthesized by hydroxylation of lysine in protocollagen, it is clear that hydroxylysine is glycosylated in peptide linkage. Studies on the biosynthesis of collagen in chick embryo tibiae have indicated that the glycosylation of hydroxylysine lags behind the synthesis of hydroxylysine by several minutes (Blumenkrantz et al., 1969).

The presence of a UDP-galactosyl transferase and a UDP-glucosyl transferase, which appear to be specific for hydroxylysine, has been demonstrated in the skin and cartilage of embryonic guinea pigs (Bosmann and Eylar, 1968a, 1968b) and in rat kidney cortex (Spiro and Spiro, 1968). Both enzymes seem to be bound to membrane fractions of the cells, and there is evidence suggesting that the enzymes are located in the plasma membrane (Hagopian et al., 1968). Both enzymes require Mn^{+2}, and initial information which is available about the properties of the enzymes suggests that the specificity of the enzymes from skin and cartilage differs significantly from the specificity of the enzymes from the kidney (above refs.). Evidently extensive further studies are required to explain why only certain hydroxylysine residues in collagen are glycosylated, and why certain residues contain galactose while others contain glucosylgalactose.

Morgan et al. (1970) have recently studied the amino acid sequences of glycopeptides from collagens from three sources. Their results indicated that the hydroxylysine substituted with glucosylgalactose was present in a sequence -Gly-X-Hyl(Gal-Glc)-Gly-X'-Arg-, where X is Phe, Met or Ile and X' is Ile or His. Because all these sequences contained Arg in the third position of the triplet following the glycosylated hydroxylysine, it was suggested that the invariant arrangement of hydroxylysine and arginine in the two triplets may be an absolute requirement for glycosylation.

4. Evidence for a Transport Form of Collagen

Several laboratories have recently reported that newly-synthesized polypeptide chains of collagen are larger than α-chains (Layman et al., 1971; Bellamy and Bornstein, 1971; Dehm and Prockop, 1971; Ramaley and Rosenbloom, 1971; Müller et al., 1971; Stark et al., 1971; Jimenez et al., 1971; Dehm et al., 1972). The larger form of collagen had solubility properties differing from those of the native tropocollagen molecule (Layman et al., 1971), and chromatographic behaviors of the polypeptide chains derived from the larger form of collagen were different from those of α-chains. After limited digestion with pepsin the larger chains were converted to α-chains (Layman et al., 1971; Bellamy and Bornstein, 1971; Jimenez et al., 1971). Pulse-chase experiments indicated a time-dependent conversion of the larger form to collagen comprised of α-chains thus demonstrating that the larger form is a precursor of collagen (Bellamy and Bornstein, 1971). The names procollagen (Bellamy and Bornstein, 1971) and transport form of collagen (Layman et al., 1971; Jimenez et al., 1971) have been used to describe this larger form of collagen.

The apparent molecular weight of the polypeptide chains of the transport form of collagen has been reported to be about 120,000 when estimated by sodium dodecyl sulfate-acrylamide gel electrophoresis (Bellamy and Bornstein, 1971) or by gel filtration in a sodium dodecyl sulfate-agarose column (Jimenez et al., 1971). The transport form of collagen was recently visualized in the electron microscope, and it was found to differ from tropocollagen molecules in that the transport form had an extension of about 13 nm at the amino-terminal end (Dehm et al., 1972). The molecular weight of the polypeptide chains of the transport form, as estimated by electron microscopy, was about 105,000 to 110,000 which is a slightly lower value than that estimated by other methods (above). Experiments with ^{14}C-cystine suggested that the extension present in the transport form contained cystine, and that the transport form was in part stabilized by disulfide bonds (Dehm et al., 1972).

Two functions have been suggested for the transport form of collagen (above refs., and Speakman, 1971). The extensions of the α-chains may aggregate rapidly and specifically with the extensions of two other α-chains so that three appropriate polypeptide chains are brought together. This mechanism could facilitate rapid formation of the triple helical molecule, and the specificity of the extensions could explain why two α1-chains and one α2-chain form the molecule. The other function of the extensions of the α-chains may be to prevent the aggregation of collagen molecules into insoluble fibres before collagen has been secreted from the cell and has moved to an appropriate distance from the cell. It seems likely that an enzyme which cleaves the extensions of the α-chains is present in the extracellular matrix, and after the cleavage, the tropocollagen molecules can aggregate into the extracellular collagen fibres.

5. Secretion of Collagen into the Extracellular Matrix

Early studies suggested that collagen is secreted from the fibroblasts by reverse pinocytosis of the Golgi vacuoles (Revel and Hay, 1963). However, more recent experiments suggested that collagen is either secreted directly from the cisternae of the endoplasmic reticulum (Ross and Benditt, 1965) or moves from ribosomal complexes to the ground cytoplasm and then through the plasma membrane (Salpeter, 1968; Cooper and Prockop, 1968).

Autoradiographic studies with chick embryo tibiae *in vitro* indicated that when the hydroxylations of proline and lysine were inhibited by anaerobic conditions or by chelating ferrous iron, protocollagen accumulated in the cells synthesizing collagen, and was either not secreted or was secreted at a markedly reduced rate (Juva et al., 1966). Control experiments indicated that these conditions did not affect the secretion of collagen which had already been hydroxylated, and thus these studies suggested that either the hydroxyproline and/or hydroxylysine residues or the carbohydrates linked to hydroxylysine are required for the secretion of the molecules. These observations have later been confirmed with tendon cells incubated *in vitro* (Dehm and Prockop, 1971) and with fibroblasts in culture (Ramaley and Rosenbloom, 1971; Switzer and Summer, 1971). One study suggested that rapidly dividing fibroblasts secreted protocollagen into the medium (Gribble et al., 1969), but the size of the molecules was not studied, and more recent studies suggested that the secreted material may have been degradation products of accumulated protocollagen (Ramaley and Rosenbloom, 1971).

Studies with analogues of proline and lysine have further demonstrated the special features involved in the secretion of collagen. Four analogues of proline were shown to be incorporated into collagen and other proteins in place of proline, and the collagen containing the analogue was found to remain in the cells or to be secreted at a negligible rate (Takeuchi et al., 1969; Rosenbloom and Prockop, 1970, 1971). The observations made with the proline analogue *cis*-hydroxyproline were of special interest, because the collagen contained normal amounts of lysine and hydroxylysine, and the only apparent difference from normal collagen was that about half the positions normally occupied by proline and *trans*-hydroxyproline were substituted by *cis*-hydroxyproline (Rosenbloom and Prockop, 1971). The results suggested that the collagen was not secreted either because the presence of *cis*-hydroxyproline interfered with the secretion or because the *trans*-hydroxyproline content was below the level necessary for the secretion. Studies with a lysine analogue 4,5-dehydrolysine indicated that collagen containing normal amounts of hydroxyproline but reduced amounts of hydroxylysine and glycosylated hydroxylysine was not secreted either (Rosenbloom and Christner, 1971). It thus seems that the presence of both hydroxyproline and hydroxylysine or glycosylated hydroxylysine in the molecule may be necessary for the secretion of collagen, and the cell must have some fine sensing mechanism or barrier which determines that only appropriate molecules are extruded (Grant and Prockop, 1972).

6. Factors Affecting Collagen Biosynthesis

As indicated in this review, the biosynthesis of collagen fibres is a complex process which involves several discrete steps. Accordingly, it seems probable that a number of factors can affect the rate of collagen biosynthesis. Some of these factors are unspecific in that conditions affecting ribosomal protein synthesis in general are probably accompanied by changes in the rate of collagen synthesis. In addition, many hormones influence the metabolism of collagen, but their exact mode of action is not known, and it is also not known to what extent their effects are specific for collagen metabolism (see Kivirikko, 1970).

Factors affecting the unique steps in the biosynthetic pathway of collagen fibres can be expected to have more specific effects. As discussed above, the synthesis of both hydroxyproline and hydroxylysine in collagen requires molecular oxygen, ferrous iron, α-ketoglutarate and perhaps ascorbate. Changes in the intracellular concentrations of these substances will thus influence the rate of collagen biosynthesis. For instance, it has been shown that chelation of ferrous iron by α',α-dipyridyl inhibits the hydroxylation reactions (Hurych and Chvapil, 1965; Juva et al., 1966), and more recently two drugs, tetracycline (Halme et al, 1969) and D-penicillamine (Uitto, 1969), were shown to inhibit collagen biosynthesis in bones in organ culture, partly because they chelated ferrous iron. Scurvy is known to be associated with defective collagen metabolism, and although ascorbate is not entirely specific as a cofactor in experiments with purified enzymes, an accumulation of protocollagen has been found in the granulomas of scorbutic guinea pigs during collagen biosynthesis *in vitro* (Gottlieb et al., 1966) and in embryonic bones cultured in the absence of ascorbic acid (Jeffrey and Martin, 1966; Halme et al, 1969). These observations may explain at least in part the defect in collagen metabolism in scurvy. The fact that α-ketoglutarate is required in the hydroxylation reactions may indicate an important connection of collagen biosynthesis with the energy metabolism of the cell.

The levels of several enzymes, such as protocollagen proline and lysine hydroxylases, the glycosylating enzymes, and the postulated enzyme cleaving the transport form of collagen may affect collagen biosynthesis. As summarized in this review, the level of protocollagen proline hydroxylase activity in the tissues has been found to be altered in several conditions, but at the present time little is known about changes in the levels of the other enzymes. Furthermore, the significance of the increases in the activities of these enzymes is not known as yet, because no clear evidence has been presented showing that these enzymes are rate-limiting enzymes of the biosynthetic pathway.

Other factors which influence specifically the biosynthesis of collagen fibres include factors affecting the secretion of collagen from the cell, and factors affecting the cross-linking of collagen. As discussed above, incorporation of several analogues of proline and lysine into collagen results in a markedly reduced rate of collagen secretion. It seems possible, therefore, that administration of the analogues can be used specifically to control the rate of collagen formation in patients with clinical conditions in which overproduction of collagen causes harmful effects (see Grant and Prockop, 1972). Attempts to influence the biosynthesis of collagen have also been made by using drugs affecting the cross-linking of collagen (see Grant and Prockop, 1972). Because the process of cross-linking is outside the scope of this review, these experiments have not been discussed here.

Many of the special features in the biosynthesis of collagen are only partly understood at the present time, and much of the basic information about the various steps in the biosynthetic pathway has been obtained only during the last few years. Consequently, it is not surprising that relatively little is known about factors affecting the rate of collagen biosynthesis and about the specific control of collagen metabolism. It can be expected, however, that such information will now gradually increase. This information should make it possible to formulate hypotheses as to the causes of various diseases affecting collagen metabolism and to develop methods for specific control of collagen biosynthesis and for treatment of conditions with altered collagen metabolism.

References

Bachra, B.N., van der Eb, A.J.: Site of proline hydroxylation during collagen synthesis in mouse fibroblasts. Biochemistry 9, 3001 - 3011 (1970).

Bellamy, G., Bornstein, P.: Evidence for procollagen, a biosynthetic precursor of collagen. Proc. nat. Acad. Sci. (Wash.) 68, 1138 - 1142 (1971).

Bhatnagar, R.S., Kivirikko, K.I., Rosenbloom, J., Prockop, D.J.: Transfer of puromycin-containing polypeptides through the plasma membrane of cartilage cells synthesizing collagen. Proc. nat. Acad. Sci. (Wash.) 58, 248 - 255 (1967a).

Bhatnagar, R.S., Prockop, D.J.: Dissociation of the synthesis of sulphated mucopolysaccharides and the synthesis of collagen in embryonic cartilage. Biochim. biophys. Acta (Amst.) 130, 383 - 392 (1966).

Bhatnagar, R.S., Prockop, D.J., Rosenbloom, J.: Intracellular pool of unhydroxylated polypeptide precursors of collagen. Science 158, 492 - 494 (1967b).

Bhatnagar, R.S., Rosenbloom, J., Kivirikko, K.I., Prockop, D.J.: Effect of cycloheximide on collagen biosynthesis as evidence for a post-ribosomal site for the hydroxylation of proline. Biochim. biophys. Acta (Amst.) 149, 273 - 281 (1967c).

Blumenkrantz, N., Rosenbloom, J., Prockop, D.J.: Sequential steps in the synthesis of hydroxylysine and in the glycosylation of hydroxylysine during the biosynthesis of collagen. Biochim. biophys. Acta (Amst.) 192, 81 - 89 (1969).

Bosmann, H.B., Eylar, E.H.: Attachment of carbohydrate to collagen. Isolation, purification and properties of the glucosyl transferase. Biochem. biophys. Res. Commun. 30, 89 - 94 (1968a).

Bosmann, H.B., Eylar, E.H.: Glycoprotein biosynthesis: The biosynthesis of the hydroxylysine-galactose linkage in collagen. Biochem. biophys. Res. Commun. 33, 340 - 346 (1968b).

Butler, W.T., Cunningham, L.: Evidence for the linkage of a disaccharide to hydroxylysine in tropocollagen. J. biol. Chem. 241, 3882 - 3888 (1966).

Cardinale, G.J., Rhoads, R.E., Udenfriend, S.: Simultaneous incorporation of ^{18}O into succinate and hydroxyproline catalyzed by collagen proline hydroxylase. Biochem. biophys. Res. Commun. 43, 537 - 543 (1971).

Chvapil, M., Hurych, J.: Control of collagen biosynthesis. Int. Rev. Connect. Tissue Res. 4, 67 - 196 (1968).

Comstock, J.P., Udenfriend, S.: Effect of lactate on collagen proline hydroxylase activity in cultured L-929 fibroblasts. Proc. nat. Acad. Sci. (Wash.) 66, 552 - 557 (1970).

Cooper, G.W., Prockop, D.J.: Intracellular accumulation of protocollagen and extrusion of collagen by embryonic cartilage cells. J. Cell Biol. 38, 523 - 537 (1968).
Dehm, P., Jimenez, S.A., Olsen, B., Prockop, D.J.: A transport form of collagen from embryonic tendon. Electron microscopic demonstration of an NH_2-terminal extension and evidence suggesting the presence of cystine in the molecule. Proc. nat. Acad. Sci. (Wash.) 69, 60 - 64 (1972).
Dehm, P., Prockop, D.J.: Synthesis and extrusion of collagen by freshly isolated cells from chick embryo tendon. Biochim. biophys. Acta (Amst.) 240, 358 - 369 (1971).
Fernández-Madrid, F.: Collagen biosynthesis - a review. Clin. Orthop. Related Res. 68, 163 - 181 (1970).
Fujimoto, D., Prockop, D.J.: Denatured collagen from the cuticle of *Ascaris lumbricoides* as a substrate for protocollagen proline hydroxylase. J. biol. Chem. 243, 4138 - 4142 (1968).
Fujimoto, D., Tamiya, N.: Incorporation of ^{18}O from air into hydroxyproline by chick embryo. Biochem. J. 84, 333 - 335 (1962).
Fuller, G.C., Langner, R.O.: Elevation of aortic proline hydroxylase: a biochemical defect in arteriosclerosis. Science 168, 987 - 989 (1970).
Gallop, P.M., Blumenfeld, O.O., Seifter, S.: Subunits and special structural features of tropocollagen. In: G.N. Ramachandran (Ed.): Treatise on Collagen. Vol. 1. Chemistry of Collagen, p. 339 - 366. London: Academic Press 1967.
Gottlieb, A.A., Kaplan, A., Udenfriend, S.: Further evidence for the accumulation of a hydroxyproline-deficient, collagenase-degradable protein during collagen biosynthesis *in vitro*. J. biol. Chem. 241, 1551 - 1555 (1966).
Gould, B.S.: Collagen biosynthesis at a ribosomal level. Int. Rev. Connect. Tissue Res. 4, 33 - 65 (1968).
Grant, M.E., Prockop, D.J.: The biosynthesis of collagen. New Engl. J. Med. 286, 194 - 196, 242 - 249, 291 - 300 (1972).
Gribble, T.J., Comstock, J.P., Udenfriend, S.: Collagen chain formation and peptidyl proline hydroxylation in monolayer tissue cultures of L-929 fibroblasts. Arch. Biochem. 129, 308 - 316 (1969).
Hagopian, A., Bosmann, H.B., Eylar, E.H.: Localization of peptidyl: N-acetyl galactosaminyl, collagen glucosyl and glycoprotein galactosyl transferases in HeLa cell membrane fractions. Arch. Biochem. 128, 387 - 396 (1968).
Halme, J.: Development of protocollagen proline hydroxylase activity in the chick embryo. Biochim. biophys. Acta (Amst.) 192, 90 - 95 (1969).
Halme, J., Jääskeläinen, M.: Protocollagen proline hydroxylase of the mouse uterus during pregnancy and post-partum involution. Biochem. J. 116, 367 - 369 (1970).
Halme, J., Kivirikko, K.I., Kaitila, I., Saxén, L.: Effect of tetracycline on collagen biosynthesis in cultured embryonic bones. Biochem. Pharmacol. 18, 827 - 836 (1969).
Halme, J., Kivirikko, K.I., Simons, K.: Isolation and partial characterization of highly purified protocollagen proline hydroxylase. Biochim. biophys. Acta (Amst.) 198, 460 - 470 (1970a).
Halme, J., Uitto, J., Kahanpää, K., Karhunen, P., Lindy, S.: Protocollagen proline hydroxylase activity in experimental pulmonary fibrosis of rats. J. Lab. clin. Med. 75, 535 - 541 (1970b).
Hausmann, E.: Cofactor requirements for the enzymatic hydroxylation of lysine in a polypeptide precursor of collagen. Biochim. biophys. Acta (Amst.) 133, 591 - 593 (1967).
Hodge, A.J.: Structure at the electron microscopic level. In: G.N. Ramachandran (Ed.): Treatise on Collagen. Vol. 1. Chemistry of Collagen, p. 185 - 206. London: Academic Press 1967.

Hurych, J., Chvapil, M.: Influence of chelating agents on the biosynthesis of collagen. Biochim. biophys. Acta (Amst.) 97, 361 - 363 (1965)
Hurych, J., Nordwig, A.: Inhibition of formation of hydroxylysine by chelating agents. Biochim. biophys. Acta (Amst.) 140, 168 - 170 (1967)
Hutton, J.J., Marglin, A., Witkop, B., Kurtz, J., Berger, A., Udenfriend, S.: Synthetic polypeptides as substrates and inhibitors of collagen proline hydroxylase. Arch. Biochem. 125, 779 - 785 (1968).
Hutton, J.J., Tappel, A.L., Udenfriend, S.: Cofactor and substrate requirements of collagen proline hydroxylase. Arch. Biochem. 118, 231 - 240 (1967).
Hutton, J.J., Udenfriend, S.: Soluble collagen proline hydroxylase and its substrates in several animal tissues. Proc. nat. Acad. Sci. (Wash. 56, 198 - 202 (1966).
Jeffrey, J.J., Martin, G.R.: The role of ascorbic acid in the biosynthesis of collagen. II. Site and nature of ascorbic acid participation. Biochim. biophys. Acta (Amst.) 121, 281 - 291 (1966).
Jimenez, S.A., Dehm, P., Prockop, D.J.: Further evidence for a transport form of collagen. Its extrusion and extracellular conversion to α-chains in embryonic tendon. FEBS Letters 17, 245 - 248 (1971).
Juva, K.: Hydroxylation of proline in the biosynthesis of collagen. An experimental study with chick embryo and granulation tissue of rat. Acta physiol. scand. Suppl. 308, 1 - 73 (1968).
Juva, K., Heikkinen, E.: The activity of protocollagen hydroxylase in rat skin at different ages. Abstr. 5th Meeting of the Federation of European Biochem. Soc. Prague, No. 561 (1968).
Juva, K., Prockop, D.J., Cooper, G.W., Lash, J.W.: Hydroxylation of proline and the intracellular accumulation of a polypeptide precursor of collagen. Science 152, 92 - 94 (1966).
Kao, K.-Y.T., Arnett, W.M., McGavack, T.H.: Effect of endometrial phase, ovariectomy, estradiol and progesterone on uterine protocollagen hydroxylase. Endocrinology 85, 1057 - 1061 (1969).
Kikuchi, Y., Fujimoto, D., Tamiya, N.: The enzymic hydroxylation of protocollagen models. Biochem. J. 115, 569 - 574 (1969).
Kivirikko, K.I.: Urinary excretion of hydroxyproline in health and disease. Int. Rev. Connect. Tissue Res. 5, 93 - 163 (1970).
Kivirikko, K.I., Bright, H.J., Prockop, D.J.: Kinetic patterns of protocollagen hydroxylase and further studies on the polypeptide substrate. Biochim. biophys. Acta (Amst.) 151, 558 - 567 (1968).
Kivirikko, K.I., Ganser, V., Engel, J., Prockop, D.J.: Comparison of poly-L-proline I and II as inhibitors of protocollagen hydroxylase. Z. physiol. Chem. 348, 1341 - 1344 (1967).
Kivirikko, K.I., Kishida, Y., Sakakibara, S., Prockop, D.J.: Hydroxylation of $(X\text{-Pro-Gly})_n$ by protocollagen proline hydroxylase. Effect of chain length, helical conformation and amino acid sequence in the substrate. Biochim. biophys. Acta (Amst.) 271, 347 - 356 (1972b).
Kivirikko, K.I., Prockop, D.J.: Partial characterization of protocollagen from embryonic cartilage. Biochem. J. 102, 432 - 442 (1967a).
Kivirikko, K.I., Prockop, D.J.: Purification and partial characterization of the enzyme for the hydroxylation of proline in protocollagen. Arch. Biochem. 118, 611 - 618 (1967b).
Kivirikko, K.I., Prockop, D.J.: Enzymatic hydroxylation of proline and lysine in protocollagen. Proc. nat. Acad. Sci. (Wash.) 57, 782 - 789 (1967c).
Kivirikko, K.I., Prockop, D.J.: Hydroxylation of proline in synthetic polypeptides with purified protocollagen hydroxylase. J. biol. Chem. 242, 4007 - 4012 (1967d).
Kivirikko, K.I., Prockop, D.J.: Partial purification and characterization of protocollagen lysine hydroxylase from chick embryos. Biochim. biophys. Acta (Amst.) 258, 366 - 379 (1972).

Kivirikko, K.I., Prockop, D.J., Lorenzi, G.P., Blout, E.R.: Oligopeptides with the sequences Ala-Pro-Gly and Gly-Pro-Gly as substrates or inhibitors for protocollagen proline hydroxylase. J. biol. Chem. 244, 2755 - 2760 (1969).
Kivirikko, K.I., Shudo, K., Sakakibara, S., Prockop, D.J.: Studies on protocollagen lysine hydroxylase. Hydroxylation of synthetic peptides and the stoichiometric decarboxylation of α-ketoglutarate. Biochemistry 11, 122 - 129 (1972a)
Layman, D.L., McGoodwin, E.B., Martin, G.R.: The nature of collagen synthesized by cultured human fibroblasts. Proc. nat. Acad. Sci. (Wash.) 68, 454 - 458 (1971).
Lazarides, E.L., Lukens, L.N.: Collagen synthesis on polysomes *in vivo* and *in vitro*. Nature New Biology 232, 37 - 40 (1971).
Lazarides, E.L., Lukens, L.N., Infante, A.A.: Collagen polysomes: site of hydroxylation of proline residues. J. molec. Biol. 58, 831 - 846 (1971).
Lukens, L.N.: The size of the polypeptide precursor of collagen hydroxyproline. Proc. nat. Acad. Sci. (Wash.) 55, 1235 - 1242 (1966).
Lukens, L.N.: The formation of collagen α-chains in the absence of proline hydroxylation. J. biol. Chem. 245, 453 - 461 (1970).
McGee, J.O'D., Langness, U., Udenfriend, S.: Immunological evidence for an inactive precursor of collagen proline hydroxylase in cultured fibroblasts. Proc. nat. Acad. Sci. (Wash.) 68, 1585 - 1589 (1971a).
McGee, J.O'D., Rhoads, R.E., Udenfriend, S.: The substrate recognition site of collagen proline hydroxylase: The hydroxylation of -X-Pro-Gly-sequences in bradykinin analogs and other peptides. Arch. Biochem. Biophys. 144, 343 - 351 (1971b).
Miller, R.L.: Purification and properties of lysyl-protocollagen hydroxylase. Fed. Proc. 30, 1195 (1971).
Miller, R.L., Udenfriend, S.: Hydroxylation of proline residues in collagen nascent chains. Arch. Biochem. 139, 104 - 113 (1970).
Morgan, P.H., Jacobs, H.G., Segrest, J.P., Cunningham, L.W.: A comparative study of glycopeptides derived from selected vertebrate collagens. J. biol. Chem. 245, 5042 - 5048 (1970).
Müller, P.K., McGoodwin, E., Martin, G.R.: Studies on protocollagen: Identification of a precursor of proto α_1. Biochem. biophys. Res. Commun. 44, 110 - 117 (1971).
Mussini, E., Hutton, J.J., Udenfriend, S.: Collagen proline hydroxylase in wound healing, granuloma formation, scurvy and growth. Science 157, 927 - 929 (1967).
Nordwig, A., Pfab, F.K.: Independence of collagen hydroxylation on the conformational state of the precursor substrate. Biochim. biophys. Acta (Amst.) 154, 603 - 605 (1968).
Nordwig, A., Kobrle, V., Pfab, F.K.: Investigations on collagen proline hydroxylation in a cell-free system. Biochim. biophys. Acta (Amst.) 147, 487 - 496 (1967).
Olsen, B.R., Berg, R.A., Kivirikko, K.I., Prockop, D.J.: Structure of protocollagen proline hydroxylase from chick embryos. Eur. J. Biochem. 35, 135 - 147 (1973).
Pänkäläinen, M., Aro, H., Simons, K., Kivirikko, K.I.: Protocollagen proline hydroxylase: molecular weight, subunits and isoelectric point. Biochim. biophys. Acta (Amst.) 221, 559 - 565 (1970).
Pänkäläinen, M., Kivirikko, K.I.: The absence of iron and certain coenzymes in highly purified protocollagen proline hydroxylase. Biochim. biophys. Acta (Amst.) 229, 504 - 508 (1971).
Popenoe, E.A., Aronson, R.B., van Slyke, D.D.: The sulphydryl nature of collagen proline hydroxylase. Arch. Biochem. 133, 286 - 292 (1969).
Prockop, D.J.: Intracellular biosynthesis of collagen and interactions of protocollagen proline hydroxylase with large polypeptides. In: E.A. Balazs (Ed.): The Chemistry and Molecular Biology of the Intercellular Matrix. Vol. 1, p. 335 - 370. London: Academic Press 1970.

Prockop, D.J.: Role of iron in the synthesis of collagen in connective tissue. Fed. Proc. 30, 984 - 990 (1971).
Prockop, D.J., Juva, K.: Synthesis of hydroxyproline in vitro by the hydroxylation of proline in a precursor of collagen. Proc. nat. Acad. Sci. (Wash.) 53, 661 - 668 (1965).
Prockop, D.J., Juva, K., Engel, J.: Enzymatic synthesis of hydroxyproline by the hydroxylation of poly (L-prolyl-glycyl-L-prolyl). Z. physiol. Chem. 348, 553 - 560 (1967).
Prockop, D.J., Kaplan, A., Udenfriend, S.: Oxygen-18 studies on the conversion of proline to collagen hydroxyproline. Arch. Biochem. 101, 499 - 503 (1963).
Prockop, D.J., Kivirikko, K.I.: Effect of polymer size on the inhibition of protocollagen proline hydroxylase by polyproline II. J. biol. Chem. 244, 4838 - 4842 (1969).
Prockop, D.J., Weinstein, E., Mulveny, T.: Hydroxylation of lysine in a polypeptide precursor of collagen. Biochem. biophys. Res. Commun. 22, 124 - 128 (1966).
Ramachandran, G.N. (Ed.): Treatise on Collagen. Vol. 1. London: Academic Press 1967.
Ramaley, P.B., Rosenbloom, J.: Inhibition of proline and lysine hydroxylation prevents normal extrusion of collagen by 3T6 fibroblasts in culture. FEBS Letters 15, 59 - 64 (1971).
Revel, J.P., Hay, E.D.: An autoradiographic and electron microscope study of collagen synthesis in differentiating cartilage. Z. Zellforsch. 61, 110 - 117 (1963).
Rhoads, R.E., Udenfriend, S.: Decarboxylation of α-ketoglutarate coupled to collagen proline hydroxylase. Proc. nat. Acad. Sci. (Wash.) 60, 1473 - 1478 (1968).
Rhoads, R.E., Udenfriend, S.: Substrate specificity of collagen proline hydroxylase: hydroxylation of a specific proline residue in bradykinin. Arch. Biochem. 133, 108 - 111 (1969).
Rhoads, R.E., Udenfriend, S.: Purification and properties of collagen proline hydroxylase from newborn rat skin. Arch. Biochem. 139, 329 - 339 (1970).
Rhoads, R.E., Udenfriend, S., Bornstein, P.: *In vitro* enzymatic hydroxylation of prolyl residues in the α1-CB2 fragment of rat collagen. J. biol. Chem. 246, 4138 - 4142 (1971).
Rosenbloom, J.: *Trans*-Hydroxyproline is not incorporated into collagen. Arch. Biochem. 142, 718 - 719 (1971).
Ross,R., Benditt, E.P.: Wound healing and collagen formation. V. Quantitative electron microscope radioautographic observations of proline-H^3 utilization by fibroblasts. J. Cell Biol. 27, 83 - 106 (1965).
Rosenbloom, J., Bhatnagar, R.S., Prockop, D.J.: Hydroxylation of proline after the release of proline-rich polypeptides from ribosomal complexes during uninhibited collagen biosynthesis. Biochim. biophys. Acta (Amst.) 149, 259 - 272 (1967).
Rosenbloom, J., Christner, P.: Effects of 4,5-dehydrolysine on collagen biosynthesis in embryonic chick tibiae. Fed. Proc. 30, 1195 (1971).
Rosenbloom, J., Prockop, D.J.: Biochemical aspects of collagen biosynthesis. In: J.E. Dunphy, W. van Winkle (Eds.): Repair and Regeneration. The Scientific Basis for Surgical Practice, p. 117 - 135. New York: McGraw-Hill 1969.
Rosenbloom, J., Prockop, D.J.: Incorporation of 3,4-dehydroproline into protocollagen and collagen. Limited hydroxylation of proline and lysine in the same polypeptide. J. biol. Chem. 245, 3361 - 3368 (1970).
Rosenbloom, J., Prockop, D.J.: Incorporation of *cis* -hydroxyproline into protocollagen and collagen. Collagen containing *cis*-hydroxyproline in place of proline and *trans*-hydroxyproline is not extruded at a normal rate. J. biol. Chem. 246, 1549 - 1555 (1971).

Salpeter, M.M.: H^3-Proline incorporation into cartilage. Electron microscope autoradiographic observations. J. Morph. 124, 387 - 442 (1968).
Speakman, P.T.: Proposed mechanism for the biological assembly of collagen triple helix. Nature 229, 241 - 243 (1971).
Spiro, R.C.: The structure of the disaccharide unit of the renal glomerular basement membrane. J. biol. Chem. 242, 4813 - 4823 (1967).
Spiro, R.G.: Characterization and quantitative determination of the hydroxylysine-linked carbohydrate units of several collagens. J. biol. Chem. 244, 602 - 612 (1969).
Spiro, R.G., Spiro, M.J.: Enzymatic synthesis of the hydroxylysine-linked disaccharide of basement membranes and collagens. Fed. Proc. 27, 345 (1968).
Stark, M., Lenaers, A., Lapiere, C., Kühn, K.: Electronoptical studies of precollagen from the skin of dermatosparaxic calves. FEBS Letters 18, 225 - 227 (1971).
Stetten, M.R.: Some aspects of metabolism of hydroxyproline studied with the aid of isotopic nitrogen. J. biol. Chem. 181, 31 - 37 (1949).
Stetten, M.R., Schoenheimer, R.: The metabolism of l(-)-proline studied with the aid of deuterium and isotopic nitrogen. J. biol. Chem. 153, 113 - 132 (1944).
Suzuki, F., Koyama, E.: Hydroxylation of proline in collagen model peptide. Biochim. biophys. Acta (Amst.) 177, 154 - 156 (1969).
Switzer, B.R., Summer, G.K.: Studies on collagen biosynthesis in human skin fibroblasts. Fed. Proc. 30, 1195 (1971).
Takeuchi, T., Prockop, D.J.: Protocollagen proline hydroxylase in normal liver and in hepatic fibrosis. Gastroenterology 56, 744 - 750 (1969).
Takeuchi, T., Rosenbloom, J., Prockop, D.J.: Biosynthesis of abnormal collagens with amino acid analogues. II. Inability of cartilage cells to extrude collagen polypeptides containing L-azetidine-2-carboxylic acid or *cis*-4-fluoro-L-proline. Biochim. biophys. Acta (Amst.) 175, 156 - 164 (1969).
Uitto, J.: Effect of D-penicillamine on collagen biosynthesis in organ culture. Biochim. biophys. Acta (Amst.) 194, 498 - 503 (1969).
Uitto, J., Halme, J., Hannuksela, M., Peltokallio, P., Kivirikko, K.I.: Protocollagen proline hydroxylase activity in the skin of normal human subjects and of patients with scleroderma. Scand. J. clin. Lab. Invest. 23, 241 - 247 (1969).
Uitto, J., Hannuksela, M., Rasmussen, O.: Protocollagen proline hydroxylase activity in scleroderma and other connective tissue disorders. Ann. clin. Res. 2, 235 - 239 (1970a).
Uitto, J., Lindy, S., Rokkanen, P., Vainio, K.: Increased protocollagen proline hydroxylase activity in synovial tissue in rheumatoid arthritis. Clin. chim. Acta 30, 741 - 744 (1970b).
Vuust, J., Piez, K.A.: Biosynthesis of the α-chains of collagen studied by pulse-labeling in culture. J. biol. Chem. 245, 6201 - 6207 (1970).
Weinstein, E., Blumenkrantz, N., Prockop, D.J.: Hydroxylation of proline and lysine in protocollagen involves two separate enzymatic sites. Biochim. biophys. Acta (Amst.) 191, 747 - 750 (1969).

Studies on Collagen Synthesis Applying Labelled Proline

K. Grasedyck, U. Wulff, D. Erl, and J. Lindner

Using labelled proline for investigations on collagen synthesis or its turnover is an effective method, making possible the collection of data by measuring the specific activity of hydroxyproline. Further information is received by fractionation of collagen. Application of these methods was aimed at receiving further information on collagen synthesis and turnover in skin and organs of the intestinal tract.

Material and Methods

Animals. 24 male white Wistar rats, 6 weeks of age, received 2 mCi ^{3}H-proline (spec. activity 266 mCi/mM)/kg intraperitoneally, and were killed after 1, 3, 7 or 14 days (Heikkinen, 1968). Tissue preparation and fractionation: 1 - 2 g of tissue of skin or different organs was used for fractionation of collagen. After homogenization neutral salt soluble collagen (NSC) was extracted four times in 0.45 M sodiumchloride at 4°C for 24 hours and separated by centrifugation, 30 min at 45,000 g. Acid soluble collagen (ASC) was extracted twice in 0.5 M acetic acid at 4°C for 24 hours. Insoluble collagen (ISC) was separated in 0.5 M acetic acid by heating to 90°C for 30 min. The supernatants of the fractions were recentrifuged, dialyzed for 48 hours against running water, lyophilized and finally hydrolyzed with 6 N hydrochloric acid for 18 hours at 110°C.

Hydroxyproline. The determination of hydroxyproline was carried out according to the methods described by Stegemann (1958) and Prockop et al. (1961). Proline was oxidized by chloramine - T and removed by extraction with toluene. After conversion of hydroxyproline to pyrrole and its separation with toluene, there was enough material for:
1. a quantitative determination of hydroxyproline by p-dimethylaminobenzaldehyde (Ehrlich's reagent). The absorbancy was measured at 560 mμ by an Unicam spectrophotometer.
2. a measurement of radioactivity in a Trib-Carb liquid scintillation spectrometer Model 3003 (Packard Instrument Co.) with PPO-POPOP-toluene solution. Quench correction was achieved using internal standards. The results were expressed as specific activity of hydroxyproline (dpm/mg hydroxyproline).

Autoradiography. The activity of the isotope applied was too low for autoradiography. Therefore some animals received 6 mCi ^{3}H-proline/kg i.p. 3 days before the sacrifice. Autoradiography was carried out by the stripping-film-technique (Kodak AR 10). The samples were exposed for 14, 21 or 28 days and stained slightly with hematoxylin. For quantitative evaluation the radioactivity was measured by a microscope photometer (MPV-Leitz; details see: Dörmer et al., 1966; Herrmann, 1970).

1. Results

Hydroxyproline. The total hydroxyproline content of the three collagen fractions (NSC, ASC, ISC) was calculated from the mean of the incorporation observed. As expected the highest content of hydroxyproline of the tissues studied was observed in the skin (Table 1); the three collagen fractions were equally distributed. A considerable amount of hydroxyproline was found in the muscularis of the gastrointestinal tract. The distribution pattern of the three collagen fractions in the mucosa was similar to that of the muscularis, i.e., about 2/3 of the hydroxyproline was found in the ISC-fraction, less than 10% in the NSC. The total amount of hydroxyproline found in the mucosa is rather low in contrary to that of the muscularis (stomach, small intestine and colon, resp, see Table 1). In the liver, in contrast to the other tissues examined so far, the highest amount of hydroxyproline found was in the ASC-fraction.

Table 1. Total hydroxyproline in 3 collagen fractions (mg hydroxyproline/g wet tissue, mean ± S.D.)

	NSC	ASC	ISC
Skin	2.59 ± 1.91	2.30 ± 1.09	3.06 ± 1.51
Liver	0.05 ± 0.03	0.48 ± 0.54	0.19 ± 0.12
Gastric mucosa	0.05 ± 0.02	0.43 ± 0.32	0.69 ± 0.45
muscularis	0.11 ± 0.08	0.54 ± 0.35	1.61 ± 1.01
Small intest. mucosa	0.13 ± 0.22	0.36 ± 0.47	0.41 ± 0.36
muscularis	0.19 ± 0.20	0.73 ± 0.51	1.23 ± 0.75
Colon mucosa	0.10 ± 0.18	0.35 ± 0.35	0.47 ± 0.36
muscularis	0.16 ± 0.20	0.87 ± 0.51	0.87 ± 0.51

Specific Activity of Hydroxyproline. Generally the highest specific activity of hydroxyproline was found 3 - 7 days after i.p. application of labelled proline. The highest activity was seen at the first day in the ASC-fraction of the skin (as described by Heikkinen, 1968). The specific activity of hydroxyproline was, after the initial increase, consistently at a very low level 14 days later.

The highest amount was found in the gastrointestinal tract, in contrast to skin and liver. The results are demonstrated in detail in Table 2.

Autoradiography. Three days after incorporation there was in skin extensive labelling of connective tissue cells and the intercellular substance. There can also be seen a distinct labelling of blood vessel walls as well as of epidermis and of hair follicles. In the liver the connective tissue of the periportal areas was labelled as well as blood vessel walls, bile ducts and parenchymal cells. In the gastrointestinal tract proline was incorporated especially into the muscularis (connective tissue cells, blood vessel walls, and smooth muscle cells). To a lesser degree proline was incorporated into the mucosa (stroma cells and intercellular substance, epithelial cells). The results of the microscope photometry are demonstrated in Table 3.

Table 2. Specific activity of hydroxyproline in 3 collagen fractions 1 - 14 days after i.p. injection of 2 mCi ^{3}H-proline/kg rat, mean ± S.D.

		1	3	7	14 d
Skin	NSC	12,104 ± 5,635	11,365 ± 4,731	10,021 ± 7,551	1,163 ± 141
	ASC	1,401 ± 957	19,102 ± 7,769	16,942 ± 6,096	
	ISC	1,867 ± 1,705	8,622 ± 2,393	17,477 ± 7,917	4,661 ± 334
Liver	NSC	2,699 ± 1,673	14,340 ± 6,289	12,288 ± 3,429	4,608 ± 3,152
	ASC	162 ± 85	10,590 ± 4,167	12,662 ± 4,657	3,557
	ISC	478	11,769 ± 6,024	11,165 ± 4,657	3,872 ± 995
Gastric mucosa	NSC	1,933 ± 536	14,394 ± 7,064		2,752 ± 834
	ASC	897 ± 406	21,236 ± 2,251	17,629 ± 9,018	
	ISC	1,779 ± 1,117	21,998 ± 4,232	16,274 ± 2,876	4,010 ± 1,897
muscularis	NSC	3,395 ± 1,394	15,929 ±17,833	12,190 ± 7,248	1,936 ± 766
	ASC	2,657 ± 743	37,780 ± 8,291	19,964 ± 5,096	
	ISC	4,107 ± 1,275	26,891 ±12,227	15,636 ± 3,784	6,328 ± 1,829
Small intest. muc.	NSC	1,825 ± 952	12,184 ± 5,379	13,823 ± 4,337	1,893 ± 395
	ASC	302 ± 188	33,208 ±25,450	16,285 ±11,195	
	ISC	747 ± 399	15,794 ± 3,660	17,955 ± 5,891	4,449 ± 1,000
muscularis	NSC	4,938 ± 2,063	14,492 ± 6,602	12,011 ± 4,184	2,420 ± 821
	ASC	3,342 ± 2,031	22,373 ±12,245	20,073 ± 5,026	
	ISC	2,922 ± 1,348	12,055 ± 3,217	11,929 ± 4,894	4,407 ± 1,142
Colon mucosa	NSC	2,583 ± 1,430	15,369 ±10,829	11,326 ± 9,820	3,665 ± 2,562
	ASC	831 ± 353	31,160 ±13,406	18,936 ± 7,691	
	ISC	3,231 ± 1,427	29,867 ±13,282	16,739 ± 6,926	4,022 ± 578
muscularis	NSC	4,952 ± 2,909	18,595 ± 7,129	15,339 ± 6,561	4,847 ± 1,487
	ASC	4,008 ± 3,325	20,925 ± 4,958	20,282 ± 5,440	
	ISC	6,216 ± 1,928	19,444 ± 5,342	23,616 ± 2,803	3,789 ± 1,655

Table 3. Labelling intensity of autoradiographs
3 days after i.p. injection of 6mCi ^{3}H-proline/kg rat, mean ± S.D.

Skin	27.5 ± 5.8
Liver	35.3 ± 5.7
Gastric mucosa	39.6 ± 4.8
muscularis	53.4 ± 9.0
Small intest. mucosa	17.3 ± 4.9
muscularis	41.9 ± 7.8
Colon mucosa	28.7 ± 4.7
muscularis	74.5 ± 19.3

2. Discussion

Proline is incorporated into several proteins. Hydroxyproline, its oxidized form, is found only in collagen. The hydroxyproline incorporated in collagen is not reutilized, but is excreted by the urine as free or bound hydroxyproline as a result of a normal collagen turnover or of increased degradation under pathological conditions.

When considering the specific activity of hydroxyproline, application of labelled proline is an excellent indicator for the metabolism of collagen. Although exact determinations of the collagen turnover are rather difficult, it is obvious that there is, even under normal conditions, a considerable proline incorporation into tissues which have a very low total collagen content. The activity of the protocollagen proline hydroxylase, the enzyme which is responsible for the hydroxylation of proline to hydroxyproline, is different in several organs. It may serve as a parameter for the collagen synthesis (Lazarides et al., 1971; Lindy et al., 1971; McGee et al., 1971; Pänkäläinen et al., 1970).

As known from studies of collagenases (Eisen et al., 1968; Harris et al., 1969; Jeffrey and Gross, 1967; Lazarus et al., 1968; Woods and Nichols, 1965) and collagenpeptidases (Gries and Grasedyck, 1969; Kirschke and Hanson, 1969; Strauch and Vencelj, 1967; Strauch et al., 1968) collagen is continuously catabolized. Thus there is a permanent collagen turnover, which is regulated by several factors and which may be influenced by therapeutics (Gerlach, 1966; Kulonen, 1971; Lindner, 1969). Furthermore it is dependent on age (Beneke, 1971). In general the fat free dry weight is used as parameter for the collagen or hydroxyproline content. The specific activity, however, does not depend on volume and weight of the tissue examined.

An exact definition of the collagen fractions is rather difficult as several methods of fractionation are described by different authors. Therefore, comparison is often almost impossible. Exact comparison is only possible in experiments in which the same procedures are applied. We observed that there was much collagen in the skin, with almost similar amounts in the three collagen fractions, NSC, ASC and ISC. Much less collagen was found in the gastrointestinal tract, in the muscularis more than in the mucosa. The lowest collagen content of the tissues examined by the methods used was found in the liver.

In the skin maximal labelling is seen already on the first day in the NSC after i.p. application of ^{3}H-proline, at about the third day in the ASC and at the seventh day in the ISC. In the other tissues the highest specific activity of hydroxyproline is found between the third and the seventh day in all three fractions at the same time. Generally the activity in these tissues is higher than in the skin. It seems obvious that the incorporation of proline is faster into organs with a lower content of connective tissue. These organs seem to have a higher collagen metabolism. One exception may be bone tissue, as reported by Gerber et al., 1960.

Autoradiography indicates the total proline incorporation. This, however, is not specific for the collagen synthesis, as mentioned above. Especially in the liver, a great deal of labelled proline is incorporated into other proteins, as described previously (Grasedyck et al., 1969). In autoradiography only the ISC and probably a certain part of the ASC show incorporation of proline. Therefore, the autoradiographic results in our procedure should correspond to the ISC three days after incorporation. But there is no sufficient correlation.

Conscious of the limits of autoradiography, however, topographic statements in tissue sections are possible, which would have been difficult or impossible using preparative methods.

Collagen is not only synthesized by fibroblasts (as pericytes, chondro- and osteoblasts) but also by smooth muscle cells. This can be stated for the blood vessel wall, the uterus and the so-called modified form of muscle cells in other organs (Haust et al., 1960; Ross and Benditt, 1965; Wissler, 1968). The same smooth muscle cells are responsible for the high collagen peptidase activity in smooth muscle tissues, especially the aorta.

This explains the high radioactive hydroxyproline content and its high turnover rate, in contrast to the normally low content of intercellular substance in the gastrointestinal muscularis. Autoradiographically, there was found a considerable labelling of the smooth muscle cells which does not differ from that of fibroblasts. Similar results are obtained in ^{35}S-sulfate autoradiographies and in studies on ^{35}S-sulfate incorporation into glycosaminoglycans (Hauss et al., 1968; Junge-Hülsing, 1963/65; Lindner, 1969).

Summary: Collagen synthesis may be investigated by measuring the incorporation of labelled proline if the specific activity of hydroxyproline is analyzed. The collagen synthesis of the gastrointestinal tract is lower than that of skin and higher than that of liver. The muscularis shows a higher incorporation rate than the mucosa, in stomach as well as in small intestine and colon. Autoradiography may serve as an additional parameter for exact localization, though it must be emphasized that autoradiography includes the total proline incorporation and not only that into collagen.

References

Beneke, G.: Altersabhängige Veränderungen des Kollagens und der Bindegewebszellen. In: H. Bredt und J.W. Rohen (Eds.): Altern und Entwicklung, S. 1 - 37. Stuttgart-New York: Schattauer 1971.

Dörmer, P., Brinkmann, W., Stieber, A., Stich, W.: Automatische Silberkornzählung in der Einzelzell-Autoradiographie. Eine neue photometrische Methode für die quantitative Autoradiographie. Klin. Wschr. 44, 477 - 482 (1966).
Eisen, A.Z., Jeffrey, J.J., Gross, J.: Human skin collagenase. Isolation and mechanism of attack on the collagen molecule. Biochim. biophys. Acta (Amst.) 151, 637 - 645 (1968).
Gerber, G.B., Gerber, G., Altmann, K.: Some interrelated aspects of proline and hydroxyproline metabolism. Nature 185, 767 - 768 (1960).
Gerlach, U.: Collagen metabolism. Bibl. anat. (Basel) 7, 496 (1966).
Grasedyck, K., Wulff, U., Rochová, I.: Autoradiographische und radiochemische Befunde zur Kollagensynthese des Magen-Darmtraktes. Verh. dtsch. Ges. Path. 53, 209 - 215 (1969).
Gries, G., Grasedyck, K.: Über spezifische Kollagenasen in Organen des Magen-Darmtraktes. Verh. dtsch. Ges. Path. 53, 205 - 209 (1969).
Harris, E.D., Cohen, G.L., Crane, S.M.: Synovial collagenase: its presence in culture from joint disease of diverse etiology. Arthr. and Rheum. 12, 92 - 102 (1969).
Hauss, W.H., Junge-Hülsing, G., Gerlach, U.: Die unspezifische Mesenchymreaktion. Zur Pathogenese der reaktiven Mesenchymerkrankungen. Stuttgart: Thieme 1968.
Haust, M.D., More, R.H., Movat, H.Z.: The role of smooth muscle cells in the fibrogenesis of atherosclerosis. Amer. J. Path. 37, 377 - 389 (1960).
Heikkinen, E.: Transformations of rat skin collagen with special reference to the ageing process. Acta physiol. scand. Suppl. 317 (1968).
Herrmann,R.: Physikalische Grundlagen der Mikroskop-Photometrie. Z. klin. Chem. 8, 321 - 328 (1970).
Jeffrey, J.J., Gross, J.: Isolation and characterization of a mammalian collagenolytic enzyme. Fed. Proc. 26, 670 (1967).
Junge-Hülsing, G.: Untersuchungen zur Pathophysiologie des Bindegewebes. Habil-Schrift Münster 1963. Bd. 24. Theoretische und klinische Medizin in Einzeldarstellungen. Heidelberg: Hüthig 1965.
Kirschke, H., Hanson, H.: Untersuchungen über den Kollagenabbau durch Rattenorgane. Differenzierung und Charakterisierung Prolinpeptidspaltender Enzyme. Hoppe-Seylers Z. physiol. Chem. 350, 1437 - 1448 (1969).
Kulonen, E.: Research work on structure and metabolism of collagen. II. Turku: Kirjapaino Polytypos 1971.
Lazarides, E.L., Lukens, L.N., Infante, A.A.: Collagen polysomes: site of hydroxylation of proline residues. J. molec. Biol. 58, 831 - 846 (1971).
Lazarus, G.S., Decker, J.L., Oliver, C.H., Daniels, J.R., Multz, C.V., Fullmer, H.M.: Collagenolytic activity of synovium in rheumatoid arthritis. N. Engl. J. Med. 279, 914 - 919 (1968).
Lindner, J.: Zur Physiologie und Pathologie der Schleimbildung des Darmes. Verh. dtsch. Ges. Path. 53, 111 - 153 (1969).
Lindner, J.: The histochemistry of atherosclerosis. In: F.G. Schettler, G.S. Boyd: Atherosclerosis, p. 73 - 140. Amsterdam: Elsevier 1969.
Lindy, S., Pedersen, F.B., Turto, H., Uitto, J.: Lactate, lactate dehydrogenase and protocollagen proline hydrolase in rat skin autograft. Hoppe-Seylers Z. physiol Chem. 352, 1113 - 1118 (1971).
McGee, J.O.D., Rhoads, R.E., Udenfriend, S.: The substrate recognition site of collagen proline hydroxylase: the hydroxylation of -X-Pro-Gly-sequences in bradykinin analogs and other peptides. Arch. Biochem. 144, 343 - 351 (1971).
Pänkäläinen, M., Aro, H., Simons, K., Kivirikko, K.I.: Protocollagen proline hydroxylase: molecular weight, subunits and isoelectric point. Biochem. biophys. Acta (Amst.) 221, 559 - 565 (1970).

Prockop, D.J., Udenfriend, S., Lindstedt, S.: A simple technique for measuring the specific activity of labelled hydroxyproline in biological materials. J. biol. Chem. 236, 1395 - 1398 (1961).
Ross, R., Benditt, E.P.: Wound healing and collagen formation. V. Quantitative electron microscope radioautographic observation of proline ^{3}H utilization by fibroblasts. J. Cell Biol. 27, 83 - 106 (1965).
Stegemann, H.: Mikrobestimmung von Hydroxyprolin mit Chloramin-T und p-Dimethylaminobenzaldehyd. Hoppe-Seylers Z. physiol. Chem. 311, 41 - 45 (1958).
Strauch, L., Vencelj, H.: Collagenases in mammalian cells. Hoppe-Seylers Z. physiol. Chem. 348, 465 - 468 (1967).
Strauch, L., Vencelj, H., Hannig, K.: Kollagenasen in Zellen höher entwickelter Tiere. Hoppe-Seylers Z. physiol. Chem. 349, 171 - 178 (1968).
Wissler, R.W.: Arterial media cell, smooth muscle or multifunctional mesenchymal cell. J. Atheroscler. Res. 8, 201 - 213 (1968).
Woods, J.F., Nichols, G.: Distribution of collagenase in rat tissues. Nature 208, 1325 - 1326 (1965).

III. Degradation

Mammalian Enzymes Degrading Glycosaminoglycans

E. Buddecke and H. Kresse

1. Glycosaminoglycan Catabolism

Although various enzymes from animal sources are able to split acid glycosaminoglycans and proteoglycans (Gibian, 1966; Woessner, 1965) and many studies on the turnover rates of glycosaminoglycans (Muir, 1964; Boström, 1966) clearly indicate a rapid glycosaminoglycan catabolism, our knowledge about the mechanism of their physiological breakdown is still limited.

This paper deals with enzymes involved in the catabolism of connective tissue glycosaminoglycans. Enzymes degrading the protein moiety and the linkage region of proteoglycans and enzymes from microorganisms or invertebrates are not considered. Although heparin belongs to the glycosaminoglycans, it is not a connective tissue polysaccharide in the same sense as e.g. chondroitin sulfate.

Numerous data on the turnover rates of hyaluronic acid and sulfated glycosaminoglycans obtained by in vivo experiments with ^{14}C-glucose, ^{14}C-acetate and ^{35}S-sulfate revealed half-life times of about 2 - 4 days depending on the type of glycosaminoglycan, on the organ and the age of the experimental animal (Schiller et al., 1956; Hauss et al., 1968). In fibroblast cultures the half-life time of intracellular dermatan sulfate was found to be 8 h (Fratantoni et al., 1968). As a general rule hyaluronic acid exhibits a much more rapid turnover than the sulfated glycosaminoglycans in the same type of tissue.

Information about the enzymatic degradation of acid glycosaminoglycans has been obtained by different methods:
(a) Purified glycosidases and sulfate hydrolases from testes, liver, spleen, kidney and many other organs, but also from connective tissue itself such as from bovine arterial tissue and rat bone, can be shown to cause depolymerization and desulfation of glycosaminoglycans and to degrade the split products up to monosaccharides. From the results of these in vitro experiments a degradation scheme may be designed for hyaluronic acid and chondroitin 4-sulfate (Fig. 1).
(b) Injection of radioactive labelled glycosaminoglycans into experimental animals and chemical analysis of the split products occuring in organs and excreted with the urine gave further information on the degradation of glycosaminoglycans. However, the physiological significance of these results remained doubtful, since in almost all cases glycosaminoglycans previously submitted to proteolytic digestion were used as substrates, but the native proteoglycans were not considered.
(c) Several systemic mucopolysaccharidoses have been identified as enzyme deficiency diseases, where normally occuring lysosomal enzymes are inactive in the afflicted individuals. The failure to

degrade glycosaminoglycans leads to an excessive intralysosomal storage and an increased excretion in the urine of mucopolysaccharides, predominantly of dermatan sulfate and/or heparan sulfate (McKusick, 1966). Fibroblasts taken from the skin of patients afflicted with mucopolysaccharidoses perpetuate the basic defect. When the deficient protein is supplied exogenously to the culture medium of fibroblasts of a particular genotype, the glycosaminoglycan metabolism can be normalized, e.g.corrected (Cantz et al., 1972). Since those proteins are identified as hydrolytic enzymes their participation in glycosaminoglycan degradation is evident.
(d) Another approach for studying the catabolism of glycosaminoglycans results from chemical analyses of the storage material in mucopolysaccharidoses. As the degradation of glycosaminoglycans is assumed to proceed by the stepwise action of hydrolytic enzymes from the non-reducing end of the carbohydrate chain, an enzymatic defect at any step would cause accumulation of the intermediate prior to this step, leaving increased amounts of the related non-reducing termini.

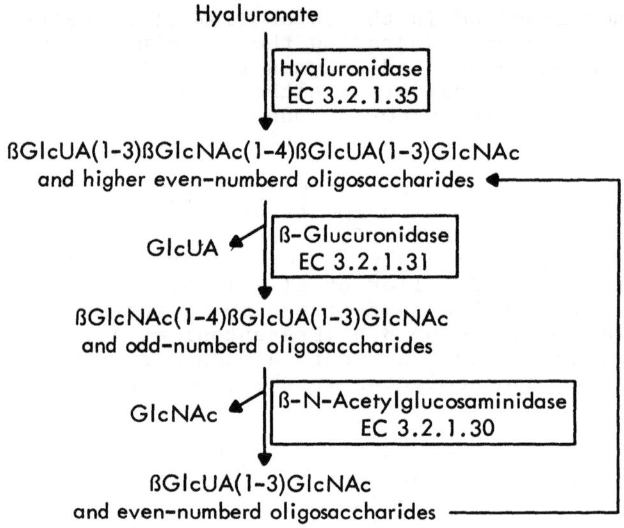

Fig. 1. In vitro degradation sequence of hyaluronate by the combined or successive action of purified connective tissue glycosidases at pH 4 - 5

2. Mammalian Hyaluronidases

Hyaluronidase (hyaluronate glycanohydrolase, EC 3.2.1.35) is widely distributed in animal tissues. Though testicular hyaluronidase is longest known and its conditions and mechanism of action have been well characterized (Meyer et al., 1960), the hyaluronidases present in other mammalian tissues do not differ from the testes enzyme with respect to their substrate specificity and the nature of their split products, except in their pH optimum. Hyaluronidases randomly attack the β-N-acetyl-hexosaminidic bonds in hyaluronate, chondroitin 4- and 6-sulfate and their desulfated derivatives, to yield even-numbered oligosaccharides with glucuronic acid residues at the nonreducing end. On digestion of chondroitin sulfate the main products or tetrasaccharides still give infrared absorption spectra for the 4- and 6-sulfate

Table 1. Properties of mammalian hyaluronidase

Source	pH-Optimum	[K_m m Mole/l] HA	CS	References
Bovine testes	6.0	5.38	-	Houck and Pearce, 1957
		0.86[a]	2.58[a]	
Rat liver	3.5	0.192	-	Aronson and Davidson, 1967
Rat bone	3.7	-	-	Vaes, 1967
Bovine spleen	3.9	2.94	0.6	Filipovic and Buddecke, 1968
Bovine aorta	4.1	4.75	1.25	Buddecke and Platt, 1965

[a]determined at pH 5.0

ester, respectively. Besides the glycosidic action, hyaluronidases catalyze the transglycosylation of oligosaccharides. This has been demonstrated by digestion of equal mixtures of hyaluronic acid and chondroitin 4- and 6-sulfate. Among the products of these reactions two hybrid oligosaccharides, composed of units derived from both polymers, have been isolated by column chromatography.

It has been shown that hyaluronidases from rat liver (Aronson and Davidson, 1965, 1967) and bone (Vaes, 1965, 1967), and from ox spleen (Filipovic and Buddecke, 1968) are localized in the lysosomes of these tissues. On comparing the properties of purified liver and testes hyaluronidases, the lysosomal enzyme showed an action pattern similar to the testes enzyme, but had a lower pH optimum at pH 3.5 and a reduced affinity for hexasaccharides. The properties of hyaluronidases from various sources are summarized in Table 1.

The action of hyaluronidase is not restricted to hyaluronate and chondroitin sulfates. Studies on the chemical heterogeneity of sulfated glycosaminoglycans demonstrated that hybrid polysaccharide chains containing two different sulfated glycosaminoglycans as part of the same polysaccharide chain may be substrates of testes hyaluronidase, provided that one of the constituting glycosaminoglycans of the hybrid polysaccharide chain is susceptible to testicular hyaluronidase. A detailed investigation of dermatan sulfate from horse aorta (Fransson and Havsmark, 1970) indicated that a large portion of it contained chondroitin 4-(6-)sulfate as well as dermatan sulfate disaccharide units in a copolymeric structure. Analogous data were presented for a proteoglycan from bovine aorta (Kresse et al., 1971) containing chondroitin sulfate and dermatan sulfate in a ratio of 75 : 25 as polysaccharide components. The susceptibility of these proteoglycans to testicular hyaluronidase is evident from the nature of their split products (see Table 2). Degradation of the polysaccharide component by hyaluronidase produced dermatan sulfate and a series of saccharide fragments ranging from disaccharides to decasaccharides. All fractions but the disaccharide fraction were found to contain iduronic acid as well as glucuronic acid indicating a copolymeric structure of chondroitin sulfate and dermatan sulfate units. Since the action of hyaluronidase is limited by the occurrence of iduronic acid residues, the oligosaccharide pattern shifts to higher molecular weight oligosaccharides; the iduronic acid/glucuronic acid ratio increased with increasing chain length (Kresse et al., 1971).

A highly purified heparan sulfate isolated from calf aorta and identified by chemical analysis, electrophoretic mobility, ion exchange column chromatography and CD-spectrum behaved as a single compound in gel chromatography (Sephadex G-25). Degradation of this heparan sulfate preparation by testicular hyaluronidase (specific activity 15 000 USP U/mg) produced a second elution peak, which contained glucosamine-N-sulfate and was identified as a heparan sulfate with the approximate size of an octasaccharide by molecular weight estimation and CD-spectra (Fig. 2, 3). Since less purified testes hyaluronidase caused no more extensive degradation of heparan sulfate, the observed breakdown is not considered to be effected by a heparan sulfate hydrolyzing enzyme present as a contaminant in the testes enzyme, but is explained by the existence of hyaluronidase sensitive linkages in the heparan sulfate. This is in accordance with the observation that high molecular weight heparan sulfate from the surface of chinese hamster cells (line CHO) was shown to be susceptible to hyaluronidase (Kraemer and Smith, 1974), interpreted as a result of a branched chain structure of heparan sulfate, the points of branching being hyaluronidase sensitive ß-N-acetylglucosaminidic bonds.

Table 2. Hydrolysis of hyaluronidase substrates by testicular and liver hyaluronidase, respectively

Hyaluronidase source	Substrates	GlcUA	% split products					References
			Di-	Tetra-	Hexa- saccharides	Octa-	Deca-	
Testes	HA	3.1	16.7	75.5	4.7	-	-	Aronson and Davidson, 1967
	C-4-S	-	8.6	39.7	27.0	24.7	-	Buddecke and Hoefele, 1966
	CS-DS-Hybrid	-	1.1	26.4	31.6	20.0	20.9[a]	Kresse et al.,1971
Liver	HA	-	16.7	56.5	26.5	-	-	Aronson and Davidson, 1967

[a]still containing higher oligosaccharides and dermatan sulfate

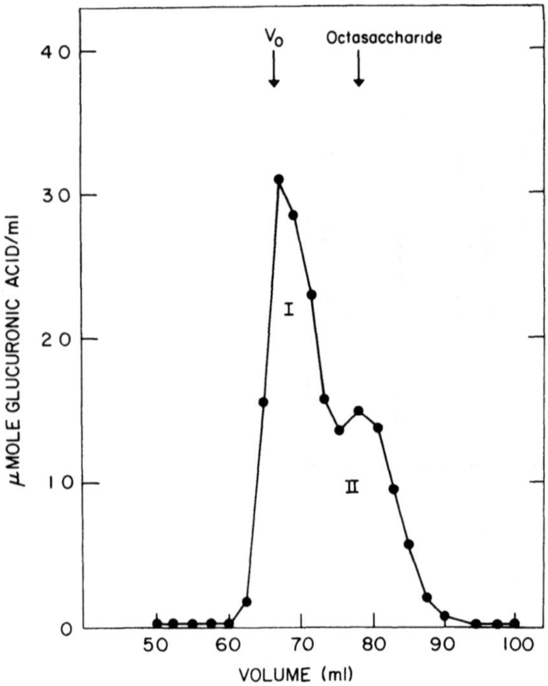

Fig. 2. Elution profile of hyaluronidase digested heparan sulfate on Sephadex G-25. 25 mg of pure heparan sulfate from calf arterial tissue were dissolved in 5.0 ml 0.1 M sodium acetate, pH 6.0, containing 0.15 M NaCl and 0.02% NaN_3 and digested 3-times with 0.34 mg each of testes hyaluronidase (AB Leo, 15 - 20 000 USP units/mg) over a period of 7 days. The digest was lyophilized, dissolved in a minimal amount of water and loaded on a 1.5 × 79 cm column of Sephadex G-25. The column was eluted with 1.0 M NaCl

Despite the clear results of these in vitro experiments, it remains doubtful whether the tissue hyaluronidase is involved in the normal degradation of dermatan sulfate and heparan sulfate, since skin fibroblasts apparently do not attack both polymers by means of hyaluronidase (Matalon and Dorfman, 1968). The existence of two alternative metabolic pathways for the degradation of dermatan sulfate could be postulated, the physiological one being independent of the presence of hyaluronidase.

3. Exoglycosidases Involved in the Degradation of Glycosaminoglycans

a. Uronide uronosohydrolases

Oligosaccharides of hyaluronate formed by the action of testes hyaluronidase are further degraded by the action of ß-glucuronidase (EC 3.2.1.31) (Linker et al., 1955, Buddecke and Hoefele, 1966), which attacks the terminal non-reducing ends of the oligosaccharides. The end products of the reaction are glucuronic acid and odd-numbered saccharide fragments. On exclusive action of arterial tissue ß-glucuronidase on tetra-, hexa- and octasaccharides of chondroitin 4-sulfate, the terminal glucuronic acid residues were split off at equal rates, the K_m value for the tetrasaccharide being 11.0 mM (Buddecke and Hoefele, 1966).

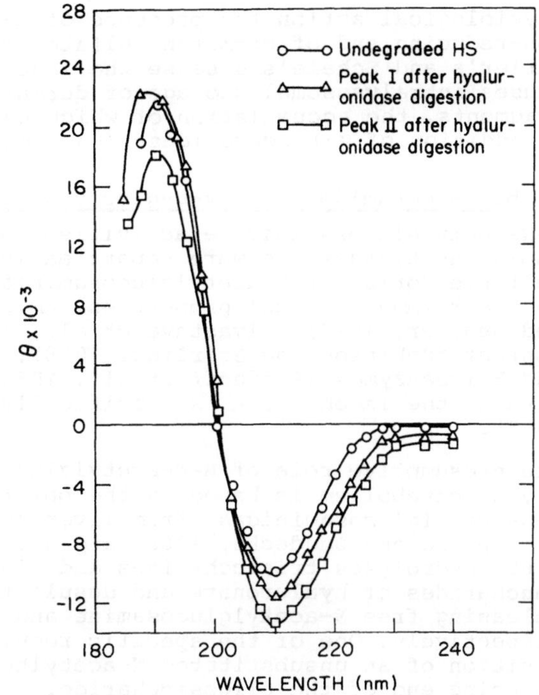

Fig. 3. CD-spectra of heparan sulfate fractions. The CD was measured at 0.5 mm pathlength for a solution of 1.0 mg/ml in the Cary Model 60 spectropolarimeter

Beside its action on hyaluronate and chondroitin 4-(6-)sulfate ß-glucuronidase is required for the hydrolysis of ß-glucuronidic linkages present in dermatan sulfate. This is concluded from the fact that cultured skin fibroblasts from patients with genetic ß-glucuronidase deficiency accumulate intracellularly excessive amounts of dermatan sulfate (Hall et al., 1972), and individuals affected with this disease excrete increased amounts of chondroitin sulfate and dermatan sulfate with their urine (Sly et al., 1972).

Although the presence of ß-glucuronidic linkages has also been established in heparan sulfate and heparin (Helting and Lindahl, 1971; Atkins and Laurent, 1973) and therefore ß-glucuronidase should be expected to be involved in their degradation, no increased heparan sulfate excretion was observed in the urine of enzyme deficient patients.

The detection of a phenyl-α-D-glucuronide hydrolysing enzyme in preparations from human urine and fibroblast extracts indicate the existence of mammalian α-glucuronidase (EC 3.2.1.?), which might be involved in the liberation of α-glycosidically bound glucuronic acid residues present in heparan sulfate and heparin. The enzyme purified from human urine was shown to be present in two multiple forms, exhibiting molecular weights of 66 000 and 37 000 and a double pH optimum (5.1 and 6.5) (v. Figura et al., unpublished).

The presence of α-glycosidically linked L-iduronic acid residues in dermatan sulfate and heparan sulfate suggests the obligate action of an α-L-iduronidase (EC 3.4.1.76) for the cleavage of this particular bond. α-L-Iduronidase activities have been demonstrated in extracts of human organs, leucocytes, cultured fibroblasts and urine (Matalon et al., 1971, Weissmann and Santiago, 1972, Bach et al., 1972, Matalon and Dorfman, 1972). α-L-Iduronidase requires for their physiological action the presence of L-iduronic acid residues at the non-reducing end of dermatan sulfate or heparan sulfate chains. In Hurler's and Scheie's disease the inactivity of α-L-iduronidase causes intralysosomal storage of dermatan sulfate and heparan sulfate fragments, the accumulation of which can be reduced in tissue culture by addition of α-L-iduronidase (Bach et al., 1972).

b. 2-Acetamido-2-deoxy-D-glucoside glucanohydrolases

ß-N-Acetylglucosaminidase activities have been detected in many mammalian tissues and were separated into two genetically distinct multiple forms - ß-N-acetylglucosaminidase A and B - which differ in their immunological properties, amino acid composition (Srivastava and Beutler, 1974; Srivastava et al., 1974 a,b) and sialic acid content (Robinson and Stirling, 1968). In Sandhoff disease the A and B isoenzymes (Sandhoff et al., 1968) are missing, in Tay-Sachs disease the isoenzyme A is specifically absent (Okada and O'Brien, 1969).

The presumptive role of ß-N-acetylglucosaminidases in the glycosaminoglycan catabolism is based on the observation that highly purified ß-N-acetylglucosaminidase from liver (Weissmann et al., 1964), spleen (Filipovic and Buddecke, 1968) and arterial wall (Buddecke and Platt, 1965) hydrolyzes trisaccharides and higher odd-numbered oligosaccharides of hyaluronate and desulfated chondroitin sulfate, releasing free N-acetylglucosamine and N-acetylgalactosamine, respectively. One of the specific requirements of the enzyme is the position of an unsubstituted N-acetylhexosamine residue at the non-reducing end of the oligosaccharide.

In Sandhoff's disease no increase in glycosaminoglycan content in liver, spleen and several other organs and a slight increase only in the concentration of uronic acid containing oligosaccharides in the urine have been reported (Applegarth and Bozoian, 1972, Strecker and Montreuil, 1971), but Sandhoff fibroblasts and to a much lesser extent Tay-Sachs fibroblasts accumulate excessive amounts of dermatan sulfate and chondroitin sulfate, thereby establishing the involvement of ß-N-acetylglucosaminidase in the catabolism of these two glycosaminoglycans (Cantz and Kresse, in press). Accordingly, extracts from Tay-Sachs as well as from Sandhoff fibroblasts are unable to liberate acetylated amino sugars from hyaluronic acid oligosaccharides or from chondroitin sulfate heptasaccharide (Thompson et al., 1973). Of the two isoenzymes the ß-N-acetylglucosaminidase A exhibits 5-times the corrective activity when compared with the isoenzyme B on the basis of their specific activities towards synthetic substrates (Cantz and Kresse, in press).

A special ß-N-acetylhexosaminidase is suggested to be obligate for the degradation of dermatan sulfate and proposed to be inactive in the Maroteaux-Lamy syndrome (Sjöberg et al., 1974).

In contrast to the N-acetyl-ß-D-glucosaminidases which splits - though at different rates - phenyl-ß-glycosides of both N-acetylglucosamine and N-acetylgalactosamine, the N-acetyl-α-glucosaminidase (EC 3.2.1.50) exhibits an absolute specificity towards the glucosamine moiety of its substrates (Werries et al., 1969).

α-N-Acetylglucosaminidase has been recognized to be essential for the degradation of heparan sulfate. This was proven by studies on the biochemical basis of Sanfilippo B disease by the following results: (a) Sanfilippo B disease is characterized by an increased intralysosomal storage and an excessive urinary excretion of heparan sulfate, (b) α-N-acetylglucosaminidase is inactive in organs and fibroblasts obtained from Sanfilippo B patients (O'Brien, 1972). (c) The impaired heparan sulfate breakdown of cultured Sanfilippo B fibroblasts is normalized by a protein identical with α-N-acetyl-glucosaminidase (v. Figura and Kresse, 1972) and (d) the heparan sulfate fragments stored in Sanfilippo B fibroblasts differ from heparan sulfate of other sources by an increased number of terminal N-acetyl-α-D-glucosamine residues (v. Figura and Kresse, unpublished).

4. Glycosaminoglycan-Sulfate Sulfohydrolases

Studies on the metabolic fate of sulfated glycosaminoglycans after injection into humans and dogs had the result that chondroitin 4-sulfate, chondroitin 6-sulfate, dermatan sulfate and heparan sulfate disappeared rapidly from the circulation, but could not be recovered quantitatively from the urine and were partially desulfated (Dziewiatkowski, 1956; Dodgson, 1966). An extensive degradation of ^{35}S-labelled heparin occurs in many organs (skin, aorta, lung, adrenal glands, stomach, heart and brain), the degradation products being dialysable and including inorganic ^{35}S-sulfate. Chondroitin 4-sulfate sulfohydrolases (EC 3.1.6.4) were purified from hog kidney lysosomes (Mehl and Jatzkewitz, 1963), from rat liver (Tudball and Davidson, 1969) and the cell-free supernatant of bovine arterial tissue homogenates (Held and Buddecke, 1967). The arterial tissue enzyme was purified about 100-fold to a specific activity of 0.5 mU/mg protein, its pH optimum was found to be 4.4 and the $K_m$value 7.35×10^{-2}M. While the arterial enzyme cleaves preferentially sulfate ester bonds of polymeric chondroitin sulfate, the liver enzyme acts on oligomeric substrates (Tudball and Davidson, 1969). The purified chondroitin sulfate sulfatases are different from aryl sulfatases (EC 3.1.6.1). The reduced aryl sulfatase B activity in cultured fibroblasts from some patients affected with the Maroteaux-Lamy syndrome and the unability of these fibroblasts to metabolize the stored lysosomal dermatan sulfate suggest the participation of this enzyme in the desulfation of dermatan sulfate (Austin, 1972).

A sulfatase, which specifically cleaves the ester sulfate of iduronic acid residues in terminal non-reducing position of polysaccharides and disulfated disaccharides, was detected in normal skin fibroblasts and has been shown the deficient enzyme in Hunter's disease (Bach et al., 1973). The inactivity of this iduronide sulfate sulfatase (EC 3.1.6?) in that disorder leads to the accumulation of heparan sulfate and dermatan sulfate, the latter of which has been demonstrated to contain a high proportion of disulfated disaccharides at the non-reducing end (Sjöberg et al., 1973).

The cleavage of sulfamidic linkages which are unique in heparin and heparan sulfate requires an enzyme which hydrolyzes this particular type of bond. A lysosomal sulfamidase (EC 3.10.1.?) is considered not only to be active towards heparin (Dietrich, 1970, Friedman and Arsenis, 1974), but also towards heparan sulfate. This is concluded from the finding that the enzyme activity is deficient in organs, leucocytes and cultured skin fibroblasts derived from patients affected with Sanfilippo A disease (Kresse and Neufeld, 1972, Matalon and Dorfman, 1973, Kresse, 1973)where the main storage product is heparan sulfate rather than heparin.

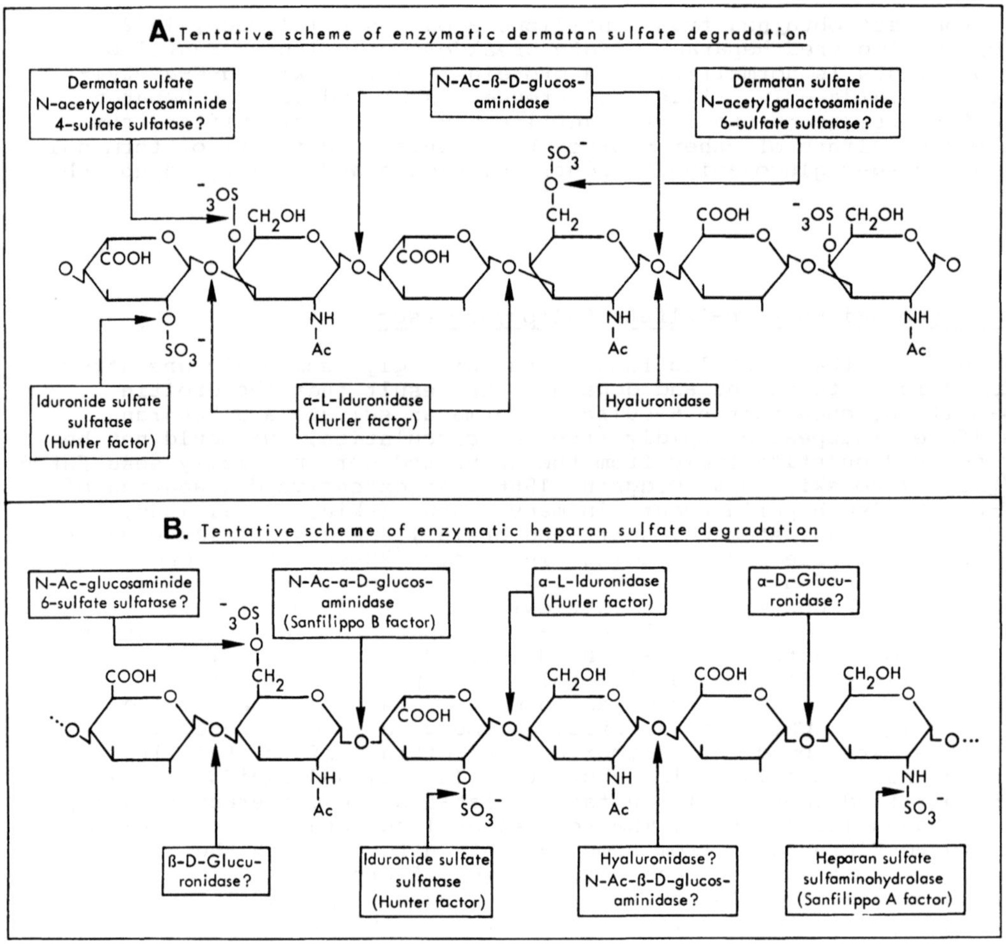

Fig. 4. Site of action of hydrolytic enzymes tentatively involved in the catabolism of dermatan sulfate (A) and heparan sulfate (B). The polysaccharide chain is stepwise degraded from the non-reducing end concerted action of glycosidases and sulfatases. An enzymatic defect at any step causes accumulation of the intermediate prior to this step

There is evidence for the existence of another sulfatase which splits the sulfate ester bonds of N-acetylglucosaminie-6-sulfate containing heparan sulfate oligosaccharides (Kresse, 1973) and a lysosomal proteochondroitin sulfate and proteokeratan sulfate sulfatase, detected in rabbit kidney (Greiling et al., 1970).

5. The Lysosomal Concept of Glycosaminoglycan Degradation

The physiological role of lysosomes in the degradative steps of glycosaminoglycan turnover is based on the following observations:
(a) Lysosomes contain a full complement of glycosidases and sulfatases

for degrading hyaluronic acid and sulfated glycosaminoglycans. In vitro experiments revealed that the combined action of the isolated enzymes virtually allows a complete degradation of hyaluronic acid, chondroitin 4-sulfate and chondroitin 6-sulfate up to their monosaccharide constituents.

(b) Hepatic lysosomes degrade highly polymerized hyaluronic acid to tetrasaccharides and several oligosaccharides as well as chondroitin 4-sulfate, chondroitin 6-sulfate and heparan sulfate under in vitro conditions.

(c) Rats treated by injection of ^{35}S-labelled chondroitin 4-sulfate and chondroitin 6-sulfate are able to concentrate this material within their liver lysosomes, from which it is completely depleted within 4 days.

(d) In mucopolysaccharidoses which are caused by a lack of glycosaminoglycan hydrolyzing enzymes partially degraded glycosaminoglycans are accumulated within the lysosomes. Upon addition of the missing enzyme to the culture medium of deficient fibroblasts from mucopolysaccharidoses patients, the stored glycosaminoglycans disappear and the glycosaminoglycan metabolism is normalized (see Table 3). This effect implies the pinocytosis of the exogenously supplied enzyme and its uptake by the lysosomes. The intralysosomal localization of a hydrolase pinocytosed by a deficient cell is indicated by the observation that after internalization the activity of this enzyme as well as its biological activity towards stored $^{35}SO_4$-labelled glycosaminoglycans persist with the same half-life time (v. Figura and Kresse, 1974).

Table 3. Effect of specific enzymes on mucopolysaccharide degradation of fibroblasts. A $^{35}SO_4$-labelled proteodermatan sulfate secreted by Sanfilippo B fibroblasts or a proteoglycan secreted by rat glia tumor cells containing ^{35}S-labelled heparan sulfate were administered to cell culture medium in the absence or presence of 50 "corrective units" of the corresponding enzymes. Units were calculated in a bioassay on the basis of the ability of the enzyme preparation to reduce the accumulation of $^{35}SO_4$-mucopolysaccharide (Cantz et al., 1972). After 24 hr, intracellularly accumulated mucopolysaccharides and degradation products in the medium were determined by measuring the radioactivity in the ethanol precipitate and in the supernatant. Degradation is expressed as the percentage of the mucopolysaccharide taken in by pinocytosis, which becomes degraded.

Genotype	Addition of enzyme	Degradation (%) (Sanfilippo B-proteoglycan)	Degradation (%) (glia cell-proteoglycan)
Normal	none	90	95
	complete mixture	91	96
Hurler	none	13	80
	α-L-Iduronidase	79	92
Hunter	none	18	78
	Iduronide sulfate sulfatase	84	93
Sanfilippo A	none	90	83
	Heparin sulfamidase	92	93
Sanfilippo B	none	89	75
	α-N-Acetylglucosaminidase	90	95

Studies on I-cell disease led to the hypothesis that at least in fibroblasts some lysosomal enzymes are primarily secreted before they reach the lysosomes of the same or neighbouring cells by adsorptive endocytosis (Hickman and Neufeld, 1972). For ß-N-acetyl-glucosaminidase the carbohydrate moiety of the enzyme molecule seems to be the specific recognition marker for the interaction with the binding site on the cell surface (Hickman et al., 1974). It seems possible that lysosomal enzymes are pinocytosed together with their physiological substrates, e.g. extracellular proteoglycans. Proteoglycans also become internalized into fibroblasts by adsorptive pinocytosis, an intact protein moiety being a prerequisite for efficient uptake. After merging of the engulfed material with lysosomes (or already within the pinosome?) degradation of glycosaminoglycans might proceed by the stepwise alternative and/or concerted action of glycosidases and sulfatases, the latter attacking sulfate ester residues at or near the non-reducing end of the polysaccharide chain. In the course of this degradation pathway cooperative effect of the involved enzymes may occur, as observed for the ß-N-acetyl-glucosaminidase-mediated activation of ß-glucuronidase (Buddecke and Hoefele, 1967).

References

Applegarth, D.A., Bozoian, G.: Mucopolysaccharide storage in organs of a patient with Sandhoff's disease. Clin. Chim. Acta 39, 269 (1972).

Aronson, M.N., jr., Davidson, E.A.: Lysosomal hyaluronidase. J. biol. Chem. 240, PC 3222 (1965).

Aronson, M.N., jr., Davidson, E.A.: Lysosomal hyaluronidase from rat liver, I. preparation. J. biol. Chem. 242, 437 (1967).

Atkins, E.D.T., Laurent, T.C.: X-ray diffraction patterns from chondroitin-4-sulphate, dermatan sulphate and heparan sulphate. Biochem. J. 133, 605 (1973).

Austin, J.: Inactivity of arylsulfatase B in the Maroteaux-Lamy syndrome. Transact. Am. Neur. Ass. 97, 29 (1972).

Bach, G., Friedman, R., Weissmann, B., Neufeld, E.F.: The defect in the Hurler and Scheie syndromes. Deficiency of α-L-iduronidase. Proc. nat. Acad. Sci. (Wash.) 69, 2048 (1972).

Bach, G., Eisenberg, F., Cantz, M., Neufeld, E.F.: The defect in the Hunter syndrome: deficiency of sulfoiduronate sulfatase. Proc. nat. Acad. Sci. 70, 2134 (1973).

Boström, H., Roden, L.: Metabolism of glycosaminoglycans. In: E.A. Balazs, R.W. Jeanloz (Ed.): The Amino Sugars, Vol. IIB, p. 77. New York: Academic press 1966.

Buddecke, E., Platt, D.: Untersuchungen zur Chemie der Arterienwand. VIII. Nachweis, Reinigung und Eigenschaften der Hyaluronidase aus der Aorta des Rindes. Hoppe-Seylers Z. physiol. Chem. 343, 61 (1965).

Buddecke, E., Hoefele, O.: Untersuchungen zur Chemie der Arterienwand. IX. Reinigung und Eigenschaften der ß-Glucuronidase aus der Aorta des Rindes. Hoppe-Seylers Z. physiol. Chem. 347, 173 (1966).

Buddecke, E., Hoefele, O.: pH-dependent activation of ß-D-glucuronidase by acetyl-ß-D-glucosaminidase. Biochim. biophys. Acta (Amst.) 132, 194 (1967).

Cantz, M., Kresse, H.: Sandhoff disease: Defective glycosaminoglycan catabolism in cultured fibroblasts and its correction by ß-N-acetylhexosaminidase. Europ. J. Biochem., in press.

Dietrich, C.P.: A heparin sulfamidase from mammalian lymphoid tissues. Can. J. Biochem. 48, 725 (1970).

Dodgson, K.S.: Sulfatases of glycosaminoglycans. In: E.A. Balazs, R.W. Jeanloz (Ed.): The Amino Sugars, Vol. IIB, p. 201. New York: Academic Press 1966.

Dziewiatkowski, D.D.: Some aspects of the metabolism of chondroitin sulfate ^{35}S in the rat. J. biol. Chem. 223, 239 (1956).

v. Figura, K., Kresse, H.: The Sanfilippo B corrective factor: A N-acetyl-α-D-glucosaminidase. Biochem. biophys. Res. Commun. 48, 262 (1972).

v. Figura, K., Kresse, H.: Quantitative aspects of pinocytosis and the intracellular fate of N-acetyl-α-D-glucosaminidase in Sanfilippo B fibroblasts. J. clin. Invest. 53, 85 (1974).

Filipovic, I., Buddecke, E.: Wechselseitige Aktivierung und intrazelluläre Lokalisation von Hyaluronidase (EC 3.2.1.35) und Kathepsin D (EC 3.4.4.23) aus Rindermilz. Hoppe-Seylers Z. physiol. Chem. 349, 553 (1968).

Fransson, L.-Å., Havsmark, B.: Structure of dermatan sulfate, VII. The copolymeric structure of dermatan sulfate from horse aorta. J. biol. Chem. 245, 4770 (1970).

Fratantoni, J.C., Hall, C.W., Neufeld, E.F.: The defect in Hurler and Hunter syndromes, faulty degradation of mucopolysaccharide. Proc. nat. Acad. Sci. (Wash.) 60, 699 (1968).

Friedman, Y., Arsenis, C.: Studies on the heparin sulphamidase activity from rat spleen. Intracellular distribution and characterization of the enzyme. Biochem. J. 139, 699 (1974).

Gibian, H.: Enzymes degrading glycosaminoglycans. In: E.A. Balazs, R.W. Jeanloz (Ed.): The Amino Sugers, Vol. IIB, p. 181. New York: Academic Press 1966.

Greiling, H., Stuhlsatz, H.W., Kisters, R.: Structure and metabolism of proteokeratan sulfate. In: E.A. Balazs (Ed.): Chemistry and Molecular Biology of Intercellular Matrix, Vol. II, p. 873. New York: Academic Press 1970.

Hall, C.W., Cantz, M., Neufeld, E.F.: A ß-glucuronidase deficiency mucopolysaccharidosis: Studies in cultured fibroblasts. Arch. Biochem. Biophys. 155, 32 (1973).

Hauss, H.W., Junge-Hülsing, G., Gerlach, U.: Die unspezifische Mesenchymreaktion. Stuttgart: Thieme 1968.

Held, E., Buddecke, E.: Untersuchungen zur Chemie der Arterienwand. XI. Nachweis, Reinigung und Eigenschaften einer Chondroitin-4-sulfatase aus der Aorta des Rindes. Hoppe-Seylers Z. physiol. Chem. 348, 1047 (1967).

Helting, T., Lindahl, U.: Occurrence and biosynthesis of ß-glucuronidic linkages in heparin. J. biol. Chem. 246, 5442 (1971).

Hickman, S., Neufeld, E.F.: A hypothesis for I-cell disease: Defective hydrolases that do not enter lysosomes. Biochem. biophys. Res. Commun. 49, 992 (1972).

Hickman, S., Shapiro, L.J., Neufeld, E.F.: A recognition marker required for uptake of a lysosomal enzyme by cultured fibroblasts. Biochem. biophys. Res. Commun. 57, 55 (1974).

Houck, J.C., Pearce, R.H.: The mechanism of hyaluronidase action. Biochim. biophys. Acta (Amst.) 25, 555 (1957).

Kraemer, P.M., Smith, D.A.: High molecular-weight heparan sulfate from the cell surface. Biochem. biophys. Res. Commun. 56, 429 (1974).

Kresse, H., Heidel, H., Buddecke, E.: Chemical and metabolic heterogeneity of a bovine aorta chondroitin sulfate-dermatan sulfate proteoglycan. Europ. J. Biochem. 22, 557 (1971).

Kresse, H., Neufeld, E.F.: Sanfilippo A corrective factor: purification and mode of action. J. biol. Chem. 247, 2164 (1972).

Linker, A., Meyer, K., Weissmann, B.: Enzymatic formation of monosaccharides from hyaluronate. J. biol. Chem. 213, 237 (1955).
Matalon, R., Dorfman, A.: The structure of acid mucopolysaccharides produced by Hurler fibroblasts in tissue culture. Proc. nat. Acad. Sci. (Wash.) 60, 179 (1968).
Matalon, R., Cifonelli. J.A., Dorfman, A.: L-Iduronidase in cultured human fibroblasts and liver. Biochem. biophys. Res. Commun. 42, 340 (1971).
Matalon, R., Dorfman, A.: Hurler's syndrome, an α-L-iduronidase deficiency. Biochem. biophys. Res. Commun. 47, 959 (1972).
Matalon, R., Dorfman, A.: Pediat. Res. 7, 384 (1973).
McKusick, V.A.: The Mucopolysaccharidoses in Heritable Disorders of Connective Tissue, p. 325. St. Louis: Mosby 1966.
Mehl, E., Jatzkewitz, H.: Über ein cerebrosidschwefelsäureesterspaltendes Enzym aus Schweineniere. Hoppe-Seylers Z. physiol. Chem. 331, 292 (1963).
Meyer, K., Hoffman, Ph., Linker, A.: Hyaluronidases. In: P.D. Boyer, H. Lardy, K. Myrbäck (Ed.): The Enzymes (Sec. Ed.), Vol. 4, p. 447. New York: Academic Press 1960.
Muir, H.: Chemistry and metabolism of connective tissue glycosaminoglycans (mucopolysaccharides). In: D.A. Hall (Ed.): International Review of Connective Tissue Research, Vol. 2, p. 101. New York: Academic Press 1964.
O'Brien, J.: Sanfilippo syndrome: Profound deficiency of alpha-acetylglucosaminidase activity in organs and skin fibroblasts from type-B patients. Proc. nat. Acad. Sci. (Wash.) 69, 1720 (1972).
Okada, S., O'Brien, J.: Tay-Sachs disease: Generalized absence of a beta-D-N-acetylhexosaminidase component. Science 165, 698 (1969).
Robinson, D., Stirling, J.L.: N-Acetyl-ß-glucosaminidases in human spleen. Biochem. J. 107, 321 (1968).
Sandhoff, K., Andreae, U., Jatzekewitz, H.: Deficient hexosaminidase activity in an exceptional case of Tay-Sachs disease with additional storage of kidney globoside in visceral organs. Life Sci. 7, 283 (1968).
Schiller, S., Mathews, M.B., Cifonelli, J.A., Dorfman, A.: The metabolism of mucopolysaccharides in animals, III. Further studies of skin utilizing ^{14}C-glucose, ^{14}C-acetate and ^{35}S-sulfate. J. biol. Chem. 218, 139 (1956).
Sjöberg, I., Fransson, L.-A., Matalon, R., Dorfman, A.: Hunter's syndrome: A deficiency of L-idurono-sulfate sulfatase. Biochem. biophys. Res. Commun. 54, 1125 (1973).
Sjöberg, I., Fransson, L.-A., Matalon, R., Dorfman, A.: Structure of dermatan sulfate accumulated in the Maroteaux-Lamy syndrome. XXIInd Colloquium on Protides of the Biological Fluids. Abstr. No. 80, Brügge, 1974.
Sly, W.S., Quinton, B.A., McAlister, W.H., Rimoin, D.L.: Beta glucuronidase deficiency: Report of clinical, radiologic, and biochemical features of a new mucopolysaccharidosis. J. Pediatr. 82, 249 (1973).
Srivastava, S.K., Beutler, E.: Studies on human ß-hexosaminidases. III. Biochemical genetics of Tay-Sachs and Sandhoff's disease. J. biol. Chem. 249, 2054 (1974).
Srivastava, S.K., Awasthi, Y.C., Yoshida, A., Beutler, E.: Studies on human ß-hexosaminidases. I. Purification and properties. J. biol. Chem. 249 2043 (1974a).

Srivastava, S.K., Yoshida, A., Awasthi, Y.C., Beutler, E.: Studies on human ß-hexosaminidases. II. Kinetic and structural properties. J. biol. Chem. 249, 2049 (1974b).
Strecker, G., Montreuil, J.: Description d'une oligosaccharidurie accompagnant une gangliosidose GM_2 a deficit total en N-acetyl-hexosaminidase. Clin. chim. Acta 33, 395 (1971).
Thompson, J.N., Stoolmiller, A.C., Matalon, R., Dorfman, A.: N-Acetyl-ß-hexosaminidase: Role in the degradation of glycosaminoglycans. Science 181, 866 (1973).
Tudball, M., Davidson, E.A.: Isolation of a novel sulphatase from rat liver. Biochim. biophys. Acta (Amst.) 171, 113 (1969).
Vaes, G.: Studies on bone enzymes. The assay of acid hydrolases and other enzymes in tissue. Biochem. J. 97, 380 (1965).
Vaes, G.: Hyaluronidase activity in lysosomes of bone tissue. Biochem. J. 103, 802 (1967).
Weissmann, B., Hadjiioannou, S., Tornheim, J.: Oligosaccharase activity of ß-N-acetyl-D-glucosaminidase of beef liver. J. biol. Chem. 239, 59 (1964).
Weissmann, B., Santiago, R.: α-L-Iduronidase in lysosomal extracts. Biochem. biophys. Res. Commun. 46, 1430 (1972).
Werries, E., Wollek, E., Gottschalk, A., Buddecke E.: Separation of N-acetyl-α-glucosaminidase and N-acetyl-α-galactosaminidase from ox spleen. Europ. J. Biochem, 10, 445 (1969).
Woessner, J.F., jr.: Acid hydrolases of connective tissue. In: D.A. Hall (Ed.): International Review of Connective Tissue Research, Vol. 3, p. 201. New York: Academic Press 1965.

The Production and Use of Water-Insoluble Derivatives of Enzymes

J. F. Kennedy

The enzymatic degradation of glycoproteins is very important from the structural point of view, not only for the purpose of providing information on the type, sequence and mode of linkage of the constituent units, but also for producing, from biologically active molecules, degraded products which retain the active site. However, one of the greatest problems in enzymatic degradation studies is that of the necessary subsequent separation of enzyme from the degradation products and any undegraded material. This difficulty can in some cases be overcome by selective absorption of either the enzyme or non-enzyme species from the incubate with e.g. ion-exchange matrices. Alternatively, if the enzyme, and any species formed from it during the incubation, has an appreciably different molecular weight from the degradation products, gel filtration will suffice. However, these separation methods are really effective only in certain specialized cases.

Enzymes which are normally used for degradation and structural studies of biological molecules can be produced in insoluble form by attachment through covalent linkage, at positions other than their active sites, to solid supports. Linked in this way, the enzymes retain their activity toward substrates but do not themselves become liberated into the solution during their action. The use of such immobilized enzymes is undoubtedly advantageous in overcoming the separation problems, since after the desired degradation the removal of enzyme is effected by simple centrifugation. It has been observed that, when coupled to insoluble carriers, enzymes are more stable than they are in the free state. The formation of the insoluble derivative therefore has the added advantage of overcoming the problems frequently arising out of loss of activity during transit and on storage.

The preparation and use of water-insoluble enzyme derivatives and related subjects has been reviewed (Crook, 1968; Goldstein and Katchalski, 1968; Kay, 1968; Guilibault, 1966; Silman and Katchalski, 1968), and there is no doubt regarding their utility. Indeed, many enzymes are commercially available in insolubilized forms. However, the lack of reports of their use by workers in laboratories other than those in which they were produced suggests that the inherrent advantages have not been fully realized. But, apart from those which are offering insolubilized enzymes as research chemicals, certain sections of industry have appreciated the potentials and are heavily committed to their development.

The group in the Department of Chemistry at the University of Birmingham is working on the production of insoluble matrices suitable for irreversible and reversible attachment of enzymes with retention of activity. Water insoluble derivatives of α-amylase, β-amylase and glucoamylase have been prepared by chemical coupling of the enzymes to a diazotized 3-(p-aminophenoxy)-2-hydroxypropyl ether of cellulose, and/or to a 2-hydroxy-3-(p-isothiocyanatophenoxy) propyl ether of cellulose (Barker et al., 1968; Barker et al., 1969a). The activity of the hydrolytic enzymes is retained on water-insolubilization, and the insoluble enzymes are relatively more resistant to heat denaturation than the free enzymes in solution.

Water insoluble derivatives of α-amylase have also been prepared by coupling to diazo, isothiocyanato and azide groups of modified cross-linked polyacrylamides, 'Enzacryl AA' and 'Enzacryl AH' (Barker et al., 1970a). Similar derivatives of β-amylase could be prepared by diazo and isothiocyanato coupling only. The stabilities of the insoluble enzyme preparations were comparable to those of the corresponding cellulose derivatives of the enzymes, and in the case of α-amylase were considerably greater than the stability of the free enzyme in solution. Comparisons of the retention of activity on re-use of these insolubilized enzymes showed that although considerable desorption of physically bound enzyme took place on initial re-use, the cellulose-bound derivatives could be subsequently re-used with comparatively little activity loss (Barker et al., 1970b). The polyacrylamide derivatives were superior in that activity retention was good on initial and repeated re-use.

An active, water-insoluble preparation of a mixture of five glycohydrolases, α- and β-galactosidase, α-L-fucosidase, α-D-mannosidase and β-2-acetamido-2-deoxy-D-glucosidase, has been obtained by coupling the free enzyme mixture, prepared from ram testis extract, with the diazotized 3-(p-aminophenoxy)-2-hydroxypropyl ether of cellulose (Barker et al., 1969b).

More recently, the chemistry of a reactive five-membered cyclic group which can be formed on cellulose has been investigated (Barker et al., 1969c, 1971a). Thus, cellulose carbonate containing 40% of the maximum possible cyclic carbonate group content was produced by the action of ethyl chloroformate in anhydrous organic solvent mixture on microcrystalline cellulose. This cellulose carbonate has been used to prepare solid phase derivatives of β-D-glucosidase (Barker et al., 1971b), a mixture of α-D-glucosidase, β-D-galactosidase, α-D-mannosidase and β-2-acetamido-2-deoxy-D-glucosidase from mouse kidney extract, and of trypsin (Barker et al., 1969d).

The cyclic carbonate group has also been generated in a different ring size, probably eight membered, in a non-enzymically degradable polymer having a simple -C-C-C- chain backbone (Barker et al., 1971c). This has been achieved by treatment of a poly (allylalcohol) with ethyl chloroformate. β-D-Glucosidase and glucoamylase (Barker et al., 1971c) and also α-amylase and trypsin (Kennedy et al., 1972) have been successfully attached to the resultant poly (allyl carbonate).

Advantages of the use of the cyclic carbonate group include the fact that it is easily generated in one step. Furthermore, the proportion of hydroxyl groups converted to carbonate groups can easily be controlled (Barker et al., 1969c, 1971a). This facet is of importance since maximum retention of enzyme specific activity on coupling to a solid phase does not necessarily follow maximum primary derivatization of the carrier.

Other later developments at Birmingham have been the preparation of two carriers amenable to regeneration or re-activation. First, the sulphhydryl groups of a cross-linked copolymer of acrylamide and N-acryloyl cysteine, 'Enzacryl Polythiol', have been used for enzyme binding, e.g. trypsin (Barker et al., 1971d). In this case, prior enrichment of the sulphohydryl group content of the enzyme is desirable and is achieved by reaction with N-acetyl-DL-homocysteine thioloactone. Subsequently, oxidative coupling with the polythiol is effected using potassium ferricyanide. When the enzyme activity has been exhausted through repeated re-use, the thiol derivatives may be displaced by washing the matrix with cysteine solution, thereby regenerating the polythiol.

Secondly, a hydrophilic copolymer of N-acryloylaminoacetaldehyde dimethyl acetal and N,N'-methylene diacrylamide, 'Enzacryl Polyacetal', on dilute acid treatment generates aldehyde groups (Epton et al., 1971). Dextranase, α-amylase, papain, trypsin and urease have been attached to this carrier. The spent enzyme can be detached in certain circumstances.

The current activities of the Birmingham group include the synthesis of polymeric organic and inorganic carriers from compounds which are easily and cheaply produced commercially. New methods being successfully investigated for the coupling of enzymes to the carrier depart dramatically from the use of organic linkages, and employ inorganic linkages and specialized microencapsulation.

Common advantageous features of all these enzyme carriers are the ease and speed with which enzymes can be attached to them. It is generally quite adequate to stir the enzyme in near neutral buffered solution with the carrier. However, the successful coupling of an enzyme to any of these insoluble supports is by no means guaranteed, since coupling may occur adjacent to or at the active site and thereby render an enzyme inactive. Nevertheless, with the choice of links which can be formed with the various derivatizable groups of enzymes, and with the various supports available, it should be possible to prepare insoluble derivatives of most enzymes used in structural analysis of glycoproteins and polysaccharides. Thereby work in the field of enzymatic degradation would be simplified.

Finally, it should be noted that the use of these carriers is not confined to the production of carrier-fixed enzymes. They are equally applicable to the production of insoluble derivatives of a whole range of molecules, e.g. insulin, albumin (Barker et al., 1971d). Indeed, cellulose carbonate has been used in the radioimmunoassay of human IgE immunoglobulin (McLauglan et al., 1971) and follicle stimulating hormone (Kennedy and Cho Tun, 1973), and in the purification of antibodies and antigens (Catty et al., 1973).

References

Barker, S.A., Chaplin, M.F., Kennedy, J.F.: Unpublished results (1969b).

Barker, S.A., Cho Tun, H., Doss, S.H., Gray, C.J., Kennedy, J.F.: Preparation of cellulose carbonate. Carbohydrate Res. 17, 471 - 474 (1971a).

Barker, S.A., Doss, S.H., Gray, C.J., Kennedy, J.F., Stacey, M., Yeo, T.H.: β-D-Glucosidase chemically bound to microcrystalline cellulose. Carbohydrate Res. 20, 1 - 7 (1971b).

Barker, S.A., Gray, C.J., Khoujah, A.M., Law, C.M.J.: Preparation and properties of a cysteine-containing polymer (Enzacryl Polythiol). Birmingham University Chemical Engineer 22, 74 - 80 (1971d).

Barker, S.A., Kennedy, J.F., Gray, C.J.: Cellulose derivative. British Patent Application 61701/69 (1969c).

Barker, S.A., Kennedy, J.F., Gray, C.J.: Enzymes. British Patent Application 61702/69 (1969d).

Barker, S.A., Kennedy, J.F., Rosevear, A.: Preparation and uses of a poly (allyl carbonate). J. Chem. Soc. C. 2726 - 2730 (1971c).
Barker, S.A., Somers, P.J., Epton, R.: Preparation and properties of α-amylase chemically coupled to microcrystalline cellulose. Carbohydrate Res. 8, 491 - 497 (1968).
Barker, S.A., Somers, P.J., Epton, R.: Preparation and stability of exo-amylolytic enzymes chemically coupled to microcrystalline cellulose. Carbohydrate Res. 9, 257 - 263 (1969a).
Barker, S.A., Somers, P.J., Epton, R.: Recovery and re-use of insoluble amylase derivatives. Carbohydrate Res. 14, 323 - 326 (1970b).
Barker, S.A., Somers, P.J., Epton, R., McLaren, J.V.: Cross-linked polyacrylamide derivatives (Enzacryls) as water-insoluble carriers of amylolytic enzymes. Carbohydrate Res. 14, 287 - 296 (1970a).
Catty, D., Kennedy, J.F., Drew, R.L., Cho Tun, H.: The application of cellulose carbonate to the preparation of water-insoluble immunoadsorbents used in the purification of antibodies to immunoglobulins. J. Immunological Methods 2, 353 - 369 (1973).
Crook, E.M.: Enzymes attached to solid matrices. Biochem. J. 107, 1P - 2P (1968).
Epton, R., McLaren, J.V., Thomas, T.H.: Enzyme insolubilisation with cross-linked polyacryloylaminoacetaldehyde dimethyl acetal. Biochem. J. 123, 21p (1971).
Goldstein, L., Katchalski, E.: Use of water-insoluble enzyme derivatives in biochemical analysis and separation. Z. anal. Chem. 243, 375 - 396 (1968).
Guilbault, G.G.: Use of enzymes in analytical chemistry. Anal. Chem. 38, 527R - 536R (1966).
Kay, G.: Insolubilised enzymes. Process Biochemistry 3, 36 - 39 (1968).
Kennedy, J.F., Barker, S.A., Rosevear, A.: The use of a poly (alhyl carbonate) for the preparation of active, water-insoluble derivatives of enzymes. J. Chem. Soc. Perkin I, 2568 - 2573 (1972).
Kennedy, J.F., Cho Tun, H.: Use of cellulose carbonate for the preparation of immunoadsorbents: the radioimmunoassay of follicle-stimulatory hormone. Carbohydrate Res. 30, 11 - 19 (1973).
McLaughlan, P., Stanworth, D.R., Kennedy, J.F., Cho Tun, H.: Use of antibody-coupled cellulose as immunosorbent in the estimation of human immunoglobulin E (γE). Nature New Biology (Lond.) 232, 245 - 247 (1971).
Silman, I.H., Katchalski, E.: Water-insoluble derivatives of enzymes, antigens and antibodies. Ann. Rev. Biochem. 35, 873 - 908 (1968).

The Enzymatic Degradation of Proteoglycans

R. I. G. Morrison

In this review article, several aspects of the enzymatic degradation of proteoglycans will be considered. After a theoretical discussion of the enzymes required to bring about the complete breakdown of a proteoglycan molecule, recent work concerning the characterization of the degradation products resulting from the action of various purified proteases, including those thought to be of importance in connective tissue catabolism, will be described. Studies concerning the degradation of proteoglycans in *in vitro* systems where cartilage is induced to break down will also be considered, and the recent use of immunoenzymic techniques to investigate the enzymes responsible for breakdown will be discussed, as well as other studies which characterize the observed degradation products.

Theoretical considerations show that a number of different enzymes would be required to effect the complete breakdown of a proteoglycan molecule. Fig. 1 is a diagrammatic representation of a proteoglycan molecule showing a number of polysaccharide chains attached to a protein core, and the points of attack of several types of enzymes are indicated.

Proteolytic attack would result in a number of products containing one or more polysaccharide chains linked to or by protein or peptide. Papain and pronase are thought to degrade the protein core extensively, leading to the appearance of single polysaccharide chains (Muir, 1958; Schubert, 1965; Morrison, 1970b). Trypsin, on the other hand, produces polysaccharide "doublets" (Schubert, 1965; Luscombe and Phelps, 1967) consisting of two polysaccharide chains linked together by polypeptide resistant to the enzyme. However, it is not these enzymes, but cathepsin D and, to a lesser extent, cathepsin B1 (for a review of the nomenclature of the tissue proteinases see Barrett and Dingle, 1971) which have been implicated in the breakdown of proteoglycans in cartilage matrix (see below). These enzymes have now been purified from several sources (Barrett, 1967, 1970), and the products of their action on proteoglycans investigated (Morrison, 1970b). Fig. 2 shows the Sepharose 6B gel chromatographic profiles of the degradation products of bovine nasal proteoglycan resulting from the action of various proteolytic enzymes.

It may be seen that exhaustive degradation of this substrate with rabbit cathepsin B1 led to products which were highly retarded on the gel, and appeared in the same position as those resulting from the action of papain or pronase, indicating that single polysaccharide chains were produced. Treatment with trypsin or pepsin resulted in a somewhat less retarded product, which it was assumed contained polysaccharide "doublets". Exhaustive treatment with cathepsin D, on the other hand, resulted in the appearence of a still less retarded (i.e. larger) product. If it is assumed that the papain and trypsin products contain one and two polysaccharide chains, respectively, and that Ve (elution volume) is proportional to log molecular weight for poly-

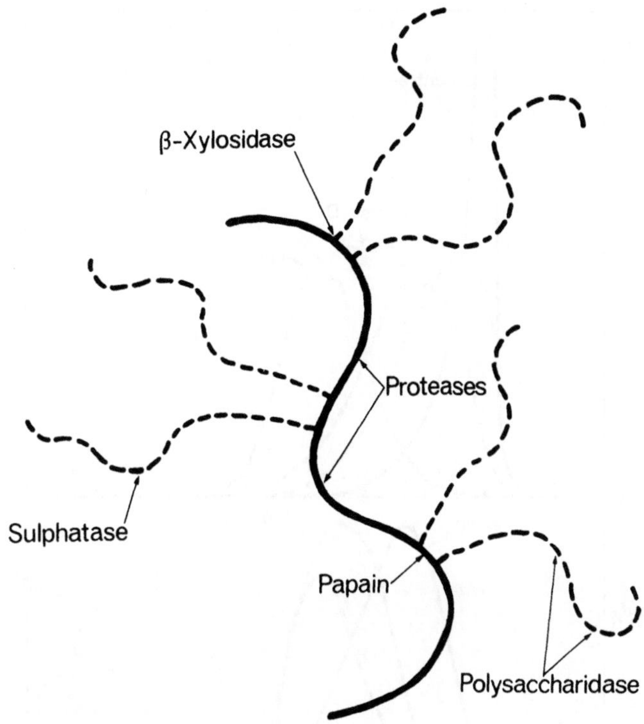

Fig. 1. Diagrammatic representation of a proteoglycan molecule showing a number of polysaccharide chains (-----) attached to a protein core (———). The points of attack of several types of enzyme are indicated

saccharide material during chromatography on agarose gels, then the cathepsin D products may be calculated to contain approximately three polysaccharide chains. These differences were not apparent when the degradation products were examined by viscometry or ultrafiltration (Morrison, 1970b). These findings are consistent with the known specificities of the enzymes in that cathepsin B1 is thought to have a specificity similar to that of papain, while cathepsin D resembles pepsin (see Keilova, 1971).

Other enzymes which would be required to effect complete degradation of the proteoglycan molecule include those attacking the protein-polysaccharide linkage region, and the polysaccharide chain itself.

β-xylosidase would theoretically hydrolyze the xylosyl-O-serine linkage, giving rise to polysaccharide chains free of protein. This action was proposed by Kent and coworkers (Fisher et al., 1966, 1967), though later work showed that β-xylosidases were apparently without action against xylosyl-O-serine linkages in intact chondroitin sulfate proteoglycans (cited by Dodgson and Lloyd, 1968). In view of these considerations, and since β-xylosidase is thought to be an exoglycosidase (see Barrett, 1969), it appears likely that the enzyme would act only at a late stage in the degradation of proteoglycans and would not cause any initial release from the matrix. This also applies to other lysosomal enzymes such as β-glucuronidase, β-N-acetyl-hexosaminidases and β-galactosidase (Barrett, 1969).

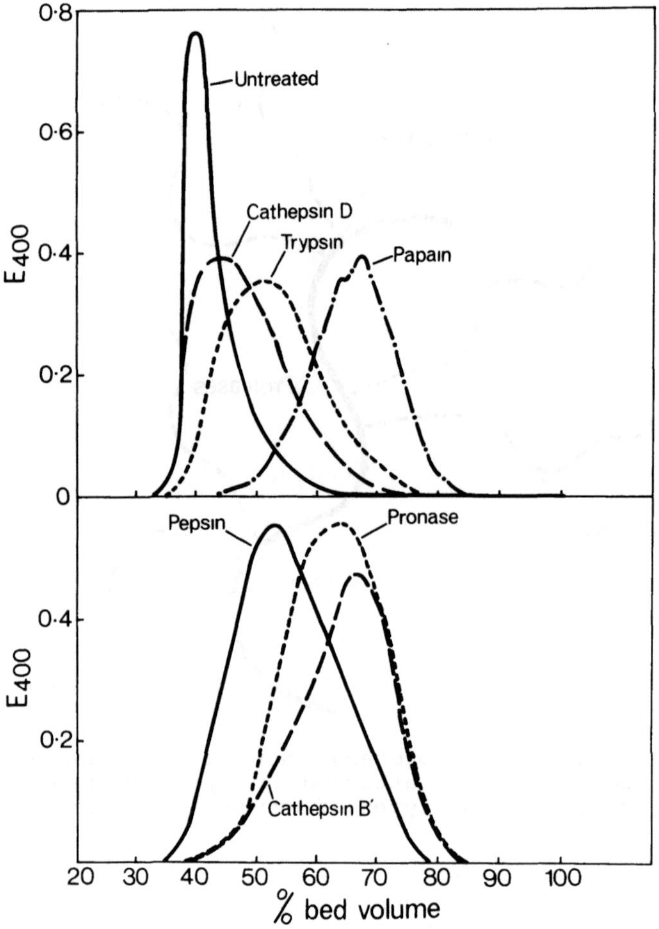

Fig. 2. Sepharose 6B gel chromatography of the degradation products resulting from the action of various proteolytic enzymes on bovine nasal proteoglycans

Enzymes exhibiting hyaluronidase activity have been demonstrated in a number of mammalian tissues, including the lysosomes of growing rat bones (Vaes, 1967), rat liver (Hutterer, 1966; Aronson and Davidson, 1967) and canine submandibular gland (Tan and Bowness, 1968), as well as in lung, liver, kidney, spleen, synovial fluid, artery, and urine (Bollet et al., 1963). As well as degrading hyaluronic acid, these enzymes have been shown to attack, more slowly, chondroitin-4-sulfate, chondroitin-6-sulfate and proteoglycan, though not dermatan sulfate, keratan sulfate and possibly heparin (see Dodgson and Lloyd, 1968; Barrett, 1969). Partial desulfation of sulfated polymers may be necessary for enzyme action.

Dodgson and Lloyd (1968) have discussed the possible action of exoglycosidases in the degradation of proteoglycans, and it appears unlikely that such enzymes contribute significantly to the initial stages in the degradation of proteoglycans (e.g. Buddecke and Platt, 1965).

However, Nishida-Fukuda and Egamit (1970) have presented evidence that degradation of keratosulfate in the liver of the marine gastropod *Charonias lampas* occurs by the concerted action of β-galactosidase, β-N-acetyl-glucosaminidase and a keratosulfatase, prior desulfation being necessary for depolymerization to occur. Greiling et al. (1971) reported an endohexosaminidase which degraded keratan sulfate in rabbit kidney lysosomes, as well as a keratan sulfate sulfohydrolase.

Although bacterial and mammalian aryl sulfatases have often been demonstrated (see Barrett, 1969), many investigations (see Dodgson and Lloyd, 1968) have failed to detect carbohydrate sulfatase activity in mammalian tissues. Recently, however, there have been reports of chondrosulfatase activity in cultured chick limb bone rudiments (Morrison, 1970a, b, 1971), bovine aorta (Held and Buddecke, 1967) and rat liver lysosomes (Tudball and Davidson, 1969).

Although all the enzymes discussed above would be necessary to bring about the complete degradation of the proteoglycan molecule, the enzymes of greatest physiological significance will be those leading to the initial breakdown and release of these molecules from the matrix of cartilage or other connective tissue. It now seems clear that proteolytic enzymes are responsible for this process in cartilage, and they have been implicated in a number of systems. Ali and coworkers (Ali, 1964; Ali et al., 1967; Ali and Evans, 1969) studied the autolytic breakdown of rabbit ear cartilage and reported that cathepsins A, B, C and D were all present in the cartilage and that cathepsin D was thought to be the most important enzyme involved in the breakdown. Whereas cathepsins B1 and D are endopeptidases which act mainly on substrates of high molecular weight, cathepsins A and C are exopeptidases and would be most likely to act on the degradation products of proteins resulting from the prior action of cathepsins B1 and D. Iodice et al. (1966) showed that cathepsins D and A had a synergistic effect in an *in vitro* test system. Woessner (1967) concluded that cathepsin D was principally responsible for the autolytic breakdown of bovine cartilage.

Lysosomal proteases have been implicated in the breakdown of cartilage which occurs during differentiation of cartilage into bone. Proteolytic activity has been demonstrated in calcifying cartilage by Jibril and Lindenbaum (1965), Hirschman and Dziewiatkowski (1966), using immunological techniques, demonstrated the loss, or at least drastic alteration, of the protein moiety of proteoglycans during or just preceding calcification. Fessel and Chrisman (1964) and Dziewiatkowski et al. (1968) described an acid protease, in cartilage extracts, which would degrade proteoglycans, and which, judging by the properties of the partially purified enzyme, was probably cathepsin D. Woessner (1967, 1971) has purified cathepsin D from several types of cartilage, including embryonic chick bones and young rabbit ear cartilage, and showed that these enzymes degraded proteoglycans, even at neutral pH. Cathepsin D was also thought to be important in remodelling during development of chick skin and during post-partum involution of the rat uterus (Woessner, 1965, see 1971).

Embryonic chick cartilage may be induced to break down during growth in organ culture in the presence of various agents such as vitamin A, sucrose and complement-sufficient antiserum, and much significant work on the mechanisms of, and enzymes responsible for, the breakdown of chick cartilage has been done using this system by Dingle and Fell and their coworkers. Early work on cartilage induced to break down by growth in the presence of vitamin A indicated that a lysosomal acid protease was responsible for degradation and release of matrix components (Lucy et al., 1961; Dingle, 1961; Fell and Dingle, 1963), and subsequent work has shown that this enzyme was cathepsin D (Dingle et al., 1971; Morrison, 1970a, 1971). In order to raise antisera for use as specific inhibitors of the enzyme, cathepsin D has been purified from the livers of chicken, rabbit and man (Barrett, 1967, 1970). Since cartilage is a somewhat inconvenient starting tissue, cathepsin D was originally purified from liver, but the chicken liver enzyme has subsequently been shown to be immunologically identical with the enzyme in embryonic cartilage, kidney, brain, heart, testes and spleen (Weston, 1969). In the human, liver cathepsin D is similarly immunologically indistinguishable from cathepsin D found in rheumatoid synovium (Barrett, 1971). The presence of several isoenzymes in the liver of these species may be demonstrated by isoelectric focussing (Barrett, 1970); the isoenzymes have similar molecular weights, pH optima and again are immunologically identical (see Barrett, 1971).

The antisera raised to cathepsin D have been shown to be specific and highly inhibitory towards the enzyme at pH5 and above, though at pH3 there was no inhibition since antigen/antibody interaction did not take place. Inhibition could be demonstrated with hemoglobin or proteoglycan as substrate, using whole antisera, IgG preparations or univalent antibody fragments produced by pepsin digestion of the IgG. Inhibition was thought to occur by a steric mechanism: whereas precipitation was caused by 3 - 4 molecules of antibody per molecule of cathepsin D, 6 - 7 molecules of antibody per molecule of enzyme were required for complete inhibition to occur.

The antisera have been used as specific inhibitors of cartilage breakdown in both autolytic and organ culture systems. When ^{35}S-labelled chick cartilage was subjected to autolysis at pH5 in the presence of the anti-chicken cathepsin D antiserum, 85 - 95 % inhibition of release of labelled material occured. Similar inhibition of the much smaller release at neutral pH was also observed, showing that cathepsin D was the enzyme principally responsible for the autolytic breakdown in this system. Similar results were obtained with rabbit cartilage.

The antiserum caused variable inhibition of breakdown of chick cartilage during growth in organ culture in the presence of vitamin A. The interpretation of the experiments was complicated by factors at present under further investigation, such as penetration of the matrix by the antibodies, their ability to affect degradation occuring within digestive vacuoles in the cells and, of course, the possible action of other enzymes. It appears that sheep anti-rabbit cathepsin D can inhibit the breakdown of hemoglobin and other substrates in digestive vacuoles of rabbit alveolar macrophages (Dingle and coworkers, unpublished results).

Another approach to investigating the enzymatic breakdown of proteoglycans has been to characterize the products appearing during the degradation of cartilage in organ culture and autolytic systems (Morrison, 1970b). Sepharose 2B gel chromatography of the ^{35}S labelled material released into the culture medium during growth of chick limb bone rudiments in the presence of vitamin A (or sucrose or complement-sufficient antiserum) showed the presence of both macromolecular material, estimated to contain 4 - 8 polysaccharide chains, and a low molecular weight component shown to be inorganic sulfate by a number of techniques (Morrison, 1970b). Only macromolecular material was released from cartilage during autolytic breakdown at pH5, though considerable release of inorganic sulfate was found at pH3. In both types of experiment, the degradation products were of larger size than those resulting from exhaustive treatment of the proteoglycan with the purified enzyme thought to be responsible for the release (cathepsin D).

These results indicate that only limited proteolytic breakdown is necessary to effect the release of the molecules from the matrix.

During culture of chick cartilage in the presence of vitamin A or complement sufficient antiserum, sulfatase activity was increased well above control levels. In the presence of sucrose, which is thought to interfere with breakdown occuring within digestive vacuoles in the cell, sulfatase activity was inhibited. These experiments suggested that the sulfatase is probably a lysosomal enzyme normally acting within the digestive vacuoles in the cell. Other reported chondrosulfatases have been lysosomal in origin (see above).

When chick cartilage was cultured in the presence of both vitamin A and sucrose together, sulfatase activity was again inhibited when compared with that found in rudiments treated with vitamin A alone, and abnormal vacuolation, caused by the endocytotic uptake of the indigestible sugar, spread throughout the explants. In the presence of sucrose alone, the abnormal vacuolation was restricted to certain areas of the cartilage where remodelling was taking place. These results strongly suggested that endocytosis was a normal and essential part of the degradation of matrix components.

These experiments have provided strong support for the "two-stage" hypothesis of matrix breakdown postulated by Dingle et al. (1969) in which an initial secretion of protease (cathepsin D) is thought to cause the partial degradation and release of proteoglycan molecules from the matrix in the immediate vicinity of the cell membrane; this material is then taken into the cell by endocytosis and its digestion continued and completed within the lysosomal system.

References

Ali, S.Y.: The degradation of cartilage matrix by an intracellular protease. Biochem. J. 93, 611 - 618 (1964).

Ali, S.Y., Evans, L.: Studies on the cathepsins in elastic cartilage. Biochem. J. 112, 427 - 433 (1969).

Ali, S.Y., Evans, L., Stainthorpe, E., Lack, C.H.: Characterisation of cathepsins in cartilage. Biochem. J. 105, 549 - 557 (1967).

Aronson, N.N., Davidson, E.A.: Lysosomal hyaluronidase from rat liver. I. Preparation. J. biol. Chem. 242, 437 - 440 (1967).

Barrett, A.J.: Lysosomal acid proteinase of rabbit liver. Biochem. J. 104, 601 - 608 (1967).

Barrett, A.J.: Properties of lysosomal enzymes. In: Lysosomes in Biology and Pathology (Ed., Dingle, J.T., Fell, H.B.), Vol. 2, p. 245. Amsterdam-London: North-Holland Publishing 1969.

Barrett, A.J.: Cathepsin D. Purification of isoenzymes from human and chicken liver. Biochem. J. 117, 601 - 607 (1970).

Barrett, A.J.: Purification and properties of cathepsin D from liver of chicken, rabbit and man. In: Tissue Proteinases (Ed. Barrett, A.J., Dingle, J.T.), p. 109 - 128. Amsterdam-London: North-Holland Publishing 1971.

Barrett, A.J., Dingle, J.T. (Ed.): Tissue Proteinases. Amsterdam-London: North-Holland Publishing 1971.

Bollet, A.J., Bonner, W.M., Nance, J.L.: The presence of hyaluronidase in various mammalian tissues. J. biol. Chem. 238, 3522 - 3527 (1963).

Buddecke, E., Platt, D.: Nachweis, Reinigung und Eigenschaften der Hyaluronidase aus der Aorta des Rindes. Hoppe-Seylers Z. physiol. Chem. 343, 61 (1965).

Dingle, J.T.: Studies on the mode of action of excess of vitamin A. 3. Release of a bound protease by the action of vitamin A. Biochem. J. 79, 509 - 512 (1961).
Dingle, J.T., Barrett, A.J., Weston, P.D.: Cathepsin D. Characteristics of immunoinhibition and the confirmation of a role in cartilage breakdown. Biochem. J. 123, 1 - 13 (1971).
Dingle, J.T., Fell, H.B., Glauert, A.M.: Endocytosis of sugars in embryonic skeletal tissues in organ culture. IV. Lysosomal and other biochemical effects. General discussion. J. Cell Sci. 4, 139 - 154 (1969).
Dodgson, K.S., Lloyd, A.G.: Metabolism of acidic glycosaminoglycans. In: Carbohydrate Metabolism and its Disorders (Ed. Dickens, F., T.J. Randle, W.J. Whelan), Vol. 1, p. 169 - 212. London: Academic Press 1968.
Dziewiatkowski, D.D., Tourtellotte, C.D., Campo, R.D.: In: The Chemical Physiology of the Mucopolysaccharides (Ed. Quintarelli, G.), p. 63. London: Churchill 1968.
Fell, H.B., Dingle, J.T.: Studies on the mode of action of vitamin A. 6. Lysosomal protease and the degradation of cartilage matrix. Biochem. J. 87, 403 - 408 (1963).
Fessel, J.M., Chrisman, O.D.: Enzymatic degradation of chondromucoprotein by cell-free extracts of human cartilage. Arthritis Rheum. 7, 398 (1964).
Fisher, D., Higham, M., Kent, P.W., Pritchard, P.: β-xylosidases of animal and other sources in relation to the degradation of chondroitin sulphate-peptide complexes. Biochem. J. 98, 46P - 47P (1966).
Fisher, D., Whitehouse, M.W., Kent, P.M.: β-xylosidase and β-galactosidase activities of mammalian connective tissues and other sources. Nature 213, 204 - 205 (1967).
Greiling, J., Stuhlsatz, H.W., Eberhard, A.: Enzymatic degradation of keratan sulphate-proteins. Z. klin. Chem. 9, 71 (1971).
Held, E., Buddecke, E.: Nachweis, Reinigung und Eigenschaften einer chondroitin-4-sulfatase aus der Aorta des Rindes. Hoppe-Seylers Z. physiol. Chem. 348, 1047 (1967).
Hirschman, A., Dziewiatkowski, D.D.: Protein polysaccharide loss during endochondral ossification: immunochemical evidence. Science 154, 393 - 395 (1966).
Hutterer, F.: Degradation of mucopolysaccharides by hepatic lysosomes. Biochem. biophys. Acta (Amst.) 115, 312 - 319 (1966).
Iodice, A.A., Leong, V., Weinstock, I.M.: Separation of cathepsins A and D of skeletal muscle. Arch. Biochem. 117, 477 - 486 (1966).
Jibril, A.O., Lindenbaum, A.: Non-dialysable sialic acid compounds in ossifying and ressing calf scapula cartilage. Biochem. biophys. Acta (Amst.) 101, 236 (1965).
Keilova, H.: On the specificity and inhibition of cathepsins D and B. In: Tissue Proteinases (Ed. Barrett, A.J., J.T. Dingle), p. 45 - 65. Amsterdam-London: North-Holland Publishing 1971.
Lucy, J.A., Dingle, J.T., Fell, H.B.: Studies on the mode of action of excess of vitamin A. 2. A possible role of intracellular proteases in the degradation of cartilage matrix. Biochem. J. 79, 500 - 508 (1961).
Luscombe, M., Phelps, C.P.: Action of degradative enzymes on the light fraction of bovine septa protein polysaccharide. Biochem. J. 103, 103 - 109 (1967).
Morrison, R.I.G.: The breakdown of proteoglycans by lysosomal enzymes and its specific inhibition by an antiserum to cathepsin D. In: The Chemistry and Molecular Biology of the Intercellular Matrix (Ed. Balazs, E.A.), Vol. 3, p. 1683 - 1706. New York: Academic Press 1970a.
Morrison, R.I.G.: Breakdown of Cartilage Proteoglycans. Ph.D. dissertation, University of Cambridge, Cambridge/England 1970b.

Morrison, R.I.G.: Studies on the enzymatic degradation of proteoglycans. Z. klin. Chem. 9, 69 - 70 (1971).
Muir, H.: The nature of the link between protein and carbohydrate of a chondroitin sulphate complex from hyaline cartilage. Biochem. J. 69, 195 - 204 (1958).
Nishida-Fukuda, M., Egami, F.: Enzymic degradation of keratosulphates. Biochem. J. 119, 39 - 47 (1970).
Schubert, M.: Protein polysaccharides, entanglement and excluded volume. In: Structure and Function of Connective and Skeletal Tissue (Eds. Fitton Jackson, S., P.D. Harkness, S.M. Partridge, G.R. Tristram), p. 124 - 131. London: Butterworths 1965.
Tan, Y.H., Bowness, J.M.: Canine submandibular-gland hyaluronidase. Identification and subcellular distribution. Biochem. J. 110, 9 (1968).
Tudball, N., Davidson, E.A.: Isolation of a novel sulphatase from rat liver. Biochim. biophys. Acta (Amst.) 171, 113 - 120 (1969).
Vaes, G.: Hyaluronidase activity in lysosomes of bone tissue. Biochem. J. 103, 802 - 804 (1967).
Weston, P.D.: A specific antiserum to lysosomal cathepsin D. Immunology 17, 421 (1969).
Woessner, J.F.: Acid hydrolases of the rat uterus in relation to pregnancy, post-partum involution and collagen breakdown. Biochem. J. 97, 855 - 866 (1965).
Woessner, J.F.: Acid cathepsins of cartilage. In: Cartilage, Degradation and Repair (Ed. Bassett, C.A.L.), p. 99 - 106. Washington: National Acad. Sci. 1967.
Woessner, J.F.: Cathepsin D: Enzymic properties and role in connective tissue breakdown. In: Tissue Proteinases (Eds. Barrett, A.J., J.T. Dingle), p. 291 - 308. Amsterdam-London: North-Holland Publishing 1971.

Lysosomal Enzymes and Proteoglycan Degradation in the Human Intervertebral Disc.

C. H. Pearson, F. Happey, A. Naylor, J. M. Osborn, and R. L. Turner

There is evidence that degradation of the proteoglycans in the intervertebral disc occurs on aging and similar changes appear to be associated with herniation or prolapse (Davidson and Woodhall, 1959; Lyons et al., 1964, 1966). It is not known whether a defect in biosynthesis or increased catabolism is responsible for the deterioration. Human discs obtained at surgery have been shown to contain β-N-acetylglucosaminidase and β-glucuronidase (Platt, 1970), which were probably derived from lysosomes but this was not proved.

In the present work the occurrence of lysosomes in different fractions of normal and prolapsed lumbar disc was investigated, using acid phosphatase activity as a guide following treatments known to rupture lysosomes. Assays were also carried out for cathepsin and acid glycosidases. In other experiments gel chromatography was employed to examine the degradation of proteoglycans in tissue slices subjected to autolytic conditions.

1. Tissue Fractionation and Enzyme Assays

The general procedure of De Duve et al. (1955) was followed, but an increased proportion of 0.25 M sucrose solution (20 - 25 mls/g of wet tissue) was required. Homogenization was carried out for 5 - 10 min only, in a VIRTIS 45 instrument at 30 - 45,000 r.p.m., cooling in ice-water. Prolapsed nucleus pulposus, obtained at laminectomy, was first rinsed with several changes of the sucrose solution to remove as much blood as possible. Centrifuging was performed at 2°C under the conditions shown in Table 1. Each sediment was washed with one or two changes of the sucrose solution, the washings being added to the previous supernatant.

Table 1. Distribution of acid glycerophosphatase activity in fractions from nucleus pulposus

g value	Time (minutes)	Fraction	% of total activity: Normal Autopsy	% of total activity: Laminectomy
1,000	10	Sediment, P_1	20 - 40	28 - 38
10-15,000	15	Sediment, P_2	45 - 60	46 - 63
38,000	60	Sediment, P_3	10 - 12	9 - 15
		Supernatant, S_4	8 - 10	7 - 12

Total enzyme activity was determined in the supernatants obtained after freezing and thawing each fraction 10 times, or after standing in 0.1 - 0.2% w/v of TRITON X - 100 (CALBIOCHEM), for 1 hour at 2°C, 0.25 M sucrose solution being present in both cases. Acid phosphatase was assayed against sodium β-glycerophosphate (BDH CHEMICALS) at pH 5 by a method similar to that of De Duve et al. (1955), suitable controls being set up to correct for contaminating inorganic phosphate. The substrates employed in assaying acid glycosidases were: phenolphthalein glucuronide, o-nitrophenyl-β-D-galactoside and N-acetyl-β-D-glucosaminide using the methods of Tishman (1967) and Sellinger et al. (1960). Turbidity-reduction tests were performed for the presence of hyaluronidase with chondroitin sulfate-A (MILES-SERAVAC) as substrate, as described by Di Ferrante (1956). Bovine haemoglobin (CALBIOCHEM) was dialyzed, before use as substrate for cathepsin (Barrett, 1967).

2. Autolysis in Vitro

Slices of nucleus pulposus (or occasionally the intermediate zone adjacent to the inner annulus fibosus) from apparently normal, autopsy specimens, were immersed in toluene-covered sodium acetate-acetic acid buffer, 0.05 M, adjusted to various pH values.

Incubation was for 24 hours at 37°C with controls of 4°C. Alternatively, a 0.1 M acetate buffer was employed, after freezing and thawing the tissue 10 times. Laminectomy specimens were rinsed initially with 0.9% NaCl solution. The volume of buffer used in the incubations varied from 1 - 30 ml/g of wet tissue. Supernatants were recovered for analysis and gel filtration by centrifuging, for up to 1 hour at 12,000 × g at 2°C for the smallest volumes.

3. Results and Discussion

In Table 1 it can be seen that the highest proportion of acid phosphatase activity is found in the sediment P_2, obtained at 10 - 15,000 × g, following removal of most of the collagen at 1,000 × g. Nearly all the activity, which showed a maximum at pH 5, was in a latent form, i.e., very little was found unless repeated freezing and thawing, or a treatment with the non-ionic detergent TRITON X - 100 was given. These procedures and also exposure to osmotic forces are known to disrupt lysosomes. The progressive release of acid phosphatase from P_2 under increasing disruptive conditions is shown in Fig. 1. The separation of several different molecular weight forms of the acid phosphatases of the disc will be reported elsewhere.

The distribution of acid phosphatase in fractions derived from laminectomy specimens did not differ significantly from normal and the average results resembled those reported by Dingle (1962) for cartilaginous rudiments (P_1 35%, P_2 51%, P_3 + P_4 16%). However, the difficulties of isolating intact lysosomes from tough connective tissues are severe, as emphasized by Woessner (1965). Even the short homogenization used here may have ruptured some lysosomes and others were probably mechanically trapped or adsorbed by the insoluble collagen, part of which was found in P_2 as well as P_1.

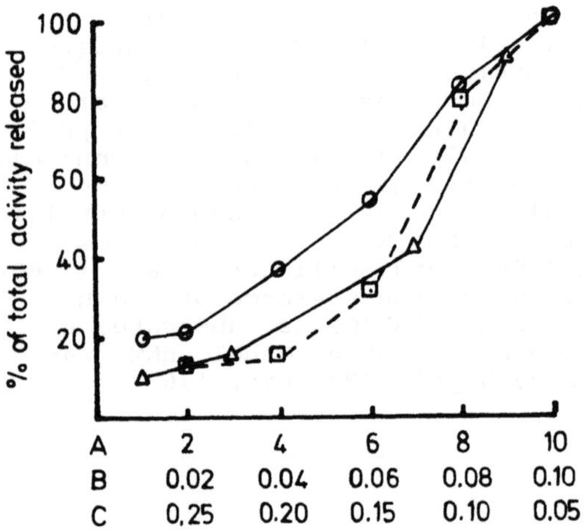

Fig. 1. Release of acid glycerophosphatase activity from fraction P_2. Scale A (Triangles) Number of times frozen and thawed Scale B (Circles) Concentration of TRITON X - 100. Scale C (Squares and broken line) molarity of sucrose

Appreciable activities of acid cathepsin β-glucuronidase, β-galactosidase and β-N-acetylglucosaminidase were found in P_2 (after activation as described for acid phosphatase) of the nucleus pulposus of autopsy and laminectomy specimens. The pH optima of the first two enzymes were established as 3.6 - 3.8 and 4.3 - 4.8, respectively. Like that of the acid phosphatase, the activities of the glycosidases varied greatly from sample to sample on a wet weight basis. Further work is needed before reliable quantitative comparison of the activities in normal and prolapsed tissue can be made. It must be inferred that these enzymes were mainly lysosomal, but in considering the general implications of lysosomes in the disc, histochemical and electron microscopical investigations are needed.

The presence of cathepsins is significant, as they are believed to initiate the degradation of proteoglycans (Morrison, 1970). On the other hand, the role of the glycosidases listed above is very uncertain. Possibly, intact chondroitin sulfate chains are susceptible to the combined action of β-glucuronidase and β-N-acetylhexosaminidase (and sulfatases), but it seems more likely that these "exoglycosidases" are important at a later stage in the degradation.

In agreement with Platt (1970) we failed to find significant hyaluronidase activity in surgical specimens and none was detected in the nucleus pulposus or the intermediate zone of normal discs. There is little definite evidence of the presence of this enzyme in most connective tissues, but irreversible inactivation, which may occur at high dilutions (Bernfield, 1966) has possibly prevented detection.

The processes by which keratan sulfate degrades are very obscure. Lysosomal β-galactosidase, which appeared to be present in the disc, has been shown to attack the molecule, but probably only at the nonreducing end and so the extent of breakdown is very small (Öckerman, 1970).

4. Autolysis

Judging by the different results obtained at 37°C and 4°C, significant autolysis occurred when two normal specimens were incubated in a very small volume of buffer and when a 36-year-old laminectomy specimen was incubated in a retalively large volume (Table 2). The remaining

Table 2. Release of protein and polysaccharides from nucleus pulposus under autolytic conditions. Protein was determined by the method of Lowry et al. (1951) and uronic acid (glucuronolactone standard) essentially as decribed by Bitter and Muir (1962). The orcinol-H_2SO_4 method was used for the determination of neutral sugars (galactose standard), calculating corrected neutral sugar and uronic acid values as given previously (Pearson et al., 1971)

Specimen Age	ml buffer/g wet tissue	pH	Temperature	mg/g wet tissue Protein	 Uronic acid	 Neutral sugars
46[a]	1	4.2	4°C	8.7	5.6	3.4
			37°C	19.0	7.6	6.5
49[a]	1	4.8	4°C	13.7	9.8	5.9
			37°C	17.8	12.6	7.0
48	1	4.2	4°C	24.6	4.2	2.5
			37°C	30.5	4.1	2.9
48[a]	30	5.0	4°C	-	8.9	3.5
			37°C	-	9.6	3.9
27	30	5.1	4°C	10.7	7.3	7.6
			37°C	10.4	7.1	8.3
36	30	5.8	4°C	10.7	1.3	1.4
			37°C	12.3	2.6	2.9
36	30	4.7	4°C	7.9	1.6	0.8
			37°C	12.5	3.4	3.6
52	30	4.4	4°C	5.7	4.3	5.7
			37°C	8.3	5.1	7.8

[a]Normal nucleus pulposus from autopsy. The remainder were obtained at laminectomy.

results were negative or doubtful, especially when the quantities released were expressed as percentages of the total present, to overcome sampling errors. Further experiments showed that the release of protein and polysaccharides at 37°C was certainly pH dependent. Maxima were found at pH 4.1 - 4.7, with indications of slightly different optima for protein, neutral sugars and uronic acid. The difficulty of demonstrating autolysis by the type of data given in Table 2 seems to be mainly related to the ease of extraction of most of the proteoglycans. The quantities, which diffuse out, even at 4°C seem to be high compared with most tissues. A tentative conclusion which was supported by the gel filtration results discussed below, was that the smallest possible volume of buffer is to be preferred, to minimize the dilution of the autolytic enzymes.

Gel filtration of supernatants from 37°C incubates or 4°C controls carried out on Sephadex G200, yielded only one major peak, at the void volume. A proteoglycan preparation extracted from discs by 3M $MgCl_2$ (Sajdera and Hascall, 1969) and purified by precipitation with cetyl pyridinium chloride, was almost completely excluded from Sepharose 2B, which has a much higher exclusion limit than Sephadex G 200. The purified preparation contained almost equal amounts of chondroitin sulfate and keratan sulfate and a high proportion of protein (uronic: neutral sugars: protein, 1:1.1:1.9). Therefore, it appeared that in this preparation both glycosaminoglycans were attached to the same protein chains. Heinegard and Gardell (1967) and Rosenberg and Schubert (1967) arrived at the same conclusion, but Gower and Pedrini (1969) obtained evidence of heterogeneity using column electrophoresis.

When no previous fractionation was carried out, 4°C controls as well as 37°C incubates gave a large, broad retarded peak on Sepharose 2B, in addition to a smaller excluded peak. The effect of increasing the volume of the buffer used in autolysis is illustrated in Fig. 2. With a very small volume of buffer, protein is degraded much more, only 11% of it appearing in the excluded peak, compared with 35% when 30 ml (or 10 ml) were used and 23% (5 ml). A similar effect on the polysaccharides has not yet been demonstrated, the proportion in the excluded peak remaining at about 20%.

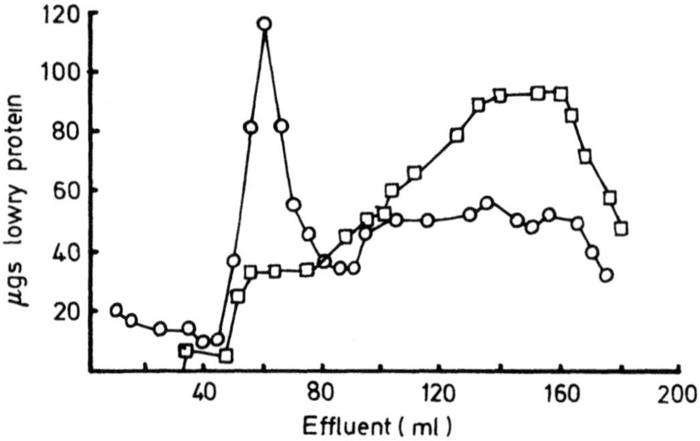

Fig. 2. Effect of volume of buffer on autolysis of tissue slices. Nucleus pulposus from autopsy was frozen and thawed 10 times before incubating for 24 hours at 37°C (pH 4.2 - 5.0). Gel filtration was carried out on a 35 ×2.5 cm column of Sepharose 2B (PHARMACIA), eluting with 0.9% NaCl solution at 1 ml/min. at room temp. (downward flow). Fractions of 5 ml were collected. Excluded peak at 45 - 80 ml.
Squares - Results using 1 ml buffer/g of wet tissue.
Circles - Using 30 ml buffer/g.
Because of the higher dilution the protein values in this chromatogram have been increased 4-fold to bring them on the indicated scale.

However, autolysis may account for the results shown in Fig. 3, in which there was a partial differentiation of chondroitin sulfate and keratan sulfate complexes in the region of the excluded peak and the immediately adjoining retarded zone. This chromatogram was unusual in that three protein peaks were clearly distinguished. There was an excess of neutral sugars over uronic acid in the region of peak 1, whereas the reverse was true in peak 2.

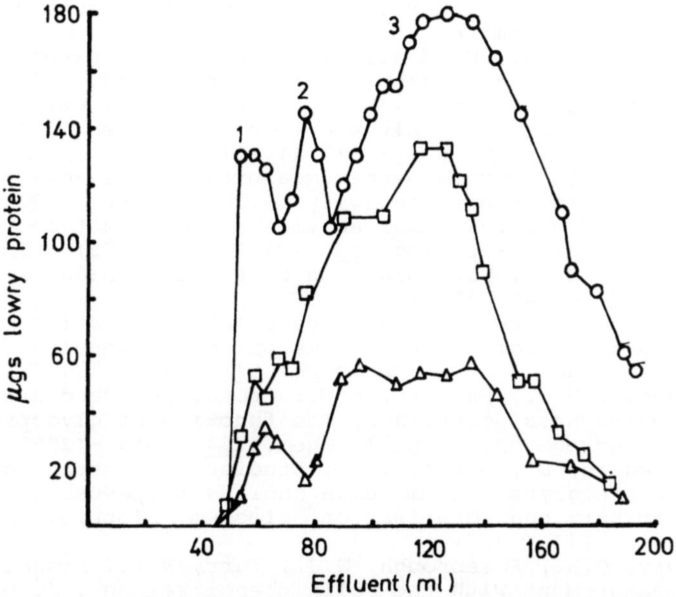

Fig. 3. Effects of autolysis on protein and polysaccharides. A specimen of intermediate zone of discs from autopsy was treated as described in Fig. 2, incubating at 37°C in 5 ml buffer/g at pH 4.7. Gel filtration as in Fig. 2. Top curve - Lowry protein. Middle curve - Uronic acid. Bottom curve - Neutral sugars (actual values reduced 2 fold)

The effects observed here are consistent with a limited proteolytic digestion, either of regions of the protein core that are relatively free of bound polysaccharides, or of the so-called glycoprotein which is involved in aggregation of the proteoglycan sub-units (Hascall and Sajdera, 1969). The changes, which were probably brought about by cathepsins, were smaller than have been demonstrated using purified substrates and concentrated enzyme preparations, particularly in relation to the polysaccharides. (Morrison, 1970). It is possible that the large proportion of keratan sulfate, which is an important feature of the disc proteoglycans, has a general protective effect.

Acknowledgements. Financial assistance was received from the Leeds Regional Hospital Board and the Laming Evans Fund of the Royal College of Surgeons of England.

References

Barrett, A.J.: Lysosomal acid proteinase of rabbit liver. Biochem. J. 104, 601 - 608 (1967).

Bernfield, P.: Activation and inhibition of enzymes by polyanions containing amino sugars. In: The Amino Sugars, Vol. IIB (Ed. Balazs, E.A., R.W. Jeanloz), p. 213. London: Academic Press 1966.

Bitter, T., Muir, H.M.: A modified uronic acid carbazole reaction. Anal. Biochem. 4, 330 - 334 (1962).

Davidson, E.A., Woodhall, B.: Biochemical alterations in herniated intervertebral discs. J. biol. Chem. 234, 2951 - 2954 (1959).

De Duve, C., Pressman, B.C., Gianetto, R., Wattiaux, R., Appelmans, F.: Intracellular distribution of enzymes in rat-liver tissues. Biochem. J. 60, 604 - 617 (1955).

Di Ferrante: Turbidometric measurements of acid mucopolysaccharides and hyaluronidase activity. J. biol. Chem. 220, 303 - 306 (1956).

Dingle, J.T.: Lysosomal enzymes and degradation of cartilage matrix. Proc. royal Soc. med. 55, 109 - 111 (1962).

Fishman, W.H.: Determination of β-glucuronidases. Meth. biochem. Anal. 15, 77 - 145 (1967).

Gower, W.E., Pedrini, V.: Age-related variations in protein-polysaccharides from human nucleus pulposus, annulus fibrosus and costal cartilage. J. Bone Jt. Surg. 51A, 1154 - 1162 (1969).

Hascall, V.C., Sajdera, S.W.: Proteinpolysaccharide complex from bovine nasal cartilage. The function of glycoprotein in the formation of aggregates. J. biol. Chem. 244, 2384 - 2396 (1969).

Heinegard, D., Gardell, S.: Studies on protein-polysaccharide complex (proteoglycan) from human nucleus pulposus. I. Isolation and preliminary characterisation. Biochem. biophys. Acta (Amst.) 148, 164 - 171 (1967).

Lowry, O.H., Rosebrough, N.J., Farr, N.L., Randall, R.J.: Protein measurement with the Folin phenol reagent. J. biol. Chem. 193, 265 (1951).

Lyons, H., Jones, E., Quinn, F.E., Sprunt, D.H.: Protein-polysaccharide complexes of normal and herniated human intervertebral discs. Proc. Soc. exp. Biol. (N.Y.) 115, 610 - 614 (1964).

Lyons, H., Jones, E., Quinn, F.E., Sprunt, D.H.: Changes in the protein-polysaccharide fractions of nucleus pulposus from human intervertebral disc with age and disc herniation. Lab. clin. Med. 68, 930 - 939 (1966).

Morrison, R.I.G.: The breakdown of proteoglycans by lysosomal enzymes and its specific inhibition by an antiserum to cathepsin D. In: Chemistry and Molecular Biology of the Intercellular Matrix, Vol. 3 (Ed. Balazs, E.A.), p. 1683. London: Academic Press 1970.

Öckerman, P.A.: Enzymic liberation of D-galactose from keratan sulphate by human tissues. Carbohydrate Res. 12, 429 - 436 (1970).

Pearson, C.H., Happey, F., Naylor, A., Turner, R.L., Palframan, J., Shentall, R.D.: Collagen and associated glycoproteins in the human intervertebral disc. Ann. Rheum. Dis. In press (1971).

Platt, D.: Altersabhängige Aktivitätsänderungen lysosomaler Enzyme (Glycosaminoglycans-Hydrolasen) in Serum und Organen des Menschen. Deutsche Med. Woch. 95, 634 - 637 (1970).

Rosenberg, L., Schubert, M.: The proteinpolysaccharides of bovine nucleus pulposus. J. biol. Chem. 242, 4691 - 4701 (1967).

Sajdera, S.W., Hascall, V.C.: Proteinpolysaccharide complex from bovine nasal cartilage. A comparison of low and high shear extraction procedures. J. biol. Chem. 244, 77 - 87 (1969).

Sellinger, O.Z., Beaufay, H., Jaques, P., Doyen, A., De Duve, C.: Intracellular distribution and properties of β-N-acetyl-glucosaminidase and β-galactosidase in rat liver. Biochem. J. 74, 450 - 456 (1960).

Woessner, J.F., Jr.: Acid hydrolases of connective tissue. In: International Review of Connective Tissue Research, Vol. 3 (Ed. Hall, D.A.), p. 201. London: Academic Press 1965.

Studies on Lysosomal Proteoglycan Degradation

H. Greiling, R. Kisters, H. W. Stuhlsatz, and A. Eberhard

1. Isolation of Proteokeratan Sulfate Degrading Enzymes from Kidney Lysosomes

After the discovery of keratan sulfate, a polysaccharide, consisting of galactose and 2-acetamido-2-deoxy-D-glucose, a bacterial enzyme was found which degrades keratan sulfate*. The existence of this enzyme in the mammalian organism is still questionable. Recently Nishida-Fukuda and Egami (1970) were able to detect a keratan sulfate degrading enzyme in liver extracts of the marine gastropode charonia lampas. They isolated a multi-enzyme system consisting of β-acetylglucosaminidase and sulfatase. Öckerman (1970) demonstrated the existence of a galactosidase in human liver and rat liver. He assumes that the β-galactose is split off from the non-reducing end of the keratan sulfate molecule.

In this paper we shall describe our studies on the isolation and characterization of a multi-enzyme complex of lysosomes of rabbit kidneys.

Tables 1 and 2 demonstrate the method of purification of keratan sulfate sulfohydrolase, of keratan sulfate glycanohydrolase and of several other lysosomal enzymes from kidney. As it is seen from the Tables, the purification of keratan sulfohydrolase and of keratan sulfate glycanohydrolase is concomitant with a purification of arylsulfatase, β-glucuronidase, β-N-acetylglucosaminidase, β-galactosidase, and α-glucosidase. Further purification was achieved by CM-Sephadex-chromatography. The final product did not show any cytochrome oxidase activity.

In the lysosomal fraction of rabbit kidney we found an enzyme which splits off sulfate of ^{35}S-labelled peptidokeratan sulfate (KPG). This keratan sulfate sulfohydrolase has a pH optimum of 4.5 (Fig. 1).

Chondroproteoglycan from cornea is also a substrate for the sulfohydrolase. For a quantitative estimation of the amount of sulfate split off, we separated the KPG from free sulfate by means of high voltage electrophoresis (at $0^{o}C$) in the presence of cetylpyridinium chloride (CPC) (Fig. 2). In the lysosomal fraction of rabbit kidney we have also detected a glycanohydrolase with a pH optimum of 3.5 acting on purified corneal and tracheal keratan sulfate. This enzyme splits peptidokeratan sulfates of cornea and trachea into oligosaccharides accompanied by an increase of free N-acetylglucosamine groups and reducing substances.

The Michaelis constant of keratan sulfate sulfohydrolase was found to be $K_M = 1.56 \times 10^{-3}$M.

With heparin, a competitive inhibitor, the constant of inhibition was $K_I = 4.77 \times 10^{-5}$M (Fig. 3).

Table 1. Purification of keratan sulfate glycanohydrolase and keratan sulfate sulfohydrolase from rabbit kidney

Step	mg Protein	KS-sulfo-hydrolase cpm/mg Prot.	CS-sulfo-hydrolase cpm/mg Prot.	KS-glycano-hydrolase mU/mg Prot. × 10^3	CS-glycano-hydrolase mU/mg Prot. × 10^3
Homogenate	21.1	3,240	13,374	1.07	0.33
Lysosomes (after freezing and thawing)	10.2	18,470	69,570	1.72	0.65
Ammonium sulfate (20 - 60%) after dialysis	0.37	61,490	182,680	4.61	1.60
Biogel P-150-chromatography	0.43	18,260	53,070	18.50	0

Table 2. Purification of lysosomal enzymes from rabbit kidney

Step	β-N-acetyl-glucosaminidase mU/mg	Arylsulfatase mU/mg	β-galactosidase mU/mg	α-manno sidase mU/mg	α-glucosidase mU/mg	β-glucuronidase mU/mg
Homogenate	15.9	12.87	5.02	1.37	0.15	0.12
Lysosomes (after freezing and thawing)	38.1	43.66	15.94	3.75	0.36	0.43
Ammonium sulfate (20 - 60%) after dialysis	68.9	229.7	31.70	125.70	1.10	1.16
Biogel P-150-chromatography	497.7	156.7	240.6	111.6	2.15	1.36

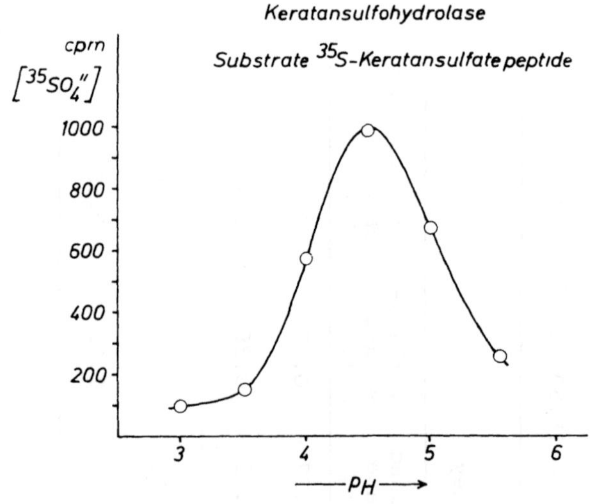

Fig. 1. Influence of pH on the action of keratan sulfate sulfohydrolase

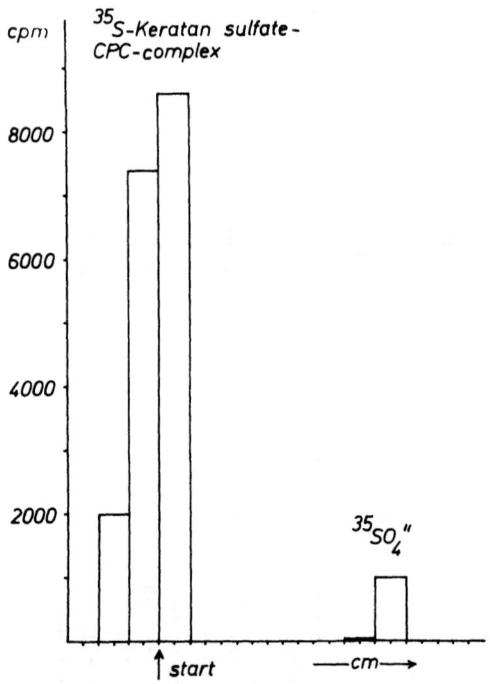

Fig. 2. Method for demonstrating the action of lysosomal keratan sulfate sulfohydrolase (high voltage electrophoresis, $0^{\circ}C$)

An inhibition was also found with sulfate (Fig. 4) phosphate, and p-chloromercuribenzoic acid.

A cleavage of 4.7% of the total activity of ^{35}S-labelled peptidokeratan sulfate was found after an incubation for 20 hours.

The Michaelis constant of keratan sulfate glycanohydrolase was found to be $K_M = 1.67 \times 10^{-3}M$ for peptidokeratan sulfate from trachea as

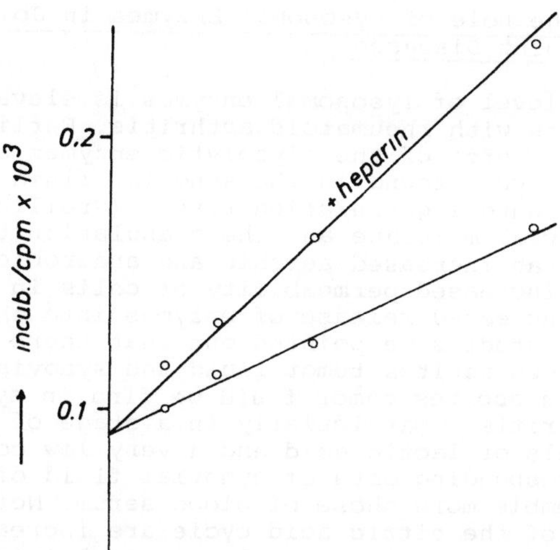

Fig. 3. Competitive inhibition of keratan sulfate sulfohydrolase with heparin

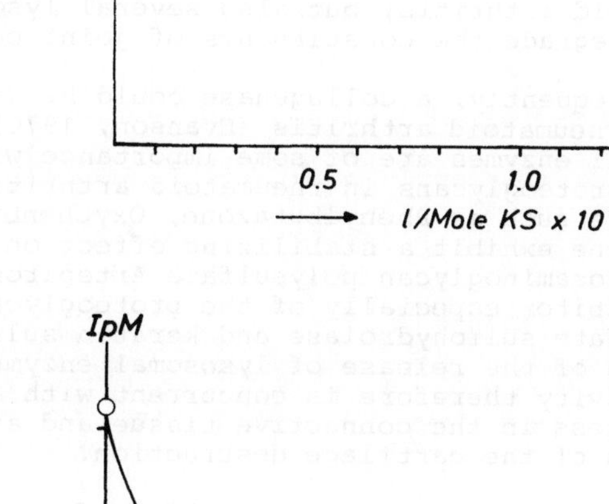

Fig. 4. Inhibition of keratan sulfate sulfohydrolase by sulfate

substrate, and $K_M = 2.44 \times 10^{-3}$M for peptidokeratan sulfate from cornea. The reactions with both substrates were competitively inhibited by heparin. The corresponding inhibitor constants were $K_I = 9.8 \times 10^{-5}$M (trachea), and $K_I = 1.65 \times 10^{-4}$M (cornea).

2. The Role of Lysosomal Enzymes in Joint Cartilage Destruction with Joint Diseases

The level of lysosomal enzymes is elevated in synovial fluid of patients with rheumatoid arthritis. Earlier experiments showed that the main source of the glycolytic enzymes and the enzymes of the citric acid cycle found in the synovial fluid is the synovial membrane and the pannous granulation tissue (Greiling et al., 1966). The inflamed synovial membrane and the granulation tissue contain cell systems which have an increased aerobic and anaerobic glycolysis. Because of the increased permeability of cells in the inflamed tissue, there is an increased release of enzymes into the extracellular fluid. In earlier studies we pointed out that there is a biochemical similarity between ascites tumor fluid and synovial fluid (Greiling et al., 1964). As in ascites tumor fluid we find in synovial fluid of rheumatoid arthritis - particularly in a stage of severe inflammation - high levels of lactic acid and a very low concentration of glucose. The corresponding data of synovial fluid of degenerative joint diseases resemble more those of blood serum. Not only the enzymes of glycolysis and of the citric acid cycle are increased in synovial fluid of rheumatoid arthritis, but also several lysosomal enzymes which are able to degrade the constituents of joint cartilage.

Consequently, a collagenase could be demonstrated in the synovial fluid of rheumatoid arthritis (Evanson, 1970). We suppose that these lysosomal enzymes are of some importance with respect to the degradation of proteoglycans in rheumatoid arthritis. Several anti-inflammatory drugs such as Phenylbutazone, Oxyphenbutazone, Chloroquine and Prednisolone exhibit a stabilizing effect on the lysosomal membrane. The glycosaminoglycan polysulfate Arteparon has proved to be a competitive inhibitor especially of the proteoglycan-degrading enzymes chondroitin sulfate sulfohydrolase and keratan sulfate sulfohydrolase. An inhibition of the release of lysosomal enzymes or an inhibition of their activity therefore is concurrent with an inhibition of the inflammatory process in the connective tissue and at the same time with an inhibition of the cartilage destruction.

The question arises, which kind of destruction is first? In the early stage of rheumatoid arthritis the concentration of collagen in the joint cartilage is unchanged, whereas a decrease in the proteoglycan content can be seen. It is therefore reasonable to presume that degradation of the proteoglycans is the first step. In the second place, an attack of the collagen matrix by collagen lysing enzymes is taking place. The determination of the lysosomal enzyme distribution pattern yielded the following order of decreasing enzyme activity in the synovial fluid of chronic joint diseases: N-acetyl-β-D-glucosaminidase, β-glucuronidase, α-mannosidase, arylsulphatase, β-galactosidase, and α-glucosidase. In comparison to rheumatoid arthritis in which the synovial fluid shows the highest enzyme activities, the lowest enzyme activities are found in the synovial fluid of degenerative joint diseases. We found about the same enzyme distribution pattern in synovial cells of a synovial fluid of a patient with rheumatoid arthritis, which consists mainly of polymorphonuclear leucocytes, as we found in the synovial fluid itself (Table 3). One therefore may suppose that the lysosomal enzymes of the synovial fluid are mainly produced by the lysosomal fraction of polymorphonuclear leucocytes. The amount of lysosomal enzyme activity is a measurement of the inflammatory process which in degenerative joint disease takes place mainly at the synovial membrane. The quantitation of the lysosomal enzyme activity, therefore, is best suited to follow the therapeutic result after intra-articular injections. Therefore a decrease of lysosomal enzyme activ-

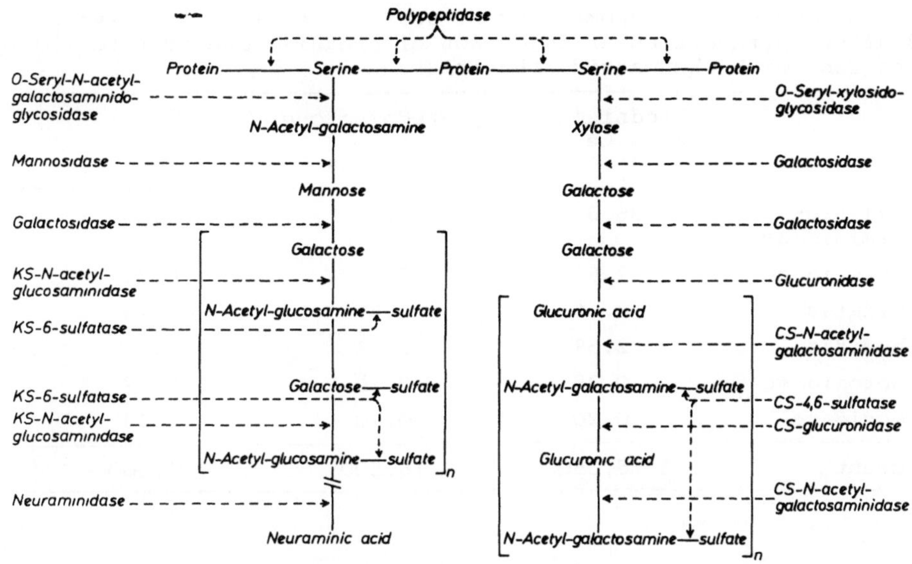

Fig. 5. Sequential degradation of proteochondroitin sulfate-keratan sulfate hybrid in human joint cartilage with a lysosomal enzyme system

Table 3. Lysosomal enzyme distribution in synovial fluid and in the extract of the cells of the same fluid (rheumatoid arthritis: female, 32 years of age)

	Synovial fluid mU/ml	Synovial cells mU/10^8 cells
β-N-acetyl-glucosaminidase	13.4	63.2
β-glucuronidase	1.53	9.1
Arylsulfatase	0.83	10.3
α-mannosidase	0.88	8.4
β-galactosidase	0.11	6.8
α-glucosidase	0.07	0.4

ities could be shown after intra-articular injection of prednisolone (Table 4). The cytostatic drug podophyllic acid ethylhydrazide also causes a decrease of lysosomal enzyme activities in the synovial fluid. In a few cases we observed an increase of lysosomal enzyme activities after a long treatment with prednisolone acetate or podophyllic acid ethylhydrazide. Perhaps a dosis dependence for the effect of anti-inflammatory or immunsuppressive drugs is the cause of this phenomenon.

Table 4. Changes in lysosomal enzyme activities (mU/ml) in synovial fluid after intra-articular injection of prednisolone acetate (50 mg) of a patient with rheumatoid arthritis

	Prednisolone acetate 11.9.1969	Prednisolone. acetate 17.9.1969	8.10.1969
N-acetal-β-D-glucosaminidase	36.3	22.1	6.25
β-glucuronidase	7.60	2.02	0.38
Arylsulfatase	3.66	0.87	0.44
α-mannosidase	2.59	0.84	0.49
β-galactosidase	0.38	0.10	0.07
α-glucosidase	0.20	0.10	0.04
Cell count	76,350	16,500	1,500

References

Evanson, J.M.: Collagen degradation by explanted rheumatoid synovial tissue. In: E.A. Balazs: Chemistry and Molecular Biology of the Intercellular Matrix, Vol. II, p. 1637. London-New York: Academic Press 1970.

Greiling, H., Engels, G., Kisters, R.: Untersuchungen über die Isoenzyme der Lactat-Dehydrogenase in der Synovialflüssigkeit. Klin. Wschr. 42, 427 - 431 (1964).

Greiling, H., Kisters, R., Engels, G.: Die Enzyme in der Synovialflüssigkeit und ihre pathophysiologische Bedeutung. Enzymologia 30, 135 - 146 (1966).

Greiling, H., Stuhlsatz, H.W., Kisters, R.: Structure and metabolism of proteokeratan sulfate. In: E.A. Balazs: Chemistry and Molecular Biology of the Intercellular Matrix, Vol. II, p. 873. London-New York: Academic Press 1970.

Nishida-Fukuda, M., Egami, F.: Enzymic degradation of keratansulphates. Biochem. J. 119, 39 (1970).

Öckerman, P.A.: Enzymic liberation of D-galactose from keratan sulfate by human tissues. Carbohydrate Res. 12, 429 (1970).

*Hirano, S., Meyer, K.: Purification and Properties of a Keratan Sulfate Hydrolyzing Enzyme, an Endo-β-Galactosidase. Conn. Tissue Res. 2, 1 (1973).

Action of Mammalian Collagenases

J. M. Evanson

It is now almost a decade since Gross and Lapiere's (1962) demonstration of the first animal collagenase, achieved by the novel technique of culturing tadpole tissues on reconstituted collagen gels. Since that time many similar enzymes of animal origin have been described and they have been summarized in several recent review articles (Gross, 1970; Evanson, 1970a; Eisen et al., 1970).

The majority of the enzymes have been the subject of study by technique similar to those originally devised by the Boston workers, and as a group, in common with the amphibian enzyme, they have the following principal properties:

1. They are difficult to demonstrate in tissue extracts and usually require culture techniques for their isolation.

2. At neutral pH and physiological ionic strength they can degrade collagen in solution at 20 - 28°C or in the reconstituted fibril state at 37°C.

3. Their initial attack on the tropocollagen monomer in solution occurs at a locus 1/4 of the way from the B (carboxy-terminal) end of the molecule.

4. They are susceptible to inhibition by serum, EDTA and cysteine.

Several of the enzymes described do not however share all these characteristics. The granulocyte collagenase (Lazarus et al., 1968) may be demonstrated directly in lysosomal fractions of white cell homogenates and appears to have little, if any, action on collagen fibrils. It is also atypical in that it is not inhibited by serum. This is also true of one of the two collagenases which can occasionally be obtained directly from synovial fluids in joint disease. Collagenase B from synovial fluid resembles and is probably identical with the enzyme derived from synovial tissue culture experiments, whereas Collagenase A is of higher molecular weight and insusceptible to serum inhibition (Harris et al., 1969).

The physiological significance of these and other differences between individual enzymes within the group is not yet clear, but the demonstration that collagenases, previously demonstrated only by culture techniques, can be identified in some direct tissue extracts or biological fluids, goes a long way towards disposing of the criticism that enzymes identifiable in tissue culture medium in vitro cannot justifiably be assigned an in vivo role. We are, however, still some way from defining this role precisely, but tissue or organ culture experiments have yielded information indicating the crucial intervention of collagenase in the degradation of tissue collagen. With cultured rheumatoid synovium, for instance, addition of agents influencing the yield of activity of collagenase also produces changes in the amount and time course of appearance of collagen breakdown products in the medium (Crossley and Evanson, 1971). Hydrocortisone which suppresses enzyme yield diminishes the amount of hydroxyproline peptides appearing in the culture fluid, whilst colchicine, which by an as yet unknown mechanism enhances collagenase yield, also increases the medium hydroxyproline. A similar correlation between collagenase production and breakdown of collagen in tissue explants also obtains with cultured fetal rat skin (Steed, 1971). In the case of the explanted mouse calvarium, addition of retinol, which promotes resorption, is associated with increased amount of collagenase, and when such retinol-induced resorption is counteracted with calcitonin, collagenase yield is suppressed (Evanson et al., 1970). In yet another situation, the carrageenin granuloma, it has been shown (Perez-Tamayo, 1970a, b) that breakdown of extracellular collagen, as judged by ultra-structural studies, correlates well with the capacity of the tissue to produce collagenase. Degradation of collagen in many different situations is thus demonstrably related to the ability of the tissue to elaborate collagenase, but what is less certain is the nature of the rate-limiting step in this process, at what stage the neutral collagenases act and their relationship to lysosomal acid hydrolases.

Since collagenase is envisaged as acting principally at an extracellular site, naturally occurring inhibitors may play a critical role in regulating its action. The majority of collagenases are inhibited by normal serum, in particular by the alpha globulin component, and it is possible to show in *in vitro* culture systems that the incorporation of inhibitory serum into the medium not only prevents direct demonstration of enzyme in the culture fluid but also suppresses the yield of collagen breakdown products. In organ culture systems the degradation of endogenous collagen within explants may thus be a consequence of removal of tissue from *in vivo* mechanisms normally restraining it. This concept has recently been strengthened by the observation of Eisen et al. (1971) that crude extracts of normal human skin, which do not show collagenase activity, yield enzymatically active fractions when examined by chromatographic techniques. The explanation for this "unmasking" of collagenase activity appears to lie in the separation of enzyme from the inhibitors α1 anti-trypsin and α2 macroglobulin. Whether or not these inhibitors are true tissue inhibitors or simply derived from plasma within the homogenized skin is not yet certain.

The delay of 24 - 48 hours in the appearance of significant enzyme activity in the culture media of rheumatoid synovial tissue, human and animal skin, and other tissues, appears to be due, at least in part, to the presence at this early stage of inhibitory substances in the fluid (Evanson, 1970b). The associated delay in the accumulation of significant quantities of hydroxyproline peptides in the medium supports the view that the operation of such inhibitors may influence collagenolysis within the tissue.

With cultures of rheumatoid synovium, maximum detectable enzyme yield and hydroxyproline peptide release roughly coincide at 4 - 6 days after the beginning of the culture period, at a time when serum and/or tissue inhibitory proteins have presumably been effectively washed out of the explants. In the case of human skin cultures, it has been shown by immunological techniques that the inhibitors α1 antitrypsin and α2 macroglobulin are present in culture medium during the first day and have disappeared by the time of maximum demonstrable enzyme activity (Eisen et al., 1971). Collagenase activity of the gastrointestinal tract has been shown to increase following surgical colonic anastomosis, and this increase occurs not only locally but also throughout the gut (Hawley et al., 1970). It has also been claimed that this postoperative enhancement of demonstrable collagenolytic activity is associated with a diminished amount of circulating inhibitor (Hawley and Faulk, 1970).

Observations such as these suggest that one mechanism by which the action of collagenases might be regulated *in vivo* is by the control of inhibitory protein concentrations. Thus the operation of such enzymes at an extracellular site would be determined not only by the rate of enzyme synthesis and release but also by the local concentration of inhibitors. It remains to be explored whether or not breakdown in such a process might account for unrestrained collagen resorption in some pathological states.

The collagen of the various connective tissues comprises several fractions of differing age and solubility, ranging from the recently synthesized, loosely aggregated fibrils to the older, highly cross-linked, insoluble component. All components are subject to resorption under various physiological or pathological conditions. Jackson (1957) and Woessner (1962), working respectively with the corrageenin granuloma and the involuting uterus, adduced evidence to suggest that catabolism of insoluble collagen involves an initial conversion to the soluble form prior to further degradation. The relationship of the neutral collagenases to this step in removal of collagen is unclear, but it is perhaps in this area that the frequently-invoked intervention of lysosomal hydrolases may be involved. Particular emphasis has in the past been placed on the role of Cathepsin D in collagenolysis (Woessner, 1965).

Our own observations, however, on purified Cathepsin D, have shown that this enzyme cannot solubilize reconstituted collagen fibrils at pH values of 4.5 and above; nor does it affect the viscosity of collagen solutions. In high concentrations, however, gelatin and the denatured products of collagenase action are broken down with peak activity in the region of pH 5.0. Thus, although evidence is accumulating that Cathepsin D is concerned with the extracellular degradation of proteoglycans (Morrison, 1970) its relation to the breakdown of native collagen *in vivo* is still a matter for speculation. Other hydrolases do appear to have collagenolytic properties at acid pH values (Bazin et al., 1966; Anderson, 1969) but since their pH activity profile is so different from that of the true neutral collagenases it is difficult to envisage their operation in the tissues at the same site. It remains to be determined whether there are two separate and distinct mechanisms for collagen breakdown in the body - the one acting principally in a neutral environment and the other in an acidic milieu within digestive vacuoles of cells - or if collagenolysis is effected by the sequential action of a variety of enzymes. But whichever scheme emerges as the true description of the process, the unique role of the true collagenases is to cleave the triple helical region of the collagen molecule at neutral pH.

References

Anderson, A.J.: Effects of lysosomal collagenolytic enzymes, anti-inflammatory drugs and other substances on some properties of insoluble collagen. Biochem. J. 113, 457 - 463 (1969).

Bazin, S., Delaunay, A., Briquelet, N., Llain, J.C.: Caracteres de cathepsines collagenolytiques presentes dans les tissue inflamme du rat. Annls. Inst. Pasteur 110, 192 - 204 (1966).

Crossley, M., Evanson, J.M.: Collagen metabolism in cultured rheumatoid synovial tissue. In preparation.
Eisen, A.Z., Bauer, E.A., Jeffrey, J.J.: Animal and human collagenases. J. Invest. Derm. 55, 359 - 373 (1970).
Eisen, A.Z., Bauer, E.A., Jeffrey, J.J.: Human skin collagenase. The role of serum alpha-globulins in the control of activity in vivo and in vitro. Proc. nat. Acad. Sci. (Wash.) 68, 248 - 251 (1971).
Evanson, J.M.: Collagen degradation by explanted rheumatoid synovial tissue. In: The Chemistry and Molecular Biology of the Intercellular Matrix (Ed. E. A. Balzs), Vol. 3, p. 1637 - 1652. London, New York: Academic Press 1970a.
Evanson, J.M.: Mammalian collagenases and their role in connective tissue breakdown. In: Tissue Proteinases (Eds. Barrett, Dingle). Amsterdam: North-Holland Publishing 1970b.
Evanson, J.M., Minkin, C., Reynolds, J.: Bone collagenase. J. Bone Jt. Surg. 52B, 182 (1970).
Gross, J.: The animal collagenases. In: The Chemistry and Molecular Biology of the Intercellular Matrix (Ed. E.A. Balazs), Vol. 3, p. 1623 - 1636. London, New York: Academic Press 1970.
Gross, J., Lapiere, C.M.: Collagenolytic activity in amphibian tissues: a tissue culture assay. Proc. nat. Acad. Sci. (Wash.) 48, 1014 - 1022 (1962).
Harris, E.C., jr., Dibona, D.R., Krane, S.M.: Collagenases in human synovial fluid. J. clin. Invest. 48, 2104 - 2113 (1969).
Jackson, D.S.: Connective tissue growth stimulated by carrageenin. The formation and removal of collagen. Biochem. J. 65, 277 - 284 (1957).
Lazarus, G.S., Brown, R.S., Daniels, J.R., Fullmer, H.M.: Human granulocyte collagenase. Science 159, 1483 - 1485 (1968).
Perez-Tamayo, R.: Collagen resorption in carrageenin granulomas. I. Collagenolytic activity in in vitro explants. Lab. Invest. 22, 137 - 141 (1970).
Perez-Tamayo, R.: Collagen resorption in carrageenin granulomas. II. Ultrastructure of collagen resorption. Lab. Invest. 22, 142 - 159 (1970).
Steed, A.J.: A study of skin collagenase and some of the factors governing its production by rat tissue in an in vitro culture system. M. Sc. Thesis. Univ. of Manchester 1971.
Woessner, J.F.: Catabolism of collagen and non-collagen protein in the rat uterus during post-partum involution. Biochem. J. 83, 304 - 314 (1962).
Woessner, J.F., jr.: Biological mechanisms of collagen resorption. In: Treatise on Collagen (Ed. Ramachandran, G.N.), p. 253 - 330. London, New York: Academic Press 1968.

The Mechanism of Collagen Decomposition in Mammals

G. Gries

Collagen, like other proteins, is apparently decomposed into low-molecular peptides or amino acids by successive degradation of various enzymes. The intact collagen molecule can be split by proteases. The known enzymes differ in various ways. The enzyme discovered by Gross and Lapière (1962) first in amphibia, called "collagenase", splits the native collagen molecule at a temperature of 20 °C at only one point, between leucine and glycine. It forms two fractions - first part corresponding to three quarters and the second part to one quarter the length of the original collagen molecule. Similar enzymes were later found also in mammals, including man, by Fullmer and Gibson (1966), Evanson et al. (1967), Aer and Kivirikko (1969) and others. These enzymes react best in a neutral to slightly alkaline medium. Hitherto they have been found primarily in the outer medium of tissue cultures and are inhibited by serum globulins (Eisen et al., 1971).

In addition to these collagenases of the Gross and Lapière type, Bazin and Delaunay (1966), Schaub (1964) and others have described enzymes which can also split collagen but which function best in an acidic medium; the researchers have classed these among the cathepsins.

Because of their effect on native collagen molecule we called the enzymes described by Gross and others "collagen proteases" to distinguish them from cathepsinoid proteases.

These proteases yield high-molecular peptides, which contain the collagen-specific apolar range in addition to the polar range. Because of their alternating amino acid sequence, the polar groups can be split by various enzymes. A number of cathepsins or cathepsinoid enzymes with an acidic pH optimum, which attack the polar ranges, have been described recently (Bazin and Delaunay, 1971; Schaub, 1964). It had been previously demonstrated that denatured collagen could be decomposed by peptide hydrolases such as trypsin and α-chymotrypsin, which operate best in a neutral or slightly alkaline medium (Grassmann et al., 1937). Since these enzymes can split neither native collagen nor the apolar range, it must be assumed that they can split the polar range of high-molecular collagen fractions.

The apolar range of high-molecular collagen peptidases can be split solely by enzymes as described by our research team. These enzymes split gly-pro-x-gly-pro-x sequences between the x and the glycine. Enzymes such as clostridiopeptidase A (EC 3.4.4.19) show the same point of attack as do these peptidases. The peptidases we have described thus have the same specificity as the bacterial collagenases.

Because of their specificity for the apolar range of collagen, we have called these peptidases "collagen peptidases" to distinguish them from those active in the polar range.

On the basis of their specificity both the bacterial collagenases (Wünsch and Heidrich, 1963) and the collagen peptidases (Strauch and Vencelj, 1967; Gries et al., 1970; Gries and Grasedyck, 1969) split synthetic substrates patterned at the apolar range Wünsch and Heidrich (1963).

Like collagen proteases, the collagen peptidases also have their pH optimum in the neutral or slightly alkaline range. The decomposition of native collagen by collagen peptidases has not yet been demonstrated to the last certainty. Denatured collagen is, however, decomposed while other denatured proteins such as hemoglobin, casein and serum albumin are not affected (Strauch, unpublished data).

Successive decomposition by enzymes at polar and apolar ranges leads in the polar range to amino acids, while in the apolar range it proceeds only as far as tripeptides of the gly-pro-x type.

Tripeptides of the gly-pro-x type can be partially further decomposed by prolinases and prolidases to amino acids (Nordwich and Dehm, 1968), but they are largely eliminated as tripeptides in the urine.

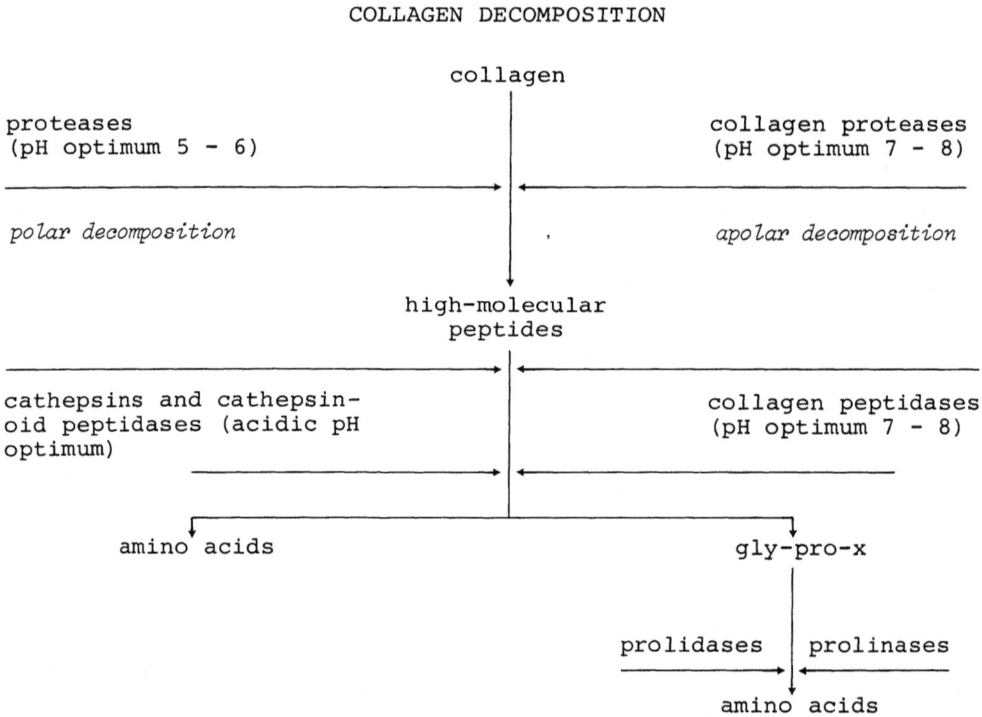

Diagram 1. Scheme of collagen decomposition

References

Aer, J., Kivirikko, K.J.: Preliminary characterization of the collagenolytic activity of red bone. Z. physiol. Chem. 350, 87 (1969).
Bazin, S., Delaunay, A.: Variations de l'activité collagenolytique dans des foyers inflammatoires en évolutions. Ann. Inst. Pasteur 110, 347 (1966).
Bazin, S., Delaunay, A.: A collagenolytic cathepsin isolated from inflamed tissues. Z. klin. Chem. 9, 66 (1971).
Eisen, A.Z., Bauer, E.A., Jeffrey, J.J.: Human skin collagenase. The role of serum alpha-globulins in the control of activity in vivo and in vitro. Proc. nat. Acad. Sci. (Wash.) 68, 248 (1971).
Evanson, J.M., Jeffrey, J.J., Krane, S.M.: Human collagenase. Identification and characterization of an enzyme from rheumatoid synovium in culture. Science 158, 499 (1967).
Fullmer, H.M., Gibson, W.: Collagenolytic activity in gingivae of man. Nature 209, 728 (1966).
Grassmann, W., Janicki, J., Schneider, F.: Über die Einwirkung von Trypsin auf Kollagen. Stiasny-Festschrift, p. 74. Darmstadt 1937.
Gries, G., Buresch, H., Strauch, L.: Collagenolytic enzymes in human serum. Experientia (Basel) 26, 31 (1970).
Gries, G., Grasedyck, K.: Über spezifische Kollagenasen in Organen des Magen-Darmtraktes. Verh. Dtsch. Ges. Path. 53, 205 (1969).
Gross, J., Lapière, Ch.M.: Collagenolytic activity in amphibian tissues. A tissue culture assay. Proc. nat. Acad. Sci. (Wash.) 48, 1014 (1962).
Nordwich, A., Dehm, P.: Breakdown of protein: Cleavage of peptides of the X-Pro-Y type by kidney peptidases. Biochim. biophys. Acta (Amst.) 160, 293 (1968).
Schaub, M.C.: Eigenschaften und intrazelluläre Verteilung eines kollagenabbauenden Kathepsins. Helv. physiol. Acta 22, 271 (1964).
Strauch, L.: Unpublished data.
Strauch, L., Vencelj, H.: Collagenases in mammalian cells. Z. physiol. Chem. 348, 465 (1967).
Wünsch, E., Heidrich, H.: Darstellung von Prolinpeptiden: III. Ein neues Substrat zur Bestimmung von Kollagenase. Z. physiol. Chem. 332, 300 (1963).

Studies on Collagen Peptidases

J. Lindner, K. Grasedyck, G. Prinz, J. Grade, and H. Kölln

Collagenolytic enzymes like collagenases and collagenpeptidases take part in the collagen turnover. As collagenases - especially collagen-proteases - split native collagen, collagenolytic activity was determined by using ^{14}C-labelled native collagen for a substrate (Woods and Nichols, 1965; Nagai et al., 1966; Jeffrey and Gross, 1967; Eisen et al., 1968; Lazarus et al., 1968; Harris et al., 1969; Keiditsch and Strauch, 1970, Vaes, 1971).

Collagenpeptidases - like clostridiopeptidase A (EC 3.4.4.19) - split polypeptides which are synthetized in analogy to the apolar regions of collagen molecules. Several synthetic polypeptides have been used for these investigations (Nagai, 1961; Hakim, 1966; Espey and Rondell, 1967; Mitchell, 1968). The activity of the collagenpeptidases which split the collagen molecule in the apolar regions can be tested by the substrate PZ-Pro-Leu-Gly-Pro-D-Arg synthetized by Wünsch and Heidrich (1963). The specifity of this substrate for the clostridiopeptidase A (EC 3.4.4.19) has been demonstrated by these authors. This polypeptide is not attacked by other enzymes (Strauch, 1970).

Using this substrate Strauch and Vencelj (1967), Strauch (1968), Strauch et al. (1968) detected enzymes in living cells of higher animals which are different from the tadpole collagenases first described by Gross and Lapière (1962) and which resemble bacterial collagenases in their specifity for the apolar regions of the collagen molecule.

Strauch (1970) showed that collagenase activity measured by ^{14}C-labelled native collagen as substrate (using the method of Nagai et al., 1966) and collagenpeptidase activity (using the PZ substrate) run parallel which means: both enzymes have the same dependence on time in the rat uterus during post partum involution.

In the last years many authors have tried to investigate the collagenpeptidase activity in tissue fluids and homogenates (Strauch and Vencelj, 1967; Strauch, 1968; Langer et al., 1968; Heidrich et al., 1969; Gries and Grasedyck, 1969; Kirschke and Hanson, 1969; Platt, 1969, 1970; Gries and Strauch, 1970; Cutroneo and Fuller, 1971; Aer, 1971; Bankowski and Galasinsky, 1971; Gries et al., 1972).

These investigations show that the collagenpeptidase activity varies in the different body fluids and tissues under various physiological and pathological conditions. Furthermore, corresponding results show that the activity of collagenpeptidases can be important for the pathogenesis and the research of several connective tissue diseases (inflammation, wound healing, rheumatoid arthritis, atherosclerosis, tumor growth, especially in the parablastomatous reactions).

Langer et al. (1968) found an increased collagenpeptidase activity in cancer tissue in comparison to the normal tissue of the same organ.

Keiditsch and Strauch (1970) demonstrated an increased enzyme activity in the invasion zone of tumors which corresponds to similar results indicating an increased collagen degradation using different methods, e.g. by the quantitative estimation of collagen degradation products: free and bound hydroxyproline (Lindner, 1957a, 1961; Gries and Lindner, 1960, 1961 and 1963).

Lindner (1957a, b, 1961, 1969a) and v. Schweinitz and Lindner (1959) demonstrated the increase of products of the intercellular substance (hydroxyproline, hexosamine, uronic acid) in the pathological status of inflammation, tumor growth and atherosclerosis, as compared to normal control tissues by means of the tissue-slide paper chromatography method.

Since the differences in concentration of free hydroxyproline in several organs correlated better with the metabolic activity of the organs than with their collagen content, it was further suggested that the concentration of free hydroxyproline in tissues is indirectly dependent on the metabolic acitivity of collagen (Chvapil and Kobrle, 1961; Woessner and Boucek, 1961; Kobrle and Chvapil, 1961a, b, 1962; Kivirikko, 1963).

Our investigations on collagenpeptidase activity of several tissues and organs confirm these observations: organs with a higher collagen content show a lower collagenpeptidase activity, organs with the lowest collagen content show usually a higher or the highest enzyme activity. This conformity of the results may indicate that the low content of collagenpeptidase in organs containing much collagen, for example skin, is not the result of absorption of enzymes, as supposed for collagenprotease. Furthermore, it can be shown that organs with a very low collagen content and high collagenpeptidase activity show a decrease of collagenpeptidase activity in diseases and at the same time an increase of the collagen content (i.e. in the liver, in the genesis of fibrosis and cirrhosis, or in arteries during the development of atherosclerosis, especially in early lesions).

The assay of collagenpeptidase activity can be used as an indicator for changes of collagen metabolism, especially in observations on the course of diseases (Grasedyck et al.: Studies on collagen synthesis with labelled proline, p. 122).

Götze et al. (1970) found an increased collagenpeptidase activity in inflamed gingiva, Grasedyck et al. (1971) in blood serum in cirrhosis of the liver. These and other investigations show that an increase of collagen turnover in inflammation, namely in the beginning of inflammation, and an increase of collagenpeptidase activity are present according to the primarily increased catabolism. The increase of collagen synthesis seems to be the second step in the disorder of collagen metabolism - for instance, in the development of liver fibrosis and cirrhosis. The increase of the collagen content is the third step in these diseases, as well as in other inflammations (Lindner et al., 1967; Lindner, 1969a, b).

We therefore studied the possibility of using collagenpeptidase activity as an indicator for collagen degradation (and metabolism) in rats during growth and in aging, as well as in several models of different diseases (in particular experimental thioacetamide intoxication, resulting in fibrosis and cirrhosis of the liver). Two further enzyme activities were investigated: (1) β-glucuronidase - a glycosidase - as indicator enzyme for the proteoglycan degradation, (2) protease, as an unspecific enzyme, also engaged in the breakdown of the intercellular components.

1. Methods

Collagenpeptidase. The tissues in question are to be homogenized in saline at 4°C, strictly avoiding any rise of temperature, because the activity of collagenpeptidase decreased rapidly when exposed to higher temperature. The synthetic PZ-Pro-Leu-Gly-Pro-D-Arg is dissolved in 0.2 mol Trisbuffer. The reaction is initiated by adding 0.25 ml tissue homogenate to 0.5 mol substrate in 0.5 ml Tris-buffer (pH 7.2). After 1 hour at 25°C it is stopped by adding 0.25 ml 10% citric acid. The pH reading 2 - 3. The intact substrate dissolves in water, while the split product PZ-Pro-Leu does not dissolve in water at low pH values. This can then be extracted in benzene. We can confirm that shaking 50 times in one minute is sufficient to transfer the split product quantitatively into the benzene phase. This procedure is followed by centrifugation.

To avoid admixtures of water, benzene is passed through filters. Then extinction can be measured at a wavelenth of 334 nm. A parallel test serves for a standard. The reaction, however, is prevented by adding citric acid at the beginning.

The nitrogen content of the homogenate is taken as a parameter. It was determined by means of Strauch's method (1965) of incineration in a thermobloc at 330°C with Nessler's reagent. The activity is expressed in CU (i.e. µmol splitted substrate/g protein/h, resp./ml blood serum/h). Three determinations are necessary.

The dilution of the tissue homogenates is a problem, which will be discussed in detail together with the results.

Blood serum is not diluted because of its very low enzyme activity. Gries and Grasedyck (1969) and Gries et al. (1972) described a second maximum of collagenpeptidases activity at pH 8.

Protease. The activity of the unspecific protease was determined according to Anson and Mirsky (1933), measuring the blue color which results from degradation products of proteins dissolved in trichloracetic acid after addition of Folin-Ciocalteau reagent. Hemoglobin of cattle, lyophilized, pure, clear of salt and denatured (Serva, Heidelberg) was used as substrate (Lit. cp. Fishman, 1970).

β-Glucuronidase. The determination of the β-glucuronidase was accomplished according to the phenolphthaleine-method (Fishman, 1970). The substrate for this method is phenolphthaleine-β-(D)-glucuronide, which was made from the cinchonidine-salt of the phenolphthaleine-glucuronide.

2. Results and Discussion

In continuation of former experiments we investigated the *collagenpeptidase activity* in tissue homogenates and also in the supernatant of centrifuged tissue homogenates as was recently done by several authors (Platt, 1969; Gries et al., 1972). The comparison between young, matured and senile rats displays no significant differences in liver, lung and blood serum (Table 1).

According to investigations on the other two enzymes, they show (a) *protease*: no alteration of the activity of the enzyme in the liver, while in lungs a significant increase can be proved with increasing

Table 1. Determination of collagenpeptidase activity (age dependent) (tissue homogenates, diluted by saline)

		13 days	4 months	3-year-old rats
Homogenate[a]	liver	12.94 ± 1.12 (16)	10.59 ± 2.76 (17)	9.86 ± 2.58 (17)
	lungs	63.57 ± 17.28 (13)	51.19 ± 20.53 (13)	75.96 ± 23.58 (10)
Supernatant[b]	liver	15.35 ± 1.23 (16)	15.14 ± 1.42 (6)	-
	lungs	62.03 ± 24.70 (11)	84.37 ± 9.79 (6)	-
Blood serum[c]		-	1.30 ± 0.44 (15)	1.33 ± 0.48 (4)

[a] 0.25 ml homogenate
[b] 0.25 ml supernatant after (10 min. 4500 rpm) centrifugation
[c] blood serum (undiluted)
Dilution of homogenates: liver 1:10, lungs 1:150; CU mean ± S.D, (n).

Table 2. Protease and β-glucuronidase activity in liver and lung (age dependent) mU/g wet weight; mean ± S.D. (n)

		13 days	4 months	3-year-old rats
Protease	liver	214.3 ± 38.1 (26)	245.0 ± 63.8 (28)	241.6 ± 109.2 (32)
	lungs	179.6 ± 31.4 (18)	309.9 ± 89.9 (20)	321.4 ± 140.0 (20)
β-Glucuroni-	liver	3.36± 0.29 (14)	2.11± 0.60 (22)	1.72± 0.62 (6)
dase	lungs	0.99± 0.15 (14)	0.82± 0.34 (18)	0.50± 0.18 (6)

age; (b) β-*glucuronidase:* a significant decrease in the liver as well as in lungs with increasing age (Table 2). Platt (1969, 1970, 1971) and Platt and Stein (1969) investigated the β-glucuronidase activity of human liver and blood serum. He found a decrease of the enzyme activity up to the second year of life, but with increasing age a constant increase of β-glucuronidase activity. In the human discus intervertebralis, however, Platt described a decrease of β-glucuronidase activity with increasing age. As mentioned above, the activities of degrading and synthetizing enzymes of connective tissues components do not run parallel to their total content. The total hexosamine content of rat liver dry weight is 0.24%, the collagen content 0.64% as compared to 0.56% resp. 11.3% of the rat lung dry weight. The collagen content of the rat liver increases postnatal until the 20th day of life, afterwards to a lesser extent up to the 22nd month of life. It increases again in the further senium. The lung collagen content (rat, guinea pig, human beings) can increase during growth and in aging, or stay constant after the end of maturation (Elster and Lowry, 1950; Neuman and Logan, 1950; Boas, 1953; Stegemann, 1958; McGavack and Kao, 1960; Schaub, 1963, etc.). Therefore, synthesis, degradation, and total content of connective tissue components must be analyzed separately in aging as well as in diseases. This will be demonstrated by the example of experimental liver cirrhosis.

We investigated the collagenpeptidase activity by using the two methods described above: the thioacetamide induced cirrhosis of the liver in rats and the simultaneous and alternative application of penicillamine and of prednisolone (Tables 3 and 4).

No influence of thioacetamide or prednisolone upon the liver can be shown in the arrangement of the experiment described in Table 3. It appeared that the collagenpeptidase activity increased in the lungs with both thioacetamide intoxication and prednisolone long-term treatment. A significant increase in chronic thioacetamide intoxication as well as in simultaneous solitary application of penicillamine, respectively, can also be stated in lungs and liver by determination of the collagenpeptidase activity in the supernatant (after 4500 rpm centrifugation, 10 min). Only in lungs treated with penicillamine is the increase not significant (Table 4). The increase in collagenpeptidase activity in lungs under long-term treatment with prednisolone corresponds to the findings concerning the rise of metabolism of the connective tissue in lungs, as shown in an increase of the collagen and proteoglycan turnover. This we found by measuring the incorporation of labelled precursors both in the synthesis of proteoglycans (by means of ^{35}S-sulfate incorporation) and in the collagen synthesis (by means of ^{3}H-proline incorporation). The enhancement of the collagen metabolism leads to an increase of the total content of collagen. This has been shown for the skeletal muscular system of the mouse (Friberg, 1957) and for stomach, liver and bones of the rat (Kowalewski, 1961) during long-term treatment with prednisolone. However, Sobel and Marmorston (1954) and Sobel et al. (1959) have proved that the collagen content in rat tissues remained relatively stable as opposed to a decrease of the hexosamine content in cortison treatment so that the hexosamine: hydroxyproline-ratio was shifted in favor of the total content of collagen. On the other hand, a decline of the total content of collagen has been stated for skin under cortisone or hydrocortisone treatment (Sakata, 1960a, b; Smith, 1962a, b; Houck, 1962).

Prednisolone appears to inhibit the activity of the collagenpeptidase activity in blood serum. This can be shown in prednisolone treatment as well as in the combination of prednisolone and thioacetamide. Not only during the combined application of thioacetamide and prednisolone

Table 3. Collagenpeptidase activity in 5-months-old rats after treatment with thioacetamide and/or prednisolone, for 15 to 20 weeks

	Controls	Thioacetamide	Thioacetamide + Prednisolone	Prednisolone
Blood serum	1.21 ± 0.20 (15)	1.81 ± 0.30 (18)	1.40 ± 0.51 (18)	1.10 ± 0.17 (18)
Liver	27.61 ± 3.59 (15)	24,22 ± 11.17 (18)	27.25 ± 4.10 (18)	27.13 ± 4.91 (18)
Lungs	51.19 ± 20.53 (13)	90.94 ± 25.24 (3)	-	92.66 ± 26.90 (9)

0.25 ml supernatant after (1o min. 4500 rpm) centrifugation
blood serum (undiluted)
Dilution of the homogenates: liver 1:25, lungs 1:150; CU mean ± S.D. (n).

Table 4. Collagenpeptidase activity in 5-months-old rats after treatment by thioacetamide for 18 weeks and/or penicillamine for 4 weeks

	Controls	Thioacetamide	Thioacetamide + Penicillamine	Penicillamine
Blood serum	1.24 ± 0.48 (6)	2.17 ± 0.36 (6)	1.95 ± 0.24 (6)	2.04 ± 0.20 (5)
Liver	15.14 ± 1.42 (6)	25.17 ± 4.84 (6)	19.32 ± 2.72 (6)	22.06 ± 3.62 (5)
Lungs	84.37 ± 9.79 (6)	116.93 ± 6.87 (5)	109.21 ± 8.49 (6)	104.26 ± 21.61 (5)

Dilution of homogenates: liver 1:10, lungs 1:150; determination in supernatant after 10 min. 4500 rpm centrifugation, CU mean ± S.D. (n).

Table 5. Protease and β-glucuronidase activity in liver and lungs of 5-months-old rats after thioacetamide for 18 weeks and/or penicillamine treatment for 4 weeks, mU/g wet weight; mean ± S.D. (n)

		Controls	Thioacetamide	Thioacetamide + Penicillamine	Penicillamine
Protease	liver	245.00 ± 63.82 (26)	619.45 ± 130.52 (12)	301.50 ± 150.18(12)	180.00 ± 31.29 (12)
	lungs	309.80 ± 89.95 (18)	307.76 ± 178.25 (8)	347.83 ± 79.91(12)	394.30 ± 54.84 (10)
β-Glucuroni-dase	liver	2.11 ± 0.60 (22)	1.83 ± 0.54 (10)	2.44 ± 0.83(12)	1.76 ± 0.51 (11)
	lungs	0.82 ± 0.34 (18)	2.73 ± 0.84 (8)	2.14 ± 0.62 (8)	2.28 ± 0.24 (8)

but also during the thioacetamide application alone the cirrhotic transformation of the liver is found.

These results are comparable to tests of other parameters of metabolism of connective tissue: 1. organ-specific differences of the influence of prednisolone (like other effectors), 2. changes of rates of metabolism of connective tissue depending on dosage and time of medication (Junge-Hülsing, 1963; Hauss et al., 1968; Lindner, 1963, 1969a, b).

In the different test series a significant increase of collagen peptidase can be shown in blood serum in the course of thioacetamide intoxication. This observation is in agreement with previous results in human blood serum in liver cirrhosis (Grasedyck et al., 1971). Other investigations on catabolic enzymes of connective tissue have been made for comparison. Thereby it was proved that the unspecific protease in lungs increased significantly with age, whereas this was not observed in liver. The activity of the β-glucuronidase, however, decreases with age in liver and lungs (Table 5).

The results show that the collagenpeptidase activity does not correlate with the activity of the unspecific protease and the β-glucuronidase which have been used as an example of enzymes degrading proteoglycans. Similar to the collagenpeptidase activity a significant increase of protease activity in the liver in chronic thioacetamide intoxication and of the β-glucuronidase in the lungs was found, But these are only single findings. A general correlation cannot be shown.

As can be gathered from our introductory remarks the determination of collagenpeptidase activity by means of the synthetic PZ substrate is an appropriate method of proving the collagenolytic activity both in body fluids and in tissue homogenates. This procedure is a suitable method for the indication of collagen degradation.

A linear dependence of the splitting of the substrate on the time of incubation was shown by Gries et al. (1972). The observation that an increased dilution of the homogenate leads to a relative increase of enzyme activity has not been clarified sufficiently. The fact, however, is that in higher dilutions the amount of the splitted substrate depends on the amount of the homogenate that has been used for the reaction. In slight dilutions linearity no longer exists. The dilution which admits linear measurement of enzyme activity differs from organ to organ. The results of the total homogenate and of the supernatant alone corresponds in the more diluted lungs but not in the less diluted liver homogenates. Additional factors - such as the unspecific absorption of the substrate by tissue proteins - may play a part. Possibly there are inhibitors against collagenpeptidases as has been shown against collagenproteases by Eisen et al. (1971). Since the collagenpeptidase according to the experiments of Gries et al. (1971) have been shown in the supernatant after centrifugation from 3,000 to 100,000 g (10 - 60') and since the determination of nitrogen as a parameter for the collagenpeptidase activity is less accurate in the total homogenate than in the supernatant, further investigations should determine the collagenpeptidase activity in the supernatant after centrifugation. The most appropriate procedure for a routine method of indicating collagen degradation, even with small amounts of tissue, is determination of enzyme activities in supernatants of tissue homogenates.

New possibilities arise from application of N-Isopropyl-Methyl-Azulen-Acetyl-Prolyl-Leucyl-Glycyl-Prolyl-Arginine-substrate (Wünsch et al., 1971), which has an absorption maximal in the longwave visible zone at about 600 nm, and which may be suitable for the histochemical determination of collagenolytic enzymes. This has not been possible with the substrates used so far.

Summary: Collagenolytic enzymes take part in the collagen turnover, Collagen degradation is synergistic in normal and pathological conditions. Investigations on collagenpeptidase activity of several tissues confirm the findings, that organs with a higher collagen content show lower activity, and vice versa. The assay on collagenpeptidase activity can be used for a test of collagen degradation. Parallel assays of β-glucuronidase, an indicator of GAG-degradation, and of an unspecific protease do not show correlation to changes of collagen-peptidase activity. The collagenolytic activity is unchanged during maturation and aging in contrast to the two additional enzymes tested. It is mostly increasing, like the other enzymes, during the development of experimental liver fibrosis as well as during long term treatment with prednisolon or penicillamine.

References

Aer, J.: Action of some hormones on the collagenolytic activities of rat bone cells. Ann. Med. exp. Biol. Fenn. 49, 9 - 12 (1971).

Bankowski, E., Galasinski, W.: Tissue collagenolytic enzymes. Postepy Biochem. 16, 553 - 563 (1953).

Boas, N: Methods for the determination of hexosamines in tissues. J. biol. Chem. 204, 553 (1953).

Chvapil, M., Kobrle, V.: Changes in proline and hydroxyproline in collagen during the development of rats. Experientia (Basel) 17, 226 - 229 (1961).

Cutroneo, K.R., Fuller, G.C.: Circadian rhythm of rat liver and lung collagenase activity. Life Sci., Pt. 2. Biochem. Gen. Molec. 10, 395 - 400 (1971).

Eisen, A.Z., Bauer, E.A., Jeffrey, J.J.: Human skin collagenase. The role of serum alpha-globulins in the control of activity in vivo and in vitro. Proc. nat. Acad. Sci. (Wash.) 68, 248 - 251 (1971).
Eisen, A.Z., Jeffrey, J.J., Gross, J.: Human skin collagenase. Isolation and mechanism of attack on the collagen molecule. Biochim. biophys. Acta (Amst.) 151, 637 - 645 (1968).
Elster, S.K., Lowry, E.L.: Collagen content of guinea pig tissues. Proc. Soc. exp. Biol. (N.Y.) 75, 127 (1950).
Espey, L.L., Rondell, P.: Estimation of mammalian collagenolytic activity with a synthetic substrate. J. appl. Physiol. 23, 757 - 761 (1967).
Fishman, W.H.: β-Glucuronidase. In: H.U. Bergmeyer (Ed.): Methoden der enzymatischen Analyse, Vol. I, p. 891 - 894. Weinheim: Verlag Chemie 1970.
Friberg, O.: Effect of adrenocortical hormones on hydroxyproline concentration of heart and skeletal muscle. An experimental investigation with mice. Ann. Med. exp. Biol. Fenn. 35, Suppl. 8, 1 - 70 (1957).
Götze, W., Gries, G., Lindner, J.: Beitrag zum Kollagenabbau im menschlichen Zahnfleisch. Dtsch. zahnärztl. Z. 25, 492 - 495 (1970).
Grasedyck, K., Ropohl, D., Szarvas, F., Lindner, J.: Kollagenpeptidaseaktivität menschlicher Seren bei Leberzirrhose. Klin. Wschr. 49, 163 - 164 (1971).
Gries, G., Grasedyck, K.: Über spezifische Kollagenasen in Organen des Magen-Darmtraktes. Verh. Dtsch. Ges. Path. 53, 205 - 209 (1969).
Gries, G., Lindner, J.: Zur Frage des Abbaues von Kollagen. Klin. Wschr. 38, 406 - 407 (1960).
Gries, G., Lindner, J.: Untersuchungen über den Kollagenabbau bei akuten Entzündungen. Z. Rheumaforsch. 20, 122 - 125 (1961).
Gries, G., Lindner, J.: Zur Frage des Kollagenabbaues bei verschiedenen Formen experimenteller Entzündungen. Z. Rheumaforsch. 22, 439 - 443 (1963).
Gries, G., Strauch, L.: Normalwerte kollagenolytischer Enzyme in menschlichem Serum. Klin. Wschr. 48, 441 - 442 (1970).
Gries, G., Strauch, L., Grasedyck, K., Lindner, J.: Die Bestimmung von Kollagenpeptidasen in tierischen Geweben. Z. Rheumaforsch. 31, 93 - 98 (1972).
Gross, J., Lapière, C.M.: Collagenolytic activity in amphibian tissues: a tissue culture assay. Proc. nat. Acad. Sci. (Wash.) 48, 1014 - 1022 (1962).
Hakim, A.W.: On the collagenolytic activity of the mammalian pancreas. Enzymologia 30, 299 - 329 (1966).
Harris, E.D., Cohen, G.L., Krane, S.M.: Synovial collagenase: its presence in culture from joint disease of diverse etiology. Arthr. and Rheum. 12, 92 - 102 (1969).
Hauss, W.H., Junge-Hülsing, G., Gerlach, U.: Die unspezifische Mesenchymreaktion. Zur Pathogenese der reaktiven Mesenchymerkrankungen. Stuttgart: Thieme 1968.
Heidrich, H.-G., Prokopová, D., Hannig, K.: The use of synthetic substrates for the determination of mammalian collagenases. Is collagenolytic activity present in mitochondria? Hoppe-Seylers Z. physiol. Chem. 350, 1430 - 1436 (1969).
Houck, J.C.: Response of the connective tissue to cortisol or salicylates. Arthr. and Rheum. 5, 111 (1962).
Jeffrey, J.J., Gross, J.: Isolation and characterization of a mammalian collagenolytic enzyme. Fed. Proc. 26, 670 (1967).
Junge-Hülsing, G.: Untersuchungen zur Pathophysiologie des Bindegewebes. Habil.-Schrift Münster 1963.
Keiditsch, E., Strauch, L.: Peptidase and collagenase activities in the invasion zones of tumors of the breast. In: E.A. Balazs (Ed.): Chemistry and Molecular Biology of the Intercellular Matrix, Vol. III, p. 1671 - 1674. London-New York: Academic Press 1970.

Kirschke, H., Hanson, H.: Untersuchungen über den Kollagenabbau durch Rattenorgane. Differenzierung und Charakterisierung Prolinpeptid-spaltender Enzyme. Hoppe-Seylers Z. physiol. Chem. 350, 1437 - 1448 (1969).

Kivirikko, K.I.: Hydroxyproline-containing fractions in normal and cortisone-treated chick embryos. Acta physiol. scand. Suppl. 219 (1963).

Kobrle, V., Chvapil, M.: Isolation of peptidase containing hydroxyproline from animal tissues. Nature (Lond.) 190, 900 - 910 (1961).

Kobrle, V., Chvapil, M.: The amount of ultrafiltrable and collagen-bound hydroxyproline in different organs of the rat during aging. Physiol. bohemoslov. 11, 243 - 248 (1962).

Kowalewski, K.: Effect of cortisone on hydroxyproline of liver, gastric tissue, skin and bone of male rats. Acta endocr. (Kbh.) 38, 427 - 431 (1961).

Langer, E., Keiditsch, E., Strauch, L., Hannig, K.: Die Kollagenase-aktivität im carcinoma solidum simplex mammae. Verh. Dtsch. Ges. Path. 52, 438 - 440 (1968).

Lazarus, G.S., Decker, J.L., Oliver, C.H., Daniels, J.R., Multz, C.V., Fullmer, H.M.: Collagenolytic activity of synovium in rheumatoid arthritis. N. Engl. J. Med. 279, 914 - 919 (1968).

Lindner, J.: Untersuchungen am Bindegewebe. Habil.-Schrift, Univ. Hamburg 1957a.

Lindner, J.: Vergleichende histochemische und papierchromatographische Untersuchungen bei Entzündungsvorgängen der Haut. Arch. klin. exp. Derm. 206, 379 - 394 (1957b).

Lindner, J.: Der cytostatische Einfluß von Bindegewebsbausteinen. Verh. Dtsch. Ges. Path. 45, 207 - 213 (1961).

Lindner, J.: Histochemie der Atherosklerose. In: G. Schettler (Hrsg.): Arteriosklerose, S. 61 - 87. Stuttgart: Thieme 1961.

Lindner, J.: Über die Synthese der organischen Interzellularsubstanz in Knorpel und Knochen. Verh. Dtsch. Ges. Path. 47, 100 - 102 (1963).

Lindner, J.: The histochemistry of atherosclerosis. In: F.G. Schettler, G.S. Boyd (Ed.): Atherosclerosis, p. 73 - 140. Amsterdam: Elsevier 1969a.

Lindner, J.: Metabolism of granulation tissue and other connective tissues. Scand. J. clin. Lab. Invest. (Suppl.) 108, 12 - 13 (1969b).

Lindner, J., Gries, G., Freytag, G., Kind, J.: Stoffwechselunter-suchungen an der atherosklerotischen Gefäßwand. Verh. Dtsch. Ges. Path. 51, 228 - 246 (1967).

McGavack, T.H., Kao, K.T.: The influence of age and sex on the soluble collagen, insoluble collagen and elastin of rat tissues. Exp. Med. Surg. 18, 104 (1960).

Mitchell, W.M.: Pseudocollagenase: a protease from clostridium histo-lyticum. Biochim. biophys. Acta (Amst.) 159, 554 - 557 (1968).

Nagai, Y.: Collagenase digestion of collagen. J. Biochem. (Tokyo) 50, 486 - 492 (1961).

Nagai, Y., Lapière, C.M., Gross, J.: Tadpole collagenase. Preparation and purification. Biochemistry 5, 3123 - 3130 (1966).

Neuman, R.E., Logan, M.A.J.: The determination of collagen and elastin in tissues. J. biol. Chem. 186, 549 - 556 (1950).

Platt, D., Stein, Z.: Untersuchungen zum katabolen Mucopolysaccharid-Protein-Stoffwechsel in menschlichen Organen. Z. klin. Chem. 7, 374 (1969).

Platt, D.: Altersabhängige Aktivitätsänderungen lysosomaler Enzyme (Glykosaminoglykan-Hydrolasen) in Serum und Organen des Menschen. Dtsch. med. Wschr. 95, 634 (1970).

Platt, D.: Biochemische Aspekte zum Altern der Interzellularsubstanz. akt. gerontol. 1, 81 (1971).

Platt, D.: Nachweis, Reinigung und Eigenschaften einer Peptidyl-Peptid-Hydrolase (Kollagenase) aus Leber und Milz des Kaninchens. Z. ges. exp. Med. 150, 185 - 193 (1969).
Platt, D.: Nachweis einer Peptidyl-Peptid-Hydrolase (Kollagenase?) in der Aortenwand des Schweins. Klin. Wschr. 48, 1420 - 1421 (1970).
Sakata, R.: Biochemical studies on the collagen metabolism. II. Collagen metabolism of the skin in some pathological states. Kumamoto med. J. 13, 41 - 53 (1960a).
Sakata, R.: Biochemical studies on the collagen metabolism. III. Collagen metabolism in healing wounds. Kumamoto med. J. 13, 54 - 72 (1960b).
Schaub, M.C.: Qualitative and quantitative changes of collagen in parenchymatous organs of the rat during aging. Gerontologia (Basel) 8, 114 (1963).
von Schweinitz, H.A., Lindner, J.: Histochemische und fermenthistochemische Untersuchungen am Geschwulstgewebe. II. Mitt.: Untersuchungen des Aszitestumor-Proteolyse. Frankf. Z. Path. 69, 667 - 678 (1959).
Smith, Q.T.: Quantitative studies on cutaneous collagen, procollagen and hexosamine in normal and cortisone treated rats. J. invest. Derm. 38, 65 - 68 (1962a).
Smith, Q.T.: Effects of cortisone administration on cutaneous collagen and hexosamine in rats of various ages. J. invest. Derm. 39, 291 - 324 (1962b).
Sobel, H., Gabay, S., Johnson, C., Hassan, B.: Carcass nitrogen and the hexosamine-collagen ratio of skin following starvation and cortisone administration in rats and guinea pigs. Metabolism 8, 180 - 184 (1959).
Sobel, H., Marmorston, J.: The effect of cortisone on the collagen and hexosamine content of the skin and femurs of one-year-old rats. Endocrinology 55, 21 - 27 (1954).
Stegemann, H.: Mikrobestimmung von Hydroxyprolin mit Chloramin-T und p-Dimethylaminobenzaldehyd. Hoppe-Seylers Z. physiol. Chem. 311, 41 - 45 (1958).
Strauch, L.: Ultramikro-Methode zur Bestimmung des Stickstoffes in biologischem Material. Z. klin. Chem. 3, 165 - 167 (1965).
Strauch, L.: Kollagenasen aus Zellen des Menschen und höher entwickelter Tiere. Mitt. Max-Planck-Ges. 1, 40 - 56 (1968).
Strauch, L.: Peptidyl-peptide hydrolase (collagenase) in cells of higher animals. In: E.A. Balazs (Ed.): Chemistry and Molecular Biology of the Intracellular Matrix, Vol. III, p. 1675 - 1682. London-New York: Academic Press 1970.
Strauch, L., Vencelj, H.: Collagenases in mammalian cells. Hoppe-Seylers Z. physiol. Chem. 348, 465 - 468 (1967).
Strauch, L., Vencelj, H., Hannig, K.: Kollagenasen in Zellen höher entwickelter Tiere. Hoppe-Seylers Z. physiol. Chem. 349, 171 - 178 (1968).
Vaes, G.: A latent collagenase released by bone and skin explants in culture. Biochem. J. 123, 23 - 24 (1971).
Woessner, J.F., Boucek, R.J.: Connective tissue development in subcutaneously implanted polyvinyl sponges. Arch. Biochem. 93, 85 - 94 (1961).
Woods, J.F., Nichols, G.: Distribution of collagenase in rat tissues. Nature 208, 1325 - 1326 (1965).
Wünsch, E., Heidrich, H.-G.: Darstellung von Prolinpeptiden. III. Ein neues Substrat zur Bestimmung der Kollagenase. Hoppe-Seylers Z. physiol. Chem. 332, 300 - 304 (1963).
Wünsch, E., Jaeger, E., Schoensteiner-Altmann, G.: Zur Bestimmung der Kollagenase mit N-((Azulen-4-yl)acetyl-) peptiden. Hoppe-Seylers Z. physiol. Chem. 352, 1560 - 1567 (1971).

Degradation of Insoluble Collagen and other Proteins by Rat Leucocytes

A. J. Anderson

Degradative enzymes in granule of tissue cells and leucocytes, active between pH values 3 and 6, digest connective tissue components intracellularly in the phagolysosome system of the cell. It is unlikely that tissue pH is normally low enough for most of these lysosomal enzymes to have significant extracellular activity, except perhaps in avascular inflamed tissue such as cartilage in rheumatoid arthritis, where large amounts of glycosaminoglycans (acid mucopolysaccharides) are present. When considering early events in inflammation, increasing attention is paid to neutral proteinases (EC 3.4.4.-) and other cellular enzymes active under neutral conditions, and which may therefore function extra-cellularly, thus initiating early inflammatory processes.

We have found that freshly harvested rat peritoneal leucocyte granules, which sediment between 400 and 15,000 g, contain two enzyme systems which degrade insoluble collagen. The first, optimally active at pH 3.5, is also in lysosome-rich fractions of rat liver (Anderson, 1969) and in paws of adjuvant-arthritic rats (Anderson, 1970). Its activity is increased by 2 mM L-cysteine. The second collagenolytic system, which bears some resemblance to certain mammalian collagenases (Evanson, 1971), acts optimally at pH 8.4 and, unlike the acid-active collagenolytic system, is absent from rat liver lysosomes and is inhibited by normal rat serum. It is unaffected by L-cysteine. Whether these collagenolytic enzymes inhabit the same, or different, granules is being investigated.

Both collagenolytic enzyme systems are inhibited by the anti-rheumatic drug phenylbutazone at 2 mM. This drug also inhibits β-glucuronidase and hyaluronidase (Anderson, 1968). These properties may contribute to the clinical effectiveness of phenylbutazone *in vivo*, by helping to prevent the enzymic breakdown of collagen, hyaluronic acid, proteoglycans (chondromucoprotein) and glycosaminoglycans which occur in synovial fluid, cartilage and other connective tissues.

We have recently detected two proteinases in rat peritoneal leucocyte granules which degrade undenatured haemoglobin at optimum pH values 8 and 4. Using the technique of differential centrifugation, or isopycnic centrifugation in sucrose-density gradients, preliminary studies indicate the former enzyme is associated with, or inhabits, denser (sp. gr. 1,300) granules than the latter enzyme (sp. gr. 1,230). The enzyme system active at pH 4 and associated with the lighter granules is probably of lysosomal origin and a mixture of cathepsins (Woessner, 1971).

In contrast, we find that enzymes with acid and alkaline phosphatase activities inhabit leucocyte granules of the same weight and density (sp. gr. 1,190 - 1,245). Acid phosphatase activity in leucocytes is probably of lysosomal origin, and we are attempting to determine if alkaline phosphatase inhibits the same, or different, granules.

Results of our work on the collagenolytic and haemoglobin splitting leucocyte granular enzyme systems tentatively support the "two-stage" theory of connective tissue destruction (Dingle, 1969). An early event in inflammation may be the partial extra-cellular digestion of connective tissue components, particularly collagen and proteoglycans, by neutral-acting enzymes discharged from tissue cells and infiltrated leucocytes. This digestion may result from aberrant leucocytes, which may have either promoted the disease process or are a result of it, and which normally do not synthesize and discharge neutral-active enzymes in great amounts. A second phase follows, when resulting connective tissue fragments are digested more extensively under acid conditions within the phagolysosomal system of leucocytes and other endocytic cells, thereby generating other inflammatory and permeability factors, some of which tend to perpetuate the inflammation and render it chronic.

Knowledge of neutral-acting enzymes (Brocklehurst, 1971), which may be associated with granules distinct from the classical lysosomes, may provide information about the early biochemical and clinical stages in rheumatoid arthritis and other inflammations, at a time when connective tissue destruction is minimal and less systemic, and therefore more amenable to curative, rather than palliative, treatment.

References

Anderson, A.J.: The effect of anti-inflammatory drugs on the enzymic activity of a rat liver granular fraction which increases vascular permeability. Biochem. Pharmacol. 17, 2253 - 2264 (1968).

Anderson, A.J.: Effects of lysosomal collagenolytic enzymes, anti-inflammatory drugs and other substances on some properties of insoluble collagen. Biochem. J. 113, 457 - 463 (1969).

Anderson, A.J.: Lysosomal enzyme activity in rats with adjuvant-induced arthritis. Ann. rheum. Dis. 29, 307 - 313 (1970).

Brocklehurst, W.E.: Role of kinins and prostaglandins in inflammation. Proc. roy. Soc. Med. 64, 4 - 6 (1971).

Dingle, J.T.: Lysosomes in Biology and Pathology (Ed. J.T. Dingle, H.B. Fell), Vol. 2. Amsterdam: North-Holland Publishing 1969.

Evanson, J.M.: Mammalian collagenases and their role in connective tissue breakdown. In: Tissue Proteinases (Ed. A.J. Barrett, J.T. Dingle), p. 327 - 345. Amsterdam: North-Holland Publishing 1971.

Woessner, J.F., jr.: Cathepsin D: enzymic properties and role in connective tissue breakdown. In: Tissue Proteinases (Ed. A.J. Barrett, J.T. Dingle) p. 291 - 311. Amsterdam: North-Holland Publishing 1971.

Enzymatic Degradation of the "Collagen-Like-Protein" in Blood Serum

D. Dabew

The protein-bound hydroxyproline in blood serum can not be enzymatically released by the usual proteolytic enzymes (trypsin, pronase). The serum content of free hydroxyproline increases under the influence of these enzymes insignificantly, due to hydrolysis of the peptide-bound amino acid. Experiments with collagenases have not led to the results expected (Leroy et al., 1964). The best results in enzymatic degradation of acid-soluble collagen, treated by collagenase, have been achieved by pronase (Fujii, 1965).

During our studies on collagen metabolism and on hydroxyproline attached to proteins in blood serum, we also examined the enzymatic degradation of hydroxyproline-containing peptides and proteins.

For our experiments we used on one hand neutral-salt-soluble collagen prepared of pig skin which was supposed to be an analogous substrate to the "Collagen-like-protein" in serum, and on the other hand blood serum from healthy blood donors. All determinations were carried out on a microliter scale. The prepared collagen and the serum samples were incubated with different enzymes, either alone or in various sequences, or combinations. There were tested: Trypsin (Trypure, Novo-industrie GmbH, Mainz), Chymotrypsin (Alphachymotrypsin, Choay AG, Paris), Papain (Merck, Darmstadt), Carboxypeptidase (Serva, Heidelberg), Aminopeptidase (M. Roth, Karlsruhe) and Collagenase (Knoll, Ludwigshafen). Further were temperature and pH varied and the influence of Ca-ions on the collagenase activity examined. The enzymatic degradation of collagen or the "Collagen-like-protein" was measured by the increase of the free hydroxyproline in the sample. The free amino acid was determined by a technique previously reported (Dabew and Struck, 1969, 1971).

The "Collagen-like-protein" in serum did not become affectable for proteolytic enzymes after thermal treatment: the corresponding samples were kept for 1 - 3 hours at 40 - 56°C, then incubated at 37°C with trypsin, chymotrypsin and papain, respectively, and finally carboxypeptidase or aminopeptidase was added. There was no increase of free hydroxyproline in the samples demonstrable during the following 24 hours.

The content of free hydroxyproline increased gradually in the samples with neutral salt soluble collagen as was well as in those of blood serum under the influence of collagenase during the time of incubation, as shown in Fig. 1

The tested collagenase preparation possessed partial peptidase activity. An optimal collagenase activity was observed at pH 6.8 - 8.0 and 37°C. The effect of addition of 0.002 M $CaCl_2$ to the samples of neutral salt soluble collagen became evident after the 8th hour: the increase of free hydroxyproline was in this case higher than in the samples without the addition of calcium.

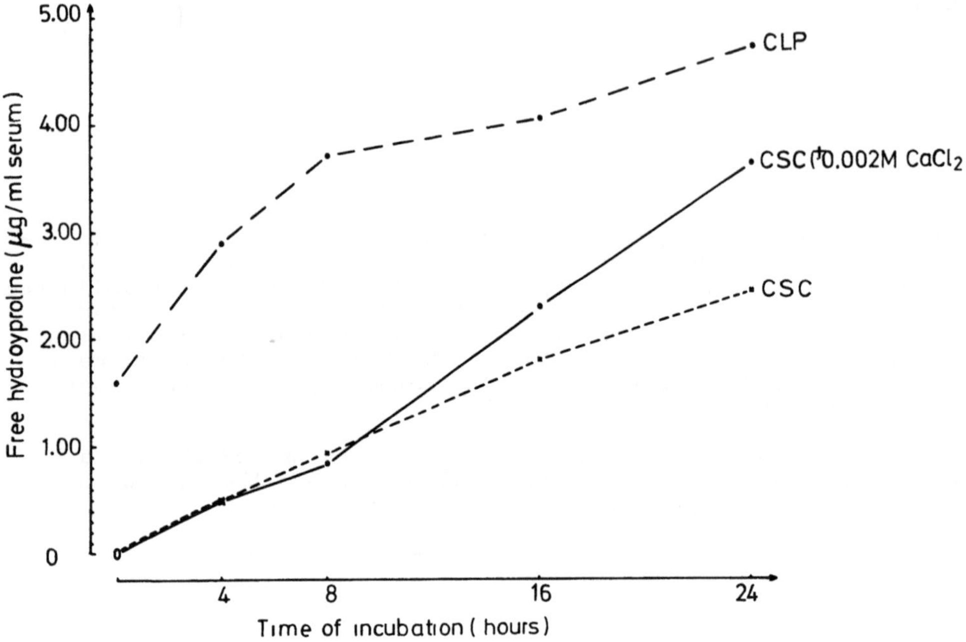

Fig. 1. Enzymic degradation of calf skin collagen (CSC) and collagen-like-protein (CLP) in serum by collagenase (Knoll AG., Ludwigshafen) at 37°C and pH 7.4

Neutral salt soluble collagen incubated for 12 hours with collagenase was degraded insignificantly by trypsin, chymotrypsin and papain, but, significantly better by pronase at 37°C and pH 6.8 - 7.4. The amount of free hydroxyproline, released by pronase, seemed to be proportional to the enzyme activity and time of incubation.

The serum concentration of free hydroxyproline increased under the influence of collagenase only up to the 24th hour; it remained unchanged during the following 24 hours. An unknown substance, which disturbed the hydroxyproline determination, was released under the influence of pronase in the serum. The absorption curve of the color complex for the proper hydroxyproline measurement in serum showed a different character after adding pronase, even without oxidation (Fig. 2).

The yield of free hydroxyproline could be increased unessentially by further addition of peptidases like carboxypeptidase and aminopeptidase.

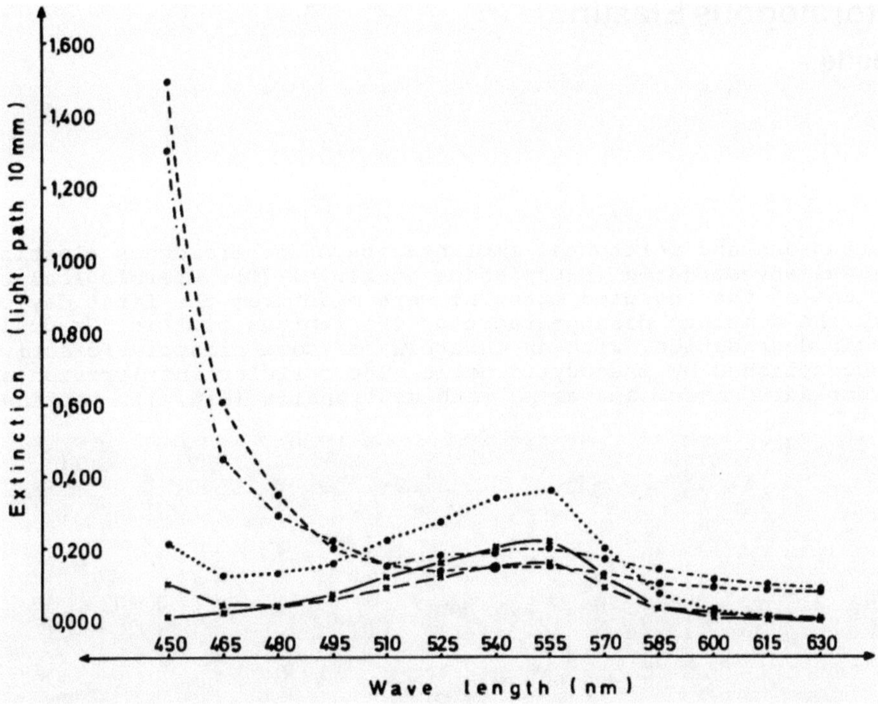

Fig. 2. Hydroxyproline absorption curves. Determination in serum before and after incubation with pronase respectively collagenase. x—x Hydroxyproline standard (3 µg/ml). x— —x Free hydroxyproline in serum. Color reaction of free hydroxyproline in serum after incubation with pronase: •----• a) blank reading (without heating), •-·-·-• b) normal assay procedure. •·······• Color reaction for hydroxyproline in serum after incubation with collagenase

References

Dabew, D., Struck, H.: Eine vereinfachte Methode zur Bestimmung des freien Hydroxyprolin im Serum. Z. klin. Chem. klin. Biochem. 7, 398 - 450 (1969).

Dabew, D., Struck, H.: Microliter determination of free hydroxyproline in blood serum. Biochem. Med. 5, 17 - 21 (1971).

Fujii, T.: The Action of "Pronase" on Acid Soluble and Insoluble Collagens. Biochimie et Physiologie du Tissu Conjonctif, p. 461. Lyon 1965.

Leroy, E.C., Kaplan, A., Udenfriend, S., Sjoerdsma, A.: A hydroxyproline-containing, collagen-like-protein in plasma and a procedure for its assay. J. biol. Chem. 239, 3350 - 3356 (1964).

Sequential Degradation and Phagocytosis of Heterologous Elastin

L. Gotte

Subcutaneous and peritoneal implantation of heterologous elastin in rats was investigated (Papajiannis et al., 1970). Histological examinations of the injected material were made from the first day onward, until the complete disappearance of the fibrous protein. Gradual elastin degradation, with an interplay of some elastolytic activity, is accomplished by phagocytic cells. The cellular infiltration around the implanted material begins with neutrophils (Fig. 1), continues

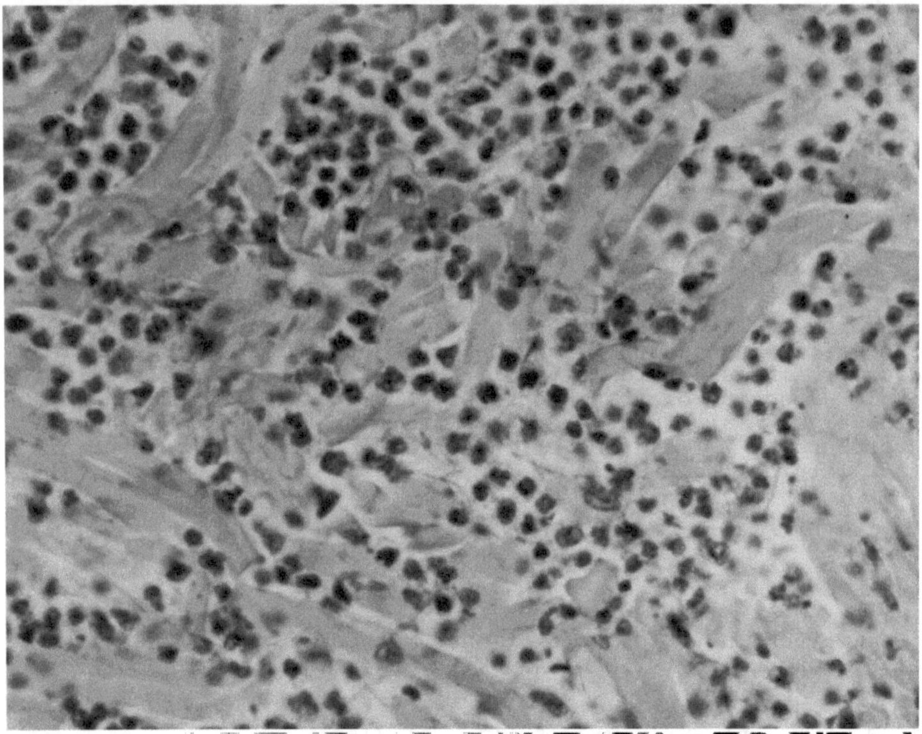

Fig. 1

with mononuclear cells, lymphocytes, macrophages and ends with the formation of giant cells (Fig. 2). In the satellite lymph nodes a gradual medullary hyperplastic activity with plasma cell proliferation, (Fig. 3) and, at a later stage, the development of germinal centers in the outer cortex were found. A parallel study of the serum electrophoretic patterns indicates first an increase of α1-globulins followed by a pronounced increase of γ-globulins.

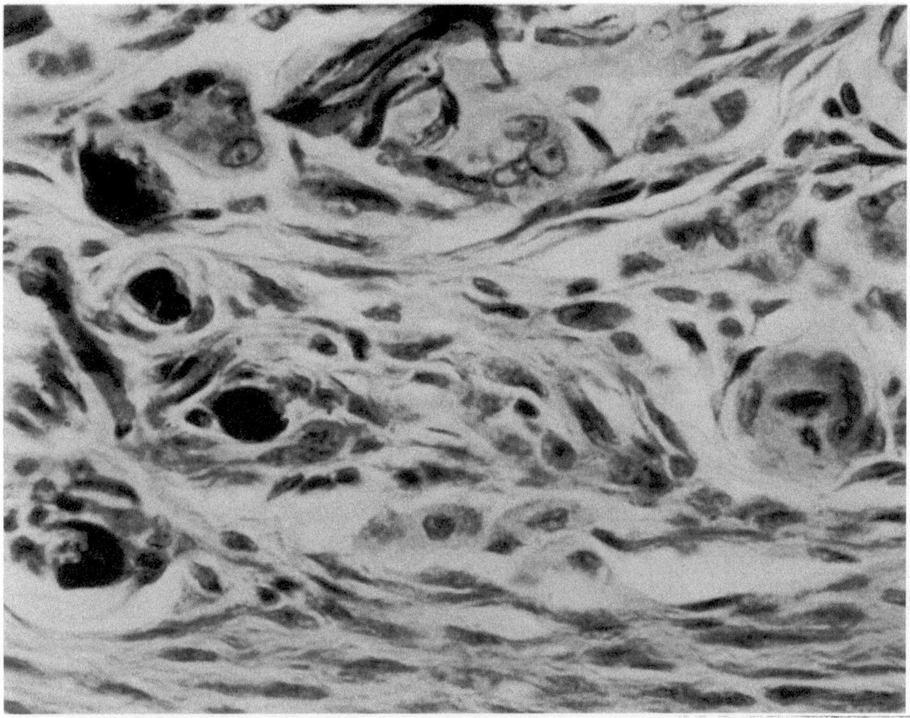

Fig. 2

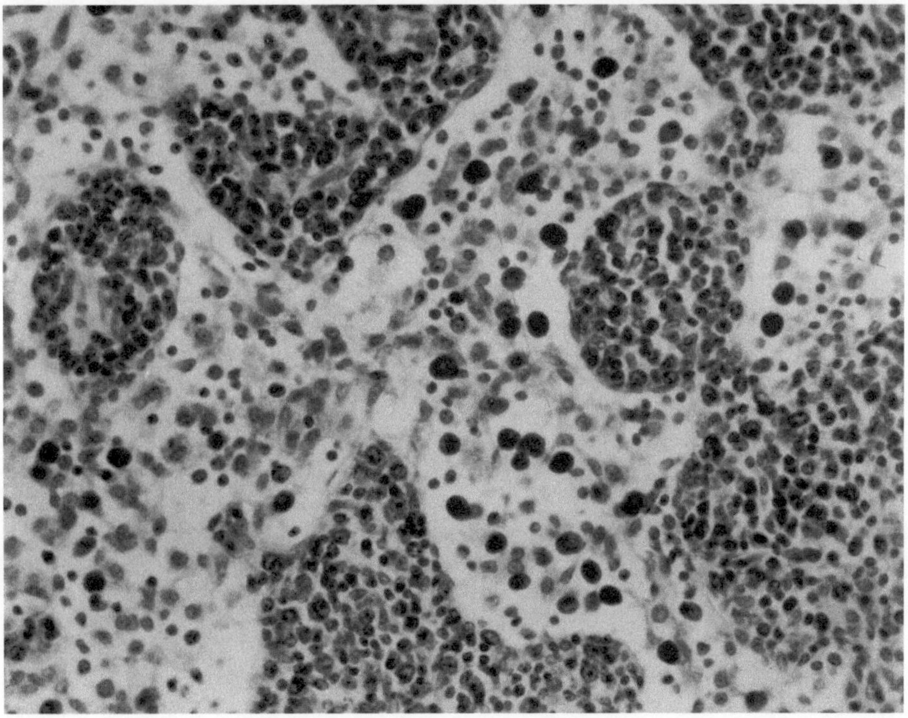

Fig. 3

References

Papajiannis, S.P., Spina, M., Gotte, L.: Arch. Path. 89, 434 (1970).

IV. Immunobiology

Immunobiology of Glycoproteins

G. Uhlenbruck, A. Rothe and O. Prokop

Glycoproteins consist of protein chains to which carbohydrate chains are attached. Both types of chains can vary in length, size and number. In ovine submaxillary mucin, for instance, there are about 800 prosthetic groups in form of a disaccharide attached to a protein chain (Graham and Gottschalk, 1960); in blood-group substances, on the other hand, there are about 110 heterosaccharide chains - branched oligosaccharides of a higher molecular weight - per protein chain (Schiffman and Marcus, 1964). It is necessary to distinguish between

1. naturally occurring glycoproteins and
2. glycoproteins made artificially by coupling oligosaccharides to
   a) natural proteins (Westphal, 1970) or
   b) synthetic polypeptide chains (Rüde, 1970).

In the latter two cases, the oligosaccharide chain may be immunogenic, but the protein part can be blocked in its immunogenicity or even be serologically inactive, for instance, with synthetic polypeptides. One must differentiate between several possible locations of the antigenic site:

a) the protein part is antigenic;
b) the carbohydrate part is antigenic;
c) both parts are antigenic (the most common form);
d) the branching point between protein and carbohydrate part is antigenic (rare).

The immunological tools which are available to elucidate the serological behavior of glycoproteins are listed in Table 1. There are the classical methods of evoking an antibody response or a specific cellular reaction and, in addition, certain antibody-like substances obtained from plants and invertebrates. The antibodies to glycoproteins - for instance blood-group substances - are usually of the IgM, IgG type, and in secretions (saliva, tears) of the IgA type (Prokop and Uhlenbruck, 1969). Only rarely are anaphylactic antibodies (IgE) against glycoproteins described (Halpern et al., 1967).

Table 1. Serological tools for investigation of glycoproteins

I. Classical immunological tools

- a) Antibody response (humoral)
- b) Delayed hypersensitivity (cellular)

II. "Para-immunological" tools

- a) Lectins from plants
- b) Agglutinins from invertebrate hemolymph
- c) Agglutinins form snails (albumin gland)

Methods: Hemagglutination inhibition
Precipitation inhibition
Immunodiffusion

Table 2. Comparative defense mechanisms: invertebrates and vertebrates

	Invertebrates	Vertebrates
Nature of mechanism	(1) Unspecific (directed against widespread receptors eg. carbohydrate prosthetic groups)	(1) Unspecific (eg. virus inhibitors) (2) Specific (directed against undividual defined receptors)
Defense system	(1) Naturally occurring (pre-existent) Not inducible No stimulation by immunogens No booster response	(1) Naturally occurring (pre-existent) (2) Inducible Stimulated by immunogens Booster response Immunological memory

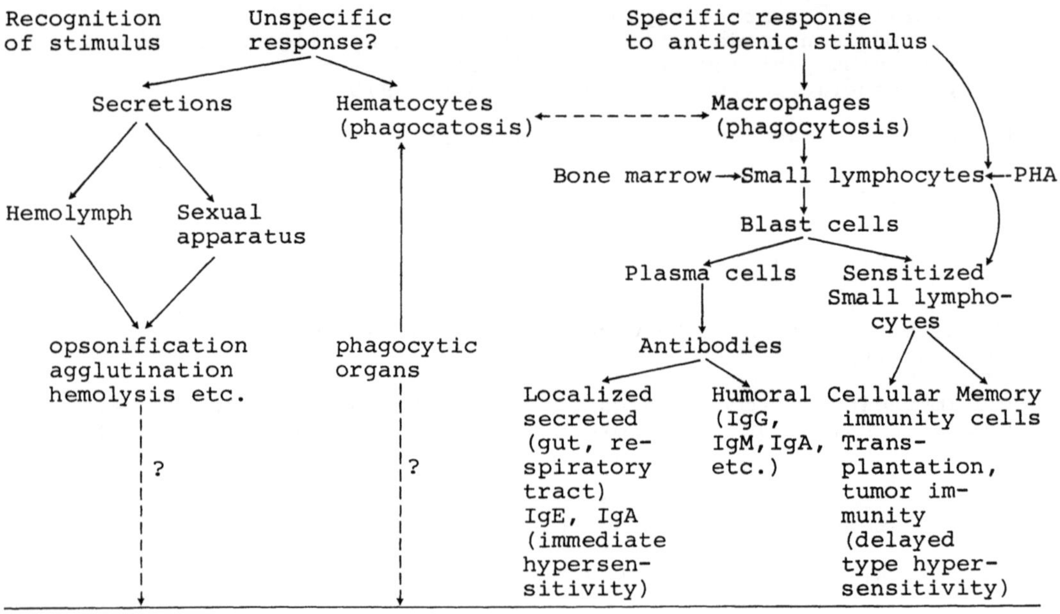

Evolution of complement and immunoglobulins	Delayed hypersensitivity	Bursa and related organs	Thymus

The different immunological systems from which these tools arise are summarized in Table 2, adapted from Pardoe and Uhlenbruck, 1970).

The biological role of the invertebrate agglutinins is not quite clear. A number of suggestions have been made (see Table 3). The literatur is fully listed elsewhere (Uhlenbruck et al., 1970).

To make it easier to discuss the immunogenicity of glycoproteins a rough classification is given in Table 4. The membrane-integrated glycoproteins (see Table 5) are of particular interest, as they carry many biological receptors of great importance: blood-group antigens, histocompatibility antigens, receptors for lectins (for instance PHA = *Phaseolus vulgaris*). We have proposed a subdivision of this group, as many of these glycoproteins behave very differently. Those from organ

Table 3. Suggested functions of invertebrate agglutinins

1. Opsoninic (lysozyme also often present)
2. Bactericidal (incorporated lectins?)
3. Parasite - immobilizing
4. Shelf-building (N-Ac-glucosamine transport)
5. Cell - lytic (anti-tumor?)
6. Cell - toxic (or anti-toxic?)
7. Carbohydrate - fixing (vitamin-like?)
8. Defense, protection (eggs)
9. Anti-viral, anti-bacteriophage
10. Anti-fungal, anti-sperm

Table 4. Classification of glycoproteins

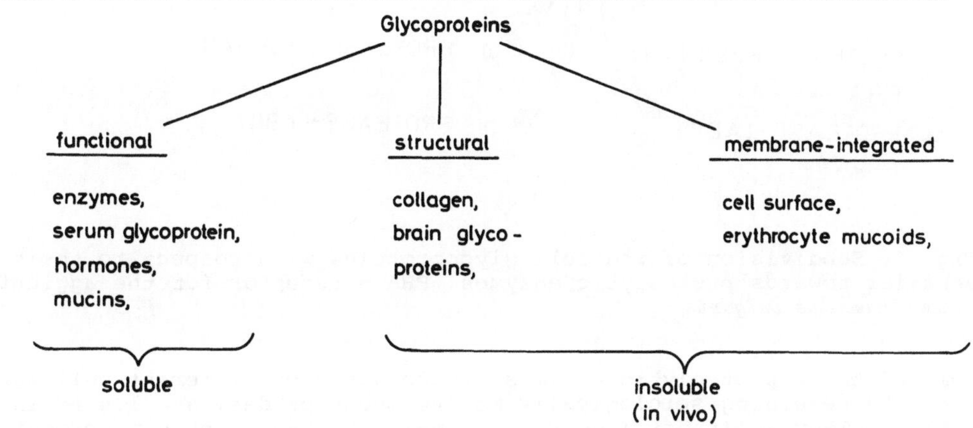

Table 5. Membrane-integrated glycoproteins

I. Red cells — Leucocytes, Lymphocytes; — Thrombocytes

II. Organ cells — Tumor cells

III. Nucleus and subcellular particles

and from tumor cells are not directly comparable (Buck et al., 1971), and glycopeptides from thrombocytes (Pepper and Jamieson, 1970) differ - at least with respect to their molecular size - from those derived from other cell surfaces (Winzler, 1970; Uhlenbruck, 1971). Furthermore, the cell surface glycoproteins can be devided into two groups with regard to their behavior towards proteolytic enzymes: the protease-releasable and the protease-resistant group, both of which are relevant for neuraminic acid (NA) containing and neutral cell surface glycoproteins. This holds also for certain serological specificities

(PHA activity) (Uhlenbruck et al., 1970), as illustrated in Fig. 1.

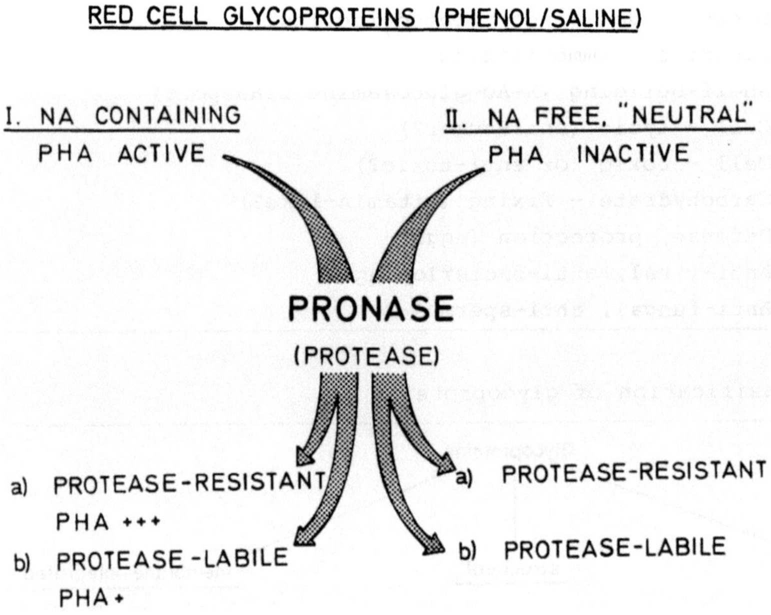

Fig. 1. Subdivision of red cell glycoproteins with respect to their behavior towards proteolytic enzymes. PHA = receptor for the agglutinin from *Phaseolus vulgaris*

Some of these proteolytic enzymes, which act upon different cell surfaces by releasing serologically active mucopeptides, are listed in Table 6. They split off (see Fig. 2) parts of the outer glycoprotein layer of the cell surface coat. This figure shows an outer layer of branched glycoproteins, which seem to be implanted into the (glyco)-lipid membrane covering the inner protein layer. All three membrane layers are connected by α-helical proteins (Zahler, 1969).

Table 6. Proteolytic enzymes with definite action on the red cell surface glycoproteins

Origin	Animal	Plant	Fungus	Bacteria
Names:	Trypsin	Agavain	Asperg. orycae	Pronase
	α, β, γ, δ-Chymotrypsins	Bromelin	Proteinase	Subtilisin
		Bromelain	Proteinase K (Sporotrichum spec.)	Thermolysin
	Pancreas-protease E_1	Papain		
	Plasmin	Ficin		

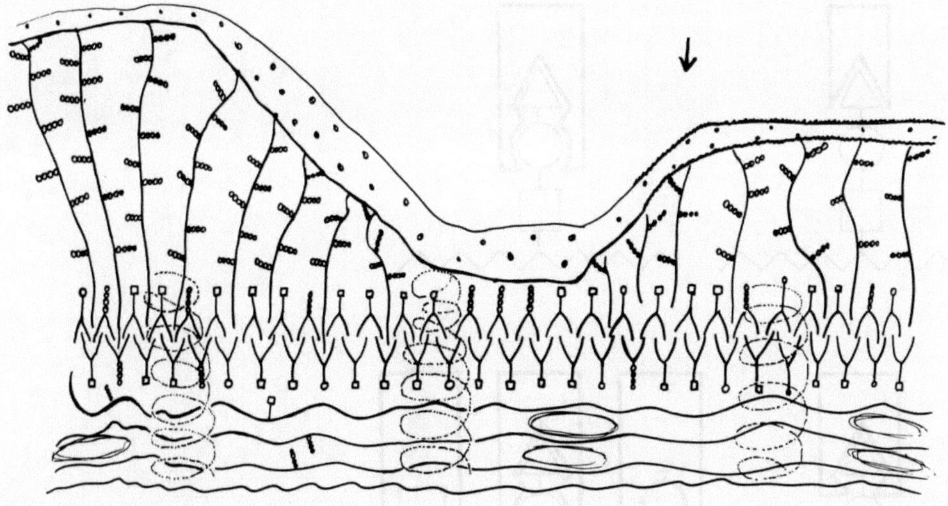

Fig. 2. Hypothetical model of a section through a red cell membrane: the upper layer consists of glycoproteins which can be partly removed (↓) by proteolytic enzymes. These glycoproteins are implanted in a lipid-glycolipid membrane backbone which is followed by a third, inner membrane containing more protein (enzymes)

The immunogenicity of the released glyco- or mucopeptides is less than that of the genuine glycoproteins whose higher molecular weight demonstrates the influence of molecular size. When a specific immune response to glycoproteins has been evoked, one can distinguish three types of reactions. Fig. 3, top left, illustrates the reaction of an antibody or antibody-like substance with a terminal sugar structure (lectin from plants, invertebrate, snail agglutinin). The sketches at bottom left and top right show a more definite specificity, extending to the anomeric linkage to the next sugar unit and the next sugar respectively. The sketch at lower right shows how an antibody reacts when directed against a trisaccharide structure: the trisaccharide unit is degraded by glycosidases or mild acid hydrolysis, so that other structures, subterminal carbohydrate determinants, become serologically available.

In Fig. 4 the upper two drawings depict the reaction of antibodies or antibody-like substances with inner-chain sugar structures, whereas the lower pictures show the immune response (antibodies) to the branching area (left) or to the protein backbone only (right).

Fig. 5 outlines the reaction type of the specific cellular immune response (delayed hypersensitivity). The top drawings show that not only the trisaccharide unit (and this even vague, see the non-reducing sugar!) is detected, but also the adjacent protein core. As indicated below this type of reaction can be prevented by a different amino-acid pattern or an elongation of the heterosaccharide chain or a sterical hindrance by closely situated heterosaccharide chains.

All types of reaction are illustrated again in Fig. 6, which presents an oversimplified model of a blood group A determinant glycoprotein. The "anti-A" from *Helix pomatia* (anti-A_{HP}) detects the terminal N-acetyl-

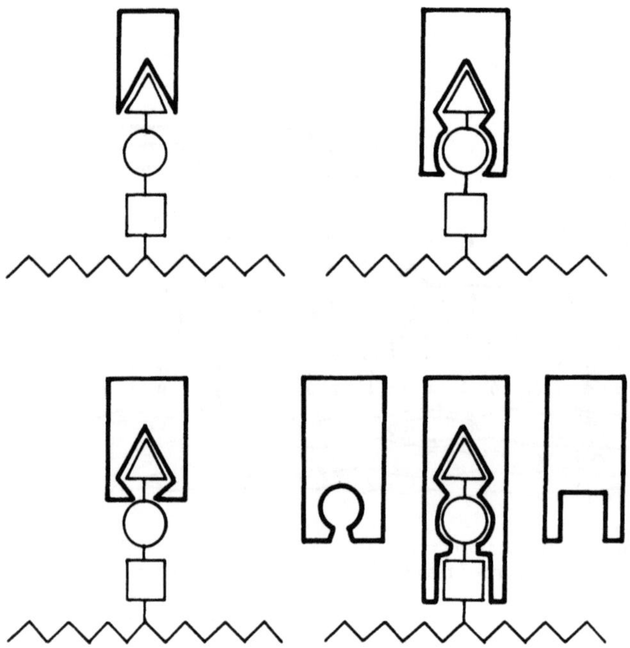

Fig. 3. Schematic representation of the reaction of different antibodies or antibody-like substances with the heterosaccharide chain of a glycoprotein

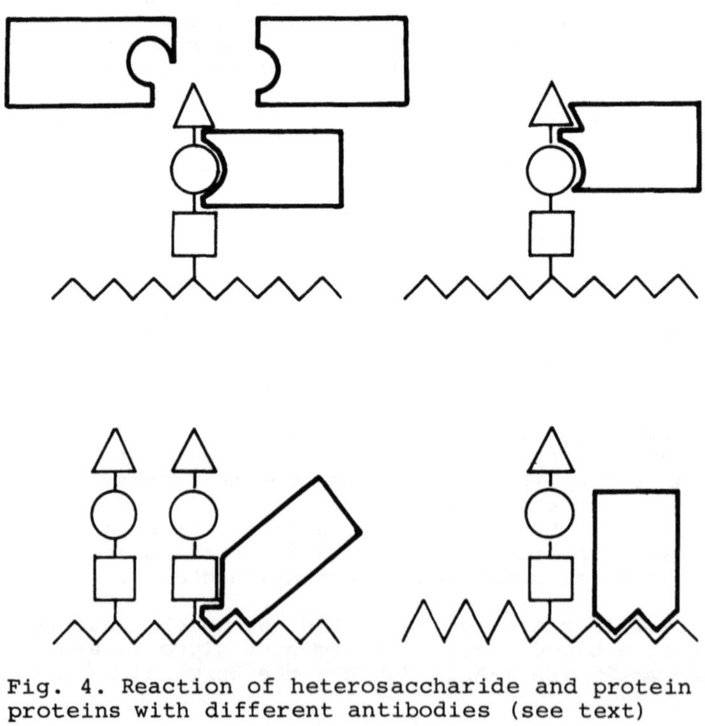

Fig. 4. Reaction of heterosaccharide and protein chains of glycoproteins with different antibodies (see text)

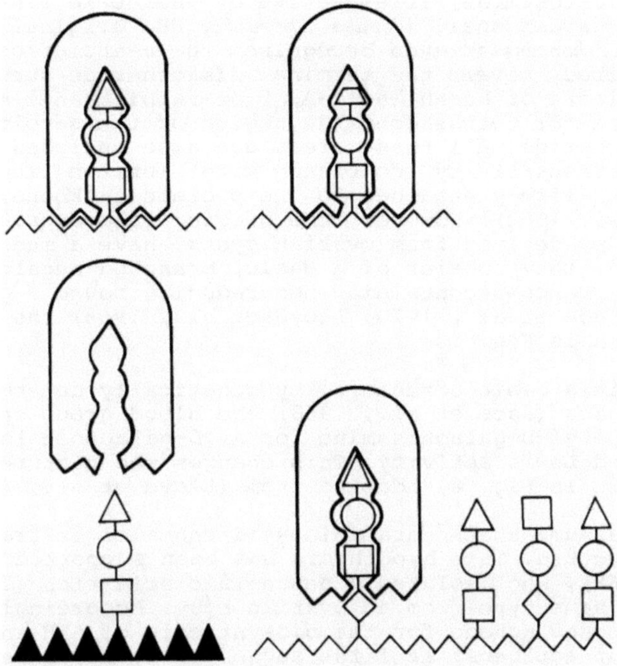

Fig. 5. Reaction of heterosaccharide and protein chains of glycoproteins in delayed-hypersensitivity response (see text)

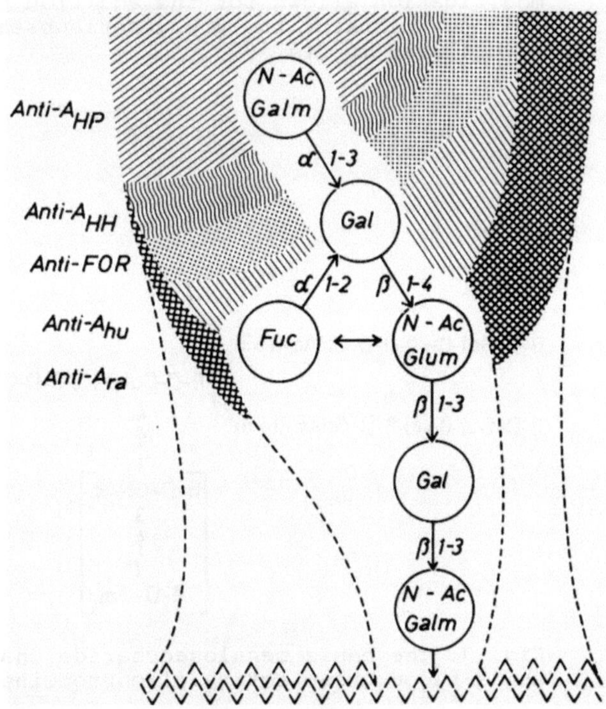

Fig. 6. Reaction of blood group A determinant with heterophile reagents, humoral antibodies and specific cellular recognition sites (see text)

D-glactosamine, irrespective of anormeric linkage. The "anti-A" from the garden snail (*Cepaea nemoralis*, HH, originally wrongly classified as *Helix hortensis*) even recognizes the α-anomeric bond, the Forssmann antibody covers the terminal disaccharide structure. Finally the animal antisera of human (anti-A_{hu}) or rabbit (anti-A_{ra}) origin also include a tri- or tetrasaccharide region of the serologically active hetero-saccharide. All these areas are also included in the delayed-type-hypersensitivity "combining site" (broken lines) which, as can be seen, firmly attaches to the protein backbone, also (Holborow and Loewi, 1962). However, the original blood group determinant glycoproteins, derived from ovarian cysts, have a much more complicated structure: they consist of a basic, branched megalosaccharide chain, with two branches containing non-reducing bound β-glycosidical D-galactose (Pardoe et al., 1970; Lloyd et al., 1968) the formula of which is given in Fig. 7.

To this basic structure, by genetically determined glycosyl-transferases (Race et al., 1968) the blood group specific sugars are added: N-acetyl-D-galactosamine for A, D-galactose for B and L-fucose for H and Lewis activity. This changes the picture given in Fig. 7 to that shown in Fig. 8, adapted from (Lloyd et al., 1968).

We assume that neuraminic acid can also be transferred to this basic structure. This hypothesis has been supported by the finding of Carlson (1968), who isolated a neuraminic acid-rich (28% NA) glycoprotein of the same type from an ovarian cyst. Accordingly, we have modified the proposed scheme for the biosynthesis of ABH and Lewis active blood group antigens, as first suggested by Morgan and Watkins (1960), starting from the Pneumococcus type XIV cross-reactive basic chain (Figs. 7 and 8) and influenced by the Le/le and Se/se secretor genes which govern the various transferases that finally lead to the expression of blood groups specificity (Kobota and Ginsburg, 1970). This modification allows for the addition of neuraminic acid (NA) to the free positions of the precursor substance, resulting in NA structures as outlined in Fig. 9.

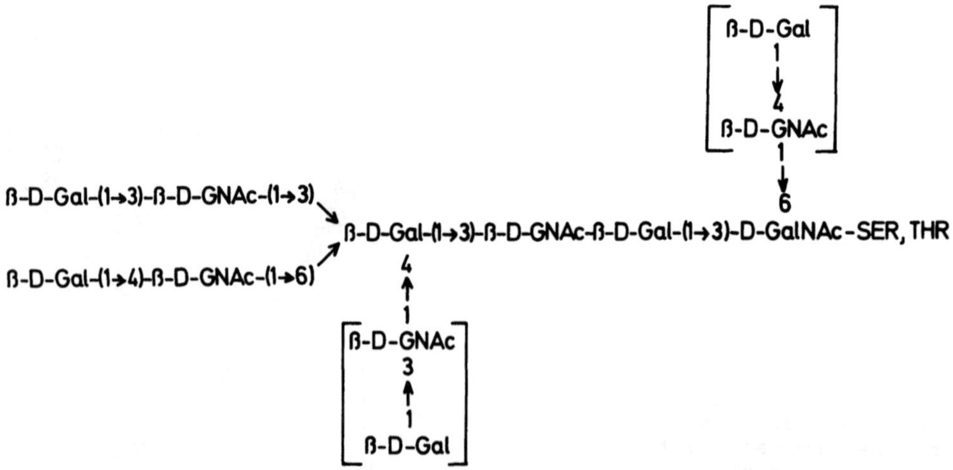

Fig. 7. The basic megalosaccharide chain of human secreted ABH and Lewis blood group active glycoproteins (Vicari and Kabat, 1970)

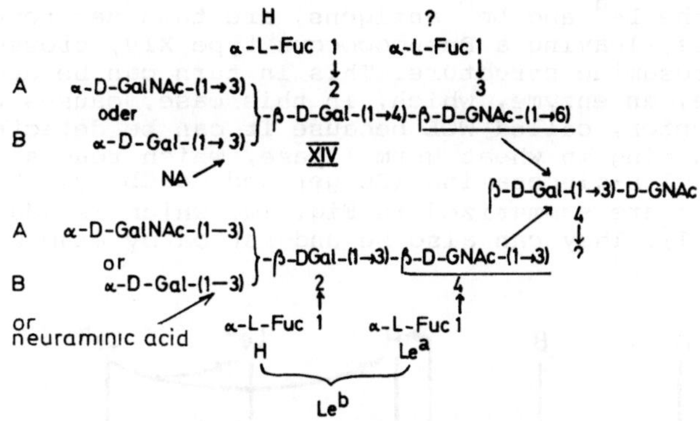

Fig. 8. The structure of ABH and Lewis active oligosaccharides. Note, specificity is accomplished by two adjacent fucose residues

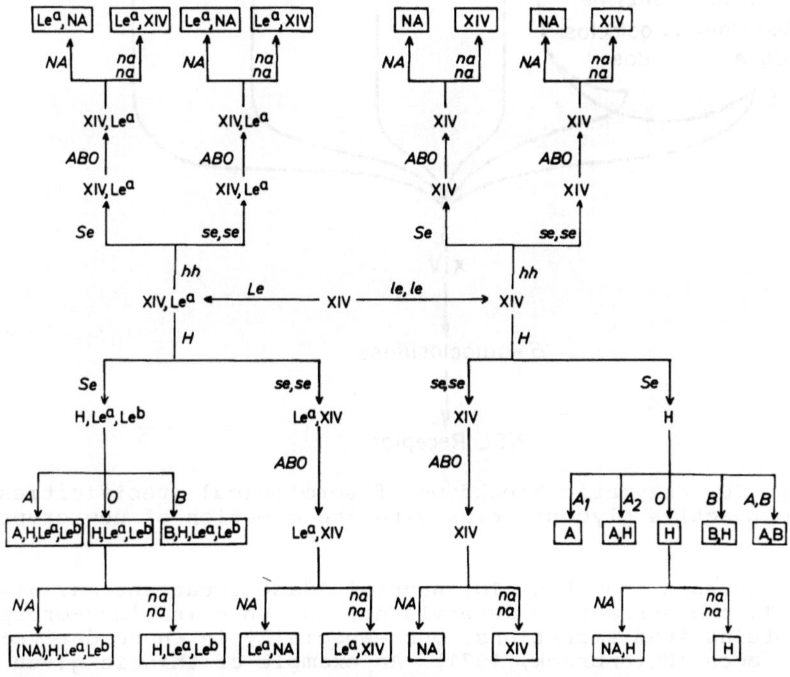

Fig. 9. Suggested pathways in the blood group ABH and Lewis systems

A detailed discussion of these possible genetic pathways has been given elsewhere (Uhlenbruck, 1969). From the genetic facts and chemical structures outlined in Figs. 7 - 9, it is obvious that the serological activities and specificities of these heterosaccharide chains in blood group active glycoproteins can be abolished by certain glocosidases which split off the terminal determinant sugar. Accordingly, A and B receptors are destroyed by α-N-acetyl-galactosaminidases or α-galactosidases, thus uncovering new H-active structures. The latter, as well as the Le^a and Le^b antigens, are then destroyed by various α-fucosidases, leaving a Pneumococcus type XIV, cross-reactive N-acetyllactosamine structure. This in turn can be degraded by a β-galactosidase, an enzyme, which, in this case, causes the appearance of another receptor, called WGL because it can be detected by an agglutinin occurring in wheat germ lipase, which reacts with the terminal N-acetyl-D-glucosamine (Burger and Goldberg, 1969). All these relationships are summarized in Fig. 10, which is adapted from Uhlenbruck (1971). They can also be understood by examining Fig. 9.

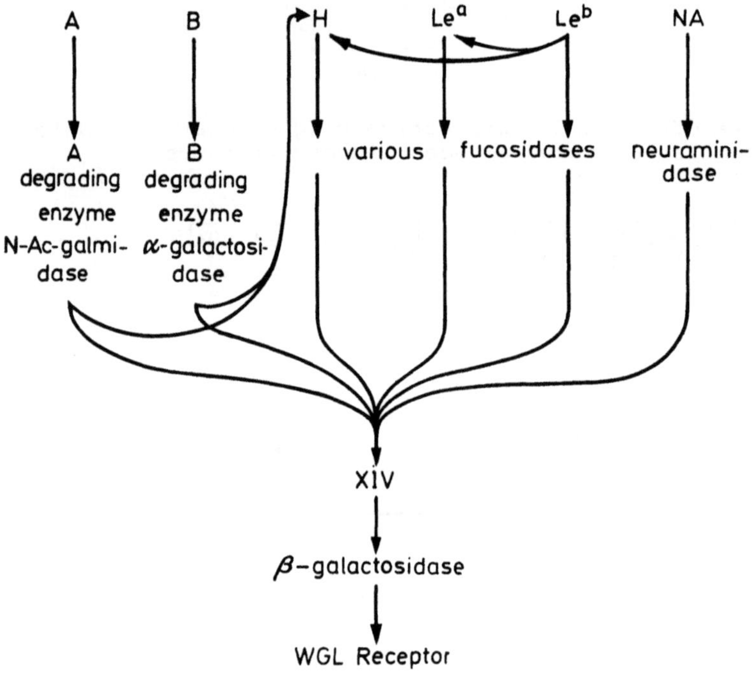

Fig. 10. Enzymatic breakdown of serological specificities in blood-group-active glycoproteins with the creation of new cryptantigens

As can seen from Fig. 10, neuraminidase treatment may also lead to a XIV cross-reactive determinant, not only in blood-group-active glycoproteins from secretions, but especially in those derived from cell surfaces (Uhlenbruck, 1971). An example of this is given in Fig. 11: bovine red cells do not cross-react with Pneumococcus type XIV antisera because neuraminic acid blocks this receptor, which occurs as F (Friedenreich antigen), so that pronase does not uncover it; however, neuraminidase (RDE) makes it accessible (left), so that one part is protease-labile and another protease-resistant (right in Fig. 11) (Pardoe and Uhlenbruck, 1969).

As can be deduced from the scheme in Fig. 10, neuraminidase has different effects on neuraminic acid-containing structures. For instance, the XIV receptor is one of the Friedenreich antigens with terminal D-galactose, others have terminal hexosamine, like the *Helix pomatia* receptor. Fig. 12 shows the receptors which are "inactivated" by neuraminidase (Pardoe et al., 1970; Springer, 1966; Roelcke et al.,

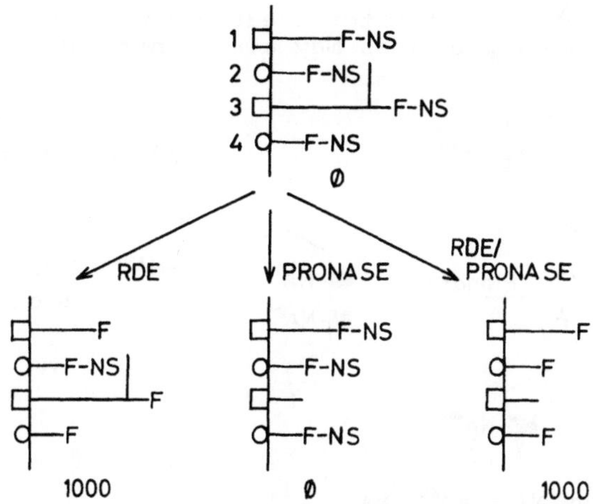

Fig. 11. Titer of bovine red cells with anti-Pneumococcus type XIV antiserum: the effect of enzyme treatment

1971): blood group receptors MN, cold agglutinin receptors HD (or Pr), invertebrate agglutinin receptors, virus receptors, and those for mycoplasma (Manchee and Taylor-Robinson, 1969). On the right of the scheme the possibility of a pseudo-inactivation of such receptors is stressed (Burger, 1968); in pseudo-Friedenreich antigens charge and sterical reasons also play a role (Currie and Bagshawe, 1968).

<u>Receptors destroyed for:</u>

Antisera: M, N, HD_1, HD_2
Agglutinins: *Limulus polyphemus*, Myxovirus, Mononucleosis and *Cepaea hortensis* (garden snail) ⟷ receptor
Mycoplasma adherence

<u>Pseudo-inactivation of receptors:</u>

charge of sterical interaction leads to optimal conformation of antigen
Example: *Triticum vulgaris* (WGL) receptor

EFFECT OF NEURAMINIDASE ON CELL ANTIGENS

<u>Receptors enhanced or de novo</u>

(Friedenreich antigens)

a) <u>with terminal D-galactose:</u>
   Vicia graminea
   Arachis hypogaea
   Ricinus communis receptor
b) <u>with terminal N-acetyl-D-galactosamine:</u>
   Helix pomatia receptor

<u>"Pseudo-Friedenreich antigens":</u>

agglutination de novo or enhanced due to

a) reduction of negative charge (repulsion of agglutinin reduced)
b) removal of sterical hindering groups

Fig. 12. "Inactivation" of cell receptors by neuraminidase

It is well known that neuraminic acid (NA) contributes to the zeta potential of the cell. Usually glycoprotein-bound NA is responsible for this, but after treatment with protease, glycolipoid-bound NA from gangliosides may also contribute to the zeta potential (Fig. 13).

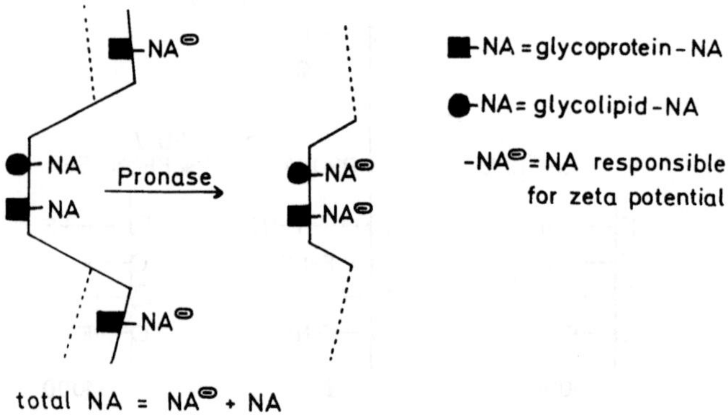

Fig. 13. Neuraminic acid and the zeta potential of cells (Rothe)

Some of the gangliosides can also be serologically active, for instance, the one called G_{XIV} which, in bovine red cells, is responsible for the anti-type-XIV cross-reaction after neuraminidase incubation (see Table 7) (Pardoe and Uhlenbruck, 1969). This means that the carrier of the NA, or of an antigen, also plays an important role, as demonstrated in Fig. 14.

Table 7. Main gangliosides in different red cells

Erythrocytes	Gangliosides		
	G_{Lact}	$G_{Lact'}$	G_{XIV}
cat	+	+	+
man,O	+	±	-
ox	+	+	+
pig	+	-	?
horse, dog	+	±	-
guinea pig	-	-	-
sheep, rabbit	-	-	-

G_{Lact} = Ceramide - Lactose - NA

$G_{Lact'}$ = Ceramide - Lactose - NA - NA

G_{XIV} = Ceramide - Lacto-N-neotetraose - NA

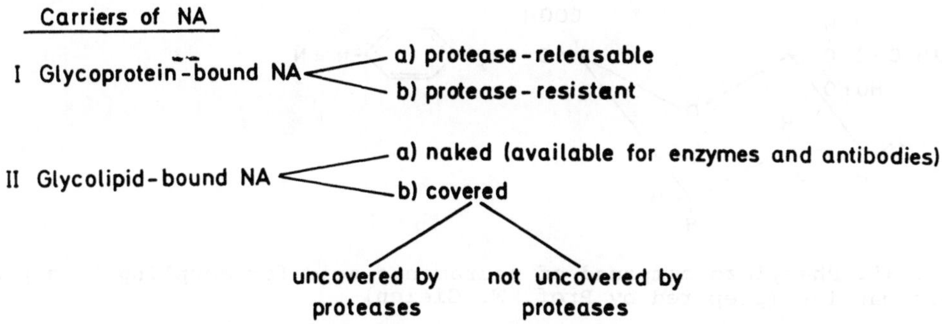

Fig. 14. Various carrier molecules of neuraminic acid (Rothe)

As already mentioned, NA can be serologically detected by certain agglutinins from snails *(Cepaea nemoralis, (Helix hortensis), Achatina granulata)* (Pardoe and Uhlenbruck, 1970; Roelcke et al., 1971; Uhlenbruck and Pardoe, 1968) and invertebrates *(Limulus polyphemus)* (Cohen, 1968). These reactions are also listed in Table 8, where some heterophile agglutinins and their specificity are presented according to Prokop et al. (1968). NA itself seems not to be antigenic, even when coupled to protein as an artificial receptor (Graf and Rapport, 1970) (Fig. 15).

Table 8. Serological specificity of some heterophile agglutinins

Name	Source	Occurrence	Specificity
Anti-A_{HP}	Helix pomatia	Albumin gland	α,β-N-Ac-D-Galm-
Anti-A_{HH}	Helix hortensis	" "	α-N-Ac-D-Galm-
Anti-A_{AG}	Achatina granulata	" "	α-N-Ac-D-Galm-
Anti-A_{OL}	Otala lactea	" "	α-N-Ac-D-Galm-
Anti-A_{OL}	" "	" "	α-L-Fucosyl-
Anti-B_{SS}	Salmo salar	Roe	α-D-Galactosyl-
Anti-B_{SI}	" irideus	"	α-D-Galactosyl-
Anti-H_{LL}	Lucioperca luciop.	"	α-L-Fucosyl-
Anti-P_{RR}	Rutilus rutilus	"	α-D-Galactosyl-
Anti-B_{RR}	" "	"	α-D-Galactosyl-
Anti-G_{SI}^{strept}	Salmo irideus	Roe	L-Rhamnosyl-(β?)
Anti-G_{LL}^{strept}	Lucioperca luciop	"	L-Rhamnosyl-
Anti-M_{AG}^{duc}	Achatina granulata	total extract	α-N-Ac-Neuraminyl-
Anti-M_{HH}^{duc}	Helix hortensis	Albumin gland	α-N-Ac-Neuraminyl-

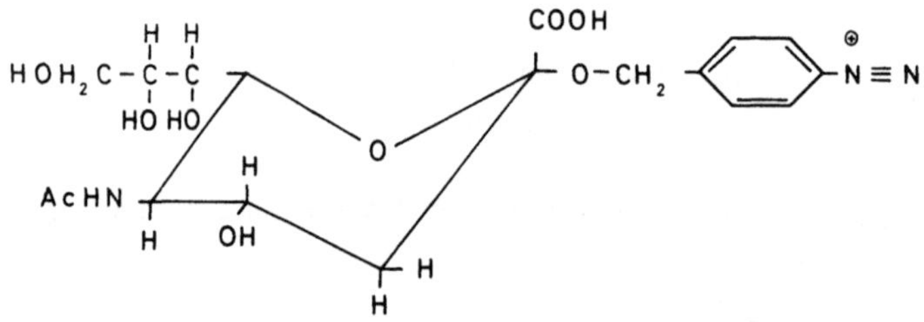

Fig. 15. Phenylazo compound of neuraminic acid for coupling to a protein carrier (prepared by Prof. W. Gielen)

The fact that *Helix pomatia* contains an agglutinin with an anti-α,β-(?) N-acetyl D-galactosamine specificity, provided this hexosamine is in terminal, non-reducing position, and that *Helix hortensis* and *Cepaea nemoralis* contain two agglutinins, namely one against terminal α-glycosidic bound N-acetyl-D-galactosamine and the other against terminal α-ketosidic bound NA, means that the snail agglutinins can have several types of reaction with cell receptors (Table 9) which must then belong to different carrier molecules (Table 10) (Prokop et al., 1968). However, there are still some contradictions and the reactions of human red cells O (or B) with anti-globoside serum and the *Helix pomatia* agglutinin are not quite clear (Table 11). The situation with rat red

Table 9. Reaction - types of Helix agglutinins

Reaction Type	Reagent		Specificity	Example
	Anti-A_{HP}	Anti-A_{HH}		
I	+	-	ß-N-Ac-Galm-	Human O-RDE-cells Tay Sachs Ganglioside Globosid I
IIa	+	+	α-N-Ac-Galm-	A substance Forssman antigen
IIb	+	+	ß-N-Ac-Galm- + α-N-Ac-NA-	Human B or O cells pronase treated
IIc	+	+	α-N-Ac-Galm- + α-N-Ac-NA-	Human A cells
III	-	+	α-N-Ac-NA-	duck erythrocytes Lymphoma cells (Burkitt)

(NA = neuraminic acid)

cells is similar: here also a glycolipid with terminal (β-linked?) N-acetyl-D-galactosamine, cytolipin R, occurs and, after trypsin treatment of the cells, is available for anti-cytolipin R sera (Graf and Rapport, 1970), but not for anti-A from snails (unpublished results).

Table 10. Partners for Helix agglutinins on red cells

End group	Detected by	Carrier	
		Glycoprotein	Glycolipid
β-N-Ac-Galm-	Anti-A_{HP}	neuraminidase treated mucoid	Globosid I (after pronase)
α-N-Ac-Galm-	Anti-A_{HP}+ Anti-A_{HH}	A active glycoprotein	A active glycolipid Forssman active glycolipid
α-N-Ac-NA-	„Anti-A_{HH}"	red cell mucoids	gangliosides

Table 11. Reaction of human red cells O with anti-globoside serum and snail agglutinin

A contradiction:

I Statement:

$$\text{N-Ac-Galm} \xrightarrow[\beta]{1\ 3} \text{Gal} \xrightarrow[\beta]{1\ 4} \text{Gal} \xrightarrow[\beta]{1\ 4} \text{Glu} \xrightarrow[\beta]{1\ 1} \text{Ceramid}$$

= Globoside I, occurs in all human red cells

II Experimental results.

Experiment 1: Anti-Globoside-antiserum. Serological reactions with

Experiment 2: Anti-β-N-Ac-Galm (Anti-A_{HP}). Serological reactions with

	Experiment 1	Experiment 2
Normal O cells:	Ø	Ø
Protease-treated mature O cells:	++	+
Protease-treated immature O cells:	+++	Ø

III Expected: identical receptor = identical serological reactions

These results, as well as those of Hammarström and Kabat (1971), favor the view that the *Helix pomatia* agglutinin detects only the α-anomeric form of N-acetyl-D-galactosamine, which is, however, supposed to be β-linked in globoside I and cytolipin R. Nevertheless - and this question is not yet solved - three points support β-specificity, too: Agglutinin from the garden snail

a) will not agglutinate neuraminidase-treated O and B cells, although it reacts strongly with human A cells which obviously have no β-specificity;

b) the inhibition by Tay-Sachs ganglioside with terminal β-linked N-acetyl-D-galactosamine (Kim et al., 1966);

c) the reaction with artificially prepared (phenylazo-protein) β-glycosidic conjugated N-acetyl-D-galactosamine (Uhlenbruck and Gielen, 1967).

The reaction of the snail agglutinins with cell surfaces has some interesting implications, as they react especially with tumor cells and also with tumor-virally transformed cells (Prokop and Uhlenbruck, 1969). Some of these cell strains are shown in Table 12. In addition,

Table 12. Tumor cells with terminal GalNAc

Tumor cells		Linkage GalNAc	Detected by agglutinin from
Mouse cells (transformed)	Polyoma virus Simian virus 40	α,β(?) GalNAc	Soybean, Soja hispida Helix pomatia?
Detroit cells		β-GalNAc	Helix pomatia
Hela cells		β-GalNAc	Helix pomatia
Zajdela hepatoma (rat)		α-GalNAc	Helix pomatia Cepaea nemoralis

transformed hamster kidney cells also react. The appearance of certain heterophile receptors on the surface of malignant transformed cells is a general phenomenon explained in different ways (Burger, 1969, 1970; Prokop and Uhlenbruck, 1969; Pardoe and Uhlenbruck, 1970; Uhlenbruck, 1971; Inbar and Sachs, 1969; Sela et al., 1970). In addition to the excellent interpretations given already by Burger and collaborators (Burger, 1969, 1970; Burger and Noonan, 1970) and by the team of Sachs (Inbar and Sachs, 1969; Sela et al., 1970), we give here our view concerning the appearance of the A-like, N-acetyl-D-galactosamine containing receptors on tumor cells.

1.) De novo:
GalNAc-Transferase → "foreign, unnatural" precursor substance

2.) Topographical reasons:
Covered GalNAc → Uncovered GalNAc receptors

3.) Incomplete conversion of GalNAc containing precursors:
Ser, Thr → GalNAc ⊣→ block

4.) Degradation of carbohydrate chains:
by neuraminidase, glycosidases:
Appearance of GalNAc structures.

Fig. 16. Appearance of "A-like" antigens on tumor cells

Fig. 16 indicates some possibilities for the establishment of N-acetyl-D-galactosamine (= GalNAc) receptors on tumor cells: the misuse of foreign precursor substances by a GalNAc transferase (Fig. 18); topographical reasons, as demonstrated in Fig. 17; incomplete conversion of a precursor substance (Fig. 18) with a different topographical incorporation; and finally, degradation of carbohydrate chains by certain glycosidases.

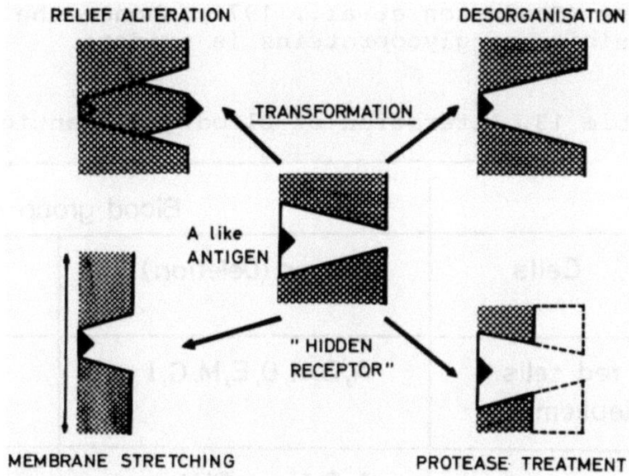

Fig. 17. Some possible explanations for the serological availability of the N-acetyl-D-galactosamine containing "A-like" antigen on tumor cells. "Transformation" can mean change of relief, disorganization of the membrane (membrane stretching, etc.) or, it can be caused artificially by proteolytic enzyme treatment

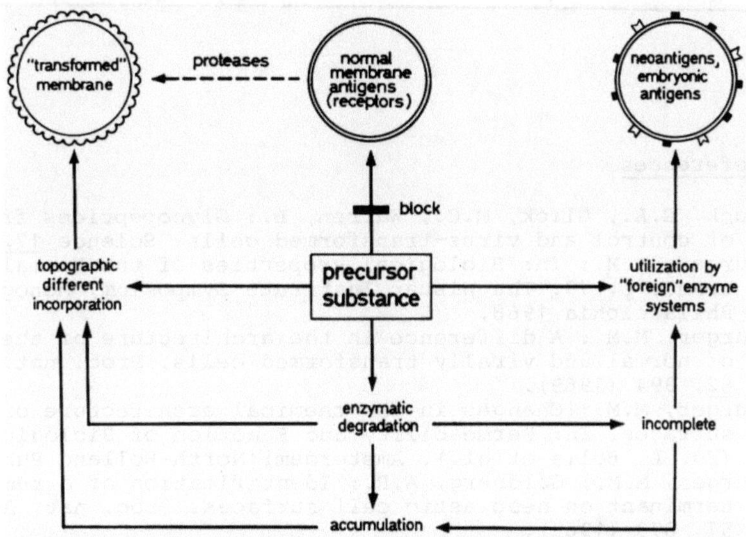

Fig. 18. Modification of the precursor substance in tumor cells: development of neo-antigens and a "transformed" membrane (for possible explanations see text)

Any of these events may lead to an alteration of blood group antigens or heterophile receptors in cancer, also in human cells, as summarized in Table 13, modified from other reviews (Majski, 1970; Uhlenbruck and Reifenberg, 1971). Like the blood group receptors, most heterophile antigens and transplantation antigens, also, are of glycoprotein nature (Sanderson et al., 1971), hence the importance of the immunogenicity of glycoproteins is evident.

Table 13. Alteration of blood group antigens in cancer

	Blood group antigens		
Cells	Loss (Deletion)	Gain	Uncovered (cryptantigens)
red cells (leukemia)	A, B, H, D, E, M, C, I	i	XIV, H
lymphocytes (leukemia)	A, B, H, PHA leucocyte specific, HL-A2	M, N HL-A 3	A like
carcinoma (gastric, cervix)	A, B, H	Le^a Le^b Le^x	H, FORSSMAN A like, WGL Con A

References

Buck, C.A., Glick, M.C., Warren, L.: Glycopeptides from the surface of control and virus-transformed cells. Science 172, 169 (1971).

Burger, M.M.: In: Biological Properties of the Mammalian Surface Membrane, p. 78. The Wistar Institute Symposium, Monograph No. 8. Philadelphia 1968.

Burger, M.M.: A difference in the architecture of the surface membrane of normal and virally transformed cells. Proc. nat. Acad. Sci. (Wash.) 62, 994 (1969).

Burger, M.M.: Changes in the chemical architecture of transformed cell surfaces. In: Permeability and Function of Biological Membranes. (Ed. L. Bolis et al.). Amsterdam: North-Holland Publishing 1970.

Burger, M.M., Goldberg, A.R.: Identification of a tumorspecific determinant on neoplastic cell surfaces. Proc. nat. Acad. Sci. (Wash.) 57, 873 (1969).

Burger, M.M., Noonan, K.D.: Restoration of normal growth by covering of agglutinin sites on tumour cell surface. Nature 228, 512 (1970).

Carlson, D.M.: Isolation of a "sialic acid rich" glycoprotein from an ovarian cyst. Fed. Proc. 27, 530 (1968).

Cohen, E.: Immunologic observations of the agglutinins of the hemolymph of *Limulus polyphemus* and *Birgus latro*. Transact. N.Y. Acad. Sci. 30, 427 (1968).

Currie, G.A., Bagshawe, K.D.: The role of sialic acid in antigenic expression: further studies of the Landschütz ascites tumour. Brit. J. Cancer 22, 843 (1968).

Graf, L., Rapport, M.M.: Immunochemical studies of organ and tumour lipids. XVIII. Cytolipin R determinants in the rat erythrocyte membrane. J. Membrane Biol. 3, 280 (1970).

Graham, E.R.B., Gottschalk, A.: The structure of prosthetic group of ovine submaxillary gland mucoprotein. Biochim. biophys. Acta (Amst.) 38, 513 (1960).
Halpern, B.N., Ky, T., Robert, B.: Clinical and immunological study of an exceptional case of reaginic type sensitation to human seminal fluid. Immunology 12, 247 (1967).
Hammarström, S., Kabat, E.A.: Studies on specificity and binding properties of the blood group A reactive hemagglutinin from *Helix pomatia*. Biochemistry 10, 1684 (1971).
Holborow, E.J., Loewi, G.: The immune response to blood group substances. Immunology 5, 278 (1962).
Inbar, M., Sachs, L.: Structural difference in sites on the surface membrane of normal and transformed cells. Nature 223, 710 (1969).
Kim, Z., Uhlenbruck, G., Prokop, O., Schlesinger, D.: Über die B-Substanz und des Anti-A von *Helix pomatia*. Z. Immunforsch. 130, 290 (1966).
Kobota, A., Ginsburg, V.: Uridine diphosphate-N-acetyl-D-galactosamine: D-galactose α-3-N-acetyl-D-galactos-aminyltransferase, a product of the gene that determines blood types A in man. J. biol. Chem. 245, 1484 (1970).
Lloyd, K.O., Kabat, E.A., Licero, E.: Immunochemical studies on blood groups XXXVIII. Structures and activities of oligosaccharides produced by alkaline degradation of blood-group Lewis^a substance. Proposed structure of the carbohydrate chains of human blood-group A, B, H, Le^a and Le^b substances. Biochemistry 7, 2976 (1968).
Majsky, A.: Serologische Modifikationen der Blutgruppenantigene menschlicher Blutzellen bei malignen Erkrankungen. Folia haemat. 94, 281 (1970).
Manchee, R.J., Taylor-Robinson, D.: Studies on the nature of receptors involved in attachment of tissue culture cells to mycoplasmas. Brit. J. Exp. Path. 50, 66 (1969).
Morgan, W.T.J.: A contribution to human biochemical genetics the chemical basis of blood group specificity. Proc. Royal Soc. B 151, 308 (1960).
Pardoe, G.I., Uhlenbruck, G.: Methods for elucidating cell surface structures. In: Organtransplantation. Stuttgart, New York: F.K. Schattauer 1969.
Pardoe, G.I., Uhlenbruck, G.: Characteristics of antigenic determinants of intact cell surfaces. J. med. Lab. Technol. 27, 249 (1970).
Pardoe, G.I., Uhlenbruck, G., Anstee, D.J., Reifenberg, U.: On the specificity of broad spectrum agglutinins. III. Heterophile agglutinins associated with anti-B-like specificity. Z. Immunforsch. 139, 468 (1970).
Pepper, D.S., Jamieson, G.A.: Isolation of a macroglycopeptide from human platelets. Biochemistry 9, 3706 (1970).
Prokop, O., Uhlenbruck, G.: N-Acetyl-D-Galaktosamin an Tumorzellmembranen: Nachweis mittels Helix-Agglutininen. Med. Welt (Stuttg.) 20, 2515 (1969).
Prokop, O., Uhlenbruck, G.: Human Blood and Serum Groups. London: McLaren 1969.
Prokop, O., Uhlenbruck, G., Köhler, W.: A new source of antibody-like substances having anti-blood group specificity. Vox Sang. (Basel) 14, 321 (1968).
Race, C., Ziderman, D., Watkins, W.M.: An α-galactosyl transferase associated with the blood group B character. Biochem. J. 107, 733 (1968).
Rüde, E.: Struktur und Immunogenität synthetischer Modellantigene. Angew. Chem. 82, 202 (1970).
Roelcke, D., Uhlenbruck, G., Metaxas, M.N.: Untersuchungen über die serologische Spezifität neuraminsäurehaltiger Rezeptoren auf Humanerythrozyten. Z. Immunforsch. 141, 141 (1971).

Sanderson, A.R., Cresswell, P., Welsh, K.I.: Involvement of carbohydrate in the immunochemical determinant area of HL-A substances. Nature 230, 8 (1971).
Schiffman, G., Marcus, D.M.: Chemistry of the ABH blood group substances. Progr. Hemat. 4, 97 (1964).
Sela, B., Lis, H., Sharon, N., Sachs, L.: Different locations of carbohydrate-containing sites in the surface membrane of normal and transformed mammalian cells. J. Membrane Biol. 3, 267 (1970).
Springer, G.F.: Beziehungen zwischen Mikroben und blutgruppen-aktiven Substanzen. Angew. Chem. 78, 967 (1966).
Uhlenbruck, G.: Das Studium der Zelloberfläche mit Hilfe von Enzymen. Mitteil. Max-Planck-Ges. 4, 227 (1965).
Uhlenbruck, G.: Serologische Eigenschaften von Ovarial-cystensekreten: ein Beitrag zur Blutgruppenkunde. Geburth. u. Frauenheilk. 29, 873 (1969).
Uhlenbruck, G.: Über Mucoide von Zelloberflächen. Chimia 25, 10 (1971).
Uhlenbruck, G.: Immunbiologie - Eine Einführung. München: W. Goldmann 1971.
Uhlenbruck, G., Gielen, W.: Serologische Eigenschaften von Ovalbumin nach Kupplung mit diazotierten p-Aminophenyl-N-acetyl-β-galaktoaminid: Ein Beitrag zur Spezifität von Anti-A_{hel}. Z. physiol. Chem. 348, 1693 (1967).
Uhlenbruck, G., Pardoe, G.I.: Serologische Besonderheiten eines heterophilen Rezeptors aus Entenblutkörperchen. Z. Naturforsch. 24b, 142 (1968).
Uhlenbruck, G., Reifenberg, U.: Tumorimmunologie: ein Problem der Membranforschung? Med. Klin. 66, 1435 (1971).
Uhlenbruck, G., Reifenberg, U., Heggen, M.: On the specificity of broad spectrum agglutinins. IV. Unvertebrate agglutinins: Current status, conception and further observations in the variation of the Hel receptor in pigs. Z. Immunforsch. 139, 486 (1970).
Uhlenbruck, G., Wintzer, G., Salfner, B., Schumacher, K., Oerkermann, H., Hirschmann, W.D., Alzer, G.: Studies on the nature of the PHA receptor. Klin. Wschr. 48, 1369 (1970).
Vicari, G., Kabat, E.A.: Structures and activities of oligosaccharides produced by alkaline degradation of a blood group substance lacking A, B, H, Le^a and Le^b specificities. Biochemistry 9, 3414 (1970).
Westphal, O.: Antigene - Spezifität und Immunogenität. In: R. Gross: Immunhämatologie, p. 29. Stuttgart, New York: Schattauer 1970.
Winzler, R.J.: Carbohydrates in cell surfaces. Int. Rev. Cytol. 29, 77 (1970).
Zahler, P.: The structure of the erythrocyte membrane. Experientia (Basel) 25, 449 (1969).

Immunological Properties of Ovine Submaxillary Glycoproteins

A Gottschalk

The glycoprotein OSM, prepared from sheep submaxillary glands in a highly purified state, has a mol. wt. of about 8×10^5 and a carbohydrate content of 50%. About eight hundred disaccharides of the structure α-N-acetylneuraminosyl-(2 → 6)N-acetylgalactosamine are linked α-glycosidically to peptide-bonded serine and threonine residues of the polypeptide (Gottschalk, 1966; Gottschalk et al., 1971). Action on OSM of neuraminidase and a specific N-acetyl-α-D-galactosaminidase which was isolated and purified, free of proteases and peptidases, from *Lumbricus terrestris* (Schauer and Gottschalk, 1968), resulted in the production of the carbohydrate-free, intact protein moiety. Ouchterlony tests with immune sera against OSM and carbohydrate-free OSM showed that the main antibody is directed against the protein moiety of OSM (see Fig. 1 and Fig. 2). Furthermore, the identity of the antibodies

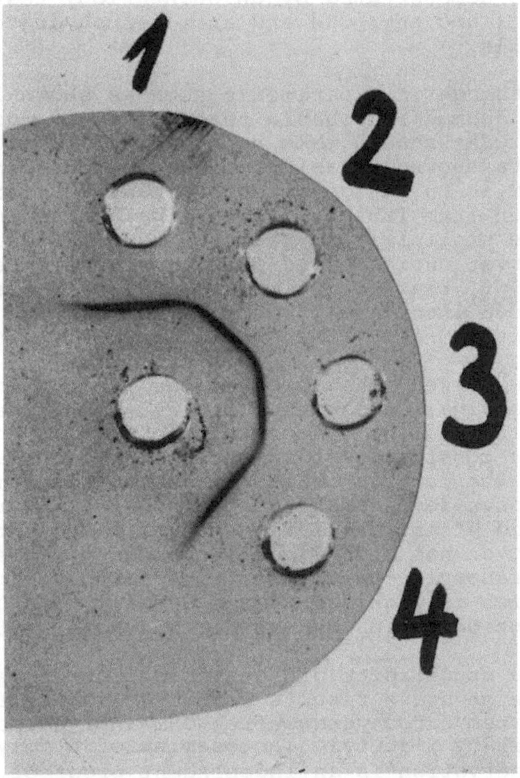

Fig. 1. Ouchterlony test with native OSM as antigen. In central well anti-OSM, in peripheral wells (1) OSM, (2) NANA-free OSM, (3) carbohydrate-free OSM, (4) OSM. (From Gottschalk et al., 1971)

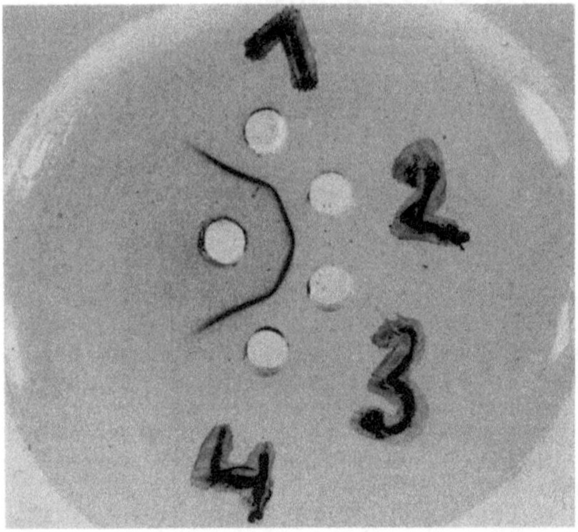

Fig. 2. Ouchterlony test with carbohydrate-free OSM as antigen. In central well anti-carbohydrate-free OSM, in peripheral wells as in Fig. 1. (From Gottschalk et al., 1971)

against the protein moiety of native OSM and carbohydrate-free OSM was demonstrated by an Ouchterlony test with native OSM in the central well and anti-OSM and anti-carbohydrate-free OSM in the peripheral wells.

Although in experiments such as shown in Fig. 1 no spur was demonstrable, it seemed improbable that the sugar moiety (50% of the molecular weight of OSM) should make no contribution to the antigenic determinants. The data shown in Table 1 leave no doubt that the carbohydrate portion of OSM is an antigenic determinant (Gottschalk et al., 1971). As may be seen from Table 1, N-acetylgalactosamine was a stronger inhibitor of the precipitation reaction than was N-acetylneuraminic acid, an observation in accord with the experience of Gielen and Uhlenbruck (1969) that no antibodies against N-acetylneuraminic acid were produced when rabbits were immunized with ovalbumin to which p-aminobenzyl N-acetylneuraminoside was coupled by diazotization.

The marked inhibition of the precipitin reaction OSM/anti-OSM by N-acetylgalactosamine (10 to 25-fold excess referred to the N-acetylgalactosamine content of the quantity of OSM used) may result from the possibility that not all N-acetylgalactosamine residues attached to the peptide chain are substituted either because of incomplete biosynthesis or because of the *in vivo* removal of N-acetylneuraminic acid after inoculation but prior to antibody formation. Such a situation has been directly proven for the disaccharide 2-O-α-D-glucopyranosyl-D-galactose, the prosthetic group of collagen (Spiro, 1969), which is linked β-glycosidically to a hydroxylysine residue of the peptide chain (Cunningham and Ford, 1968).

The specificity of the inhibitions presented in Table 1 is shown by the negative results with D-galactose and N-acetyl-glucosamine. The system carbohydrate-free OSM/anti-carbohydrate-free OSM is not inhibited by N-acetylgalactosamine or by glycopeptides (from OSM). This observation is an independent proof of the almost complete removal

Table 1. Inhibition by haptens of the precipitin reaction between immunsera against OSM and carbohydrate-free OSM and their antigens. GalNAc = N-acetylgalactosamine; GlcNAc = N-acetylglucosamine; NANA = N-acetylneuraminic acid. (From Gottschalk et al., 1971)

System	Hapten added	Excess of hapten[a] (n-fold)	Inhibition of precipitation (%)
Anti-OSM/	GalNAc	10	10 - 15
OSM	GalNAc	25	23 - 30
	GalNAc	100	20 - 30
	NANA	25	14 - 20
	NANA	100	18 - 20
	GlcNAc	10	0
	GlcNAc	25	0
	GlcNAc	100	8 - 10
	D-Galactose	25	0
	Glycopeptide	6	50 - 70
Anti-carbohydrate-free OSM/ carbohydrate-free OSM	GalNAc	see footnote[b]	2
	Glycopeptide		6

[a] The excess of added hapten is referred to the N-acetylgalactosamine content of the quantity of OSM used.

[b] In this experiment amounts of haptens were added corresponding to a 25-fold excess of GalNAc and a 6-fold excess of glycopeptides in the above assays of the system anti-OSM/OSM.

of the carbohydrate moiety of OSM.

The observation that short glycopeptides are more potent inhibitors than is N-acetylgalactosamine would indicate the presence in OSM of an antigenic determinant which results from the covalent interaction of a peptide segment containing serine or threonine with the hapten (carbohydrate) group. The high inhibitory titer of the small glycopeptides may also lend some support to the assumption that the amino acid sequence around the serine and threonine residues, to which the prosthetic groups are bound, have a fair degree of similarity.

References

Cunningham, L.W., Ford, J.D.: A comparison of glycopeptides derived from soluble and insoluble collagens. J. Biol. Chem. 243, 2390 (1968).

Gielen, W., Uhlenbruck, G.: Neuraminsäure als determinates Molekül in künstlichen Antigenen. Z. physiol. Chem. 350, 672 (1969).

Gottschalk, A., Pigman, W.: Submaxillary gland glycoproteins. In: Glycoproteins. Their Composition, Structure and Function, A. Gottschalk (Ed.), B.B.A. Library, 5, p. 434. Amsterdam: Elsevier 1966.
Gottschalk, A., Schauer, H., Uhlenbruck, G.: Immunological properties of ovine submaxillary glycoprotein. Z. physiol. Chem. 352, 117 (1971).
Schauer, H., Gottschalk, A.: Studies on glycoproteins. 17. purification of O-Seryl-N-acetyl-galactosaminide glycohydrolase D-hexosaminidase and proteases. Biochim. biophys. Acta (Amst.) 156, 304 (1968).
Spiro, R.G.: Characterization and quantitive determination of hydroxylsine-linked carbohydrate units of several collagens. J. Biol. Chem. 244, 602 (1969).

The Distribution of Antigenic Determinants in Collagen as Determined by a Radioimmunoassay

P. Bornstein, H. Lindsley, and M. Mannik

Although immunological studies of collagen have been hampered by the limited solubility and poor immunogenicity of the protein, the ability of collagen to stimulate antibody production is firmly established (Kirrane and Glynn, 1968). The collagen monomer is composed of three polypeptide chains, each with a molecular weight of about 95,000. In mammalian collagens two of the chains (α1) are identical whereas the third (α2) is different in amino acid composition (Piez et al., 1963). Further analysis of the α1 and α2 chains has utilized nonenzymatic cleavage of methionyl residues with cyanogen bromide (CNBr). Such studies have resulted in the isolation and characterization of eight fragments which constitute the α1 chain of rat skin collagen (Butler et al., 1967) and six fragments which account for rat skin collagen α2 (Fietzek and Piez, 1969). The present immunological studies were undertaken to determine whether different antigenic determinants exist on the α1 and α2 chains of rat skin collagen, to localize the determinants of these chains to specific CNBr-produced fragments, and to examine the extent to which the availability of antigenic determinants is influenced by the conformation of collagen chains.

1. Materials and Methods

The experimental procedures employed in these studies have been described in detail (Lindsley et al., 1971). Purified α1 and α2 chains from normal and lathyritic rat skin collagen were isolated (Piez et al., 1963; Bornstein and Piez, 1966), cleaved with CNBr, and the resulting fragments separated by a combination of ion-exchange and molecular sieve chromatography (Butler et al., 1967; Bornstein and Piez, 1966).

Rabbits were immunized with preparations of native collagen, α1, α2, or β12 chains which were dissolved in normal saline and mixed with an equal volume of Freund's adjuvant. The total dose of protein given to each rabbit ranged from 20 to 100 mg during the course of immunization. A goat was immunized with rabbit γG-globulin by a conventional procedure.

Isolated α1 and α2 chains were trace-labelled by the iodine monochloride method (Helmkamp et al., 1967) except that iodination was performed in 5 M guanidine HCl containing 0.2 M Tris-HCl, pH 9.0, to permit even distribution of the label. Reagents were adjusted to introduce one equivalent of iodine per 10 equivalents of α chains. Radiolabelled peptides were obtained by CNBr cleavage of iodinated chains and subsequent fractionation of the resulting peptides.

In the immunoassay a rabbit antiserum interacted with radiolabelled rat collagen chains or fragments. The resulting antigen-antibody complexes were precipitated, together with non-specific rabbit γG-globulin, by the addition of excess goat antiserum to rabbit γG-globulin. In subsequent experiments unlabelled chains or fragments were added to detect their ability to inhibit the precipitation of labelled antigen. Experiments with native collagen as an inhibitor were modified to avoid precipitation of the protein.

2. Results

a) Demonstration of Chain Specificity

α1 and α2 chains were immunologically distinct and failed to show significant cross-reactivity. When ^{125}I-α2 was used as a test antigen with antisera to native collagen or α2 chains, addition of 10 picomoles of α2 chains decreased binding of the radiolabelled antigen from 80 to 27%. In contrast, addition of as much as 100 picomoles of α1 did not inhibit binding of ^{125}I-α2 by these antisera. Reciprocal experiments were performed with an antiserum to collagen and ^{125}I-α1 as a test antigen. Addition of 10 picomoles of α1 decreased binding of ^{125}I-α1 from 72 to 34%. Addition of quantities of α2 increasing to 100 picomoles produced only a slight inhibition of binding of ^{125}I-α1 (to 60%). This inhibition can be attributed to a minor degree of contamination (less than 5%) of α2 by the β12 component which contains the α1 chain.

b) Localization of Antigenic Determinants on α1 and α2 Chain

CNBr cleaves the α1 chain of rat skin collagen into eight fragments ranging in molecular weight from 1,500 to 24,000 and the α2 chain into six fragments ranging in molecular weight from 290 to 29,500 (Butler et al., 1967; Fietzek and Piez, 1969; Piez et al., 1969; Vuust et al., 1970; Fig. 1). With ^{125}I-α1 chain as a test antigen α1 and α1-CB6

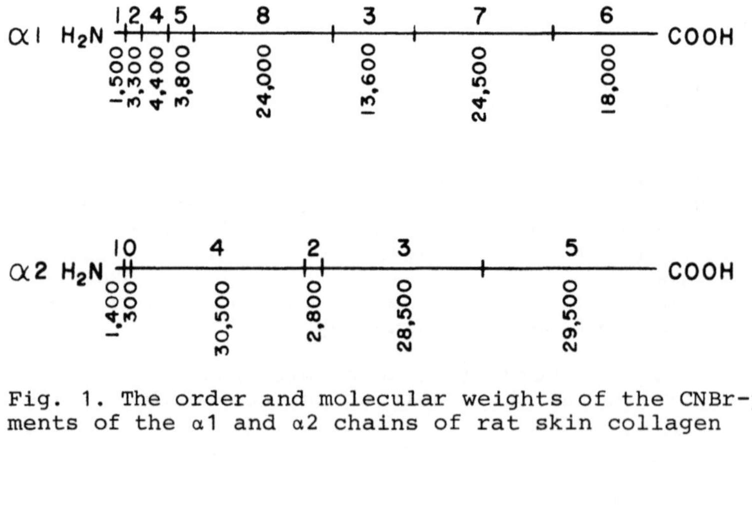

Fig. 1. The order and molecular weights of the CNBr-produced fragments of the α1 and α2 chains of rat skin collagen

proved to be the most effective inhibitors (Table 1). Lesser and progressive degrees of inhibition were seen with both α1-CB7 and a mixture of α1-CB4 and α1-CB5. Since the latter three peptides chromatographed in close proximity to α1-CB6 (Butler et al., 1967), it is possible that the observed inhibition was due to contamination with α1-CB6 rather than to the existence of separate determinants. No inhibition was seen with the four remaining peptides α1-CB1, α1-CB2, α1-CB3 and α1-CB8. These results were confirmed with ^{125}I-α1-CB6 as a test antigen. In addition, no antibodies were detected in sera from rabbits immunized with rat skin collagen, β12 or α1 chains when α1-CB1 was used as a test antigen.

Table 1. Inhibition of binding of ^{125}I-α1 and ^{125}I-α1-CB6 with α1 CNBr peptides

Inhibitor[a]	Percent ^{125}I-α1 bound	Percent ^{125}I-α1-CB6 bound
None	75	57
α1	17	9
α1-CB6	34	20
α1-CB7	50	30
α1-CB4[b] + α1-CB5	54	48
α1-CB1[c] α1-CB2 α1-CB3 α1-CB8	72	57

[a] 100 picomoles were used in all assays.

[b] Tested as a mixture.

[c] Tested individually.

With ^{125}I-α2 chain as a test antigen the α2 chain and α2-CB5 proved to be the most effective inhibitors (Table 2). Essentially no inhibition was noted with the other α2 CNBr-produced peptides. Of note is that α2-CB1 failed to inhibit precipitation of labelled α2 chains even when a large quantity of the peptide was used as an inhibitor. α2-CB1 was, however, shown to contain an antigenic determinant by direct assay with the labelled peptide. With ^{125}I-α2-CB1 as a test antigen only the α2 chain and α2-CB1 were potent inhibitors. None of the other peptides was effective at the concentrations tested (Table 2). Thus two antigenic determinants were distinguished by these experiments and were localized to α1-CB1 and α2-CB5. Comparable results were obtained in experiments with ^{14}C-α2 chains.

c) Availability of Antigenic Determinants in Native Collagen

Experiments were performed with native collagen as an inhibitor to determine whether the triple helical conformation influenced the degree to which antibodies bound to specific antigenic determinants. With both ^{125}I-α2-CB1 and ^{125}I-α2-CB5 as test antigens the α2 chain was a more effective inhibitor than equimolar quantities of the native pro-

Table 2. Inhibition of binding of ^{125}I-α2 and ^{125}I-α2-CB1 with α2 CNBr peptides

Inhibitor[a]	Percent ^{125}I-α2 bound	Percent ^{125}I-α2-CB1 bound
None	74	47
α2	11	2
α2-CB5	18	
α2-CB4	65	41 (α2-CB5, α2-CB4, α2-CB3)
α2-CB3	62	
α2-CB2	60	46
α2-CB1	70	2

[a] 100 picomoles were used in all assays

tein. Similarly, the α1 chain inhibited the precipitation of ^{125}I-α1-CB6 more effectively than did equimolar amounts of native collagen. Thus in the helical collagen molecule, the antigenic determinants of both α1 and α2 are not fully available to react with antibodies. Differences in the distribution of antibodies elicited to preparations of native and denatured collagen have also been reported by Michaeli and his associates (Michaeli et al., 1971).

3. Discussion

The studies of Schmitt, Davison and coworkers (Schmitt et al., 1964; Davison et al., 1967) indicated that the species-specific antigenic determinants of native collagen were located primarily on non-helical protease-susceptible regions of both the α1 and α2 chains of the molecule. In rat collagen such regions comprise mainly the NH_2-terminal ends of the chains (Bornstein and Piez, 1966; Bornstein et al., 1966; Kang et al., 1967). More extensive immunologic investigations by Steffen, Timpl and associates (Steffen et al., 1967, 1968; Timpl et al., 1968; Steffen and Timpl, 1970) provided evidence for both species-specific and non-species-specific or general collagen determinants. A fraction of the former determinants were pepsin-labile and presumably do not reside in the main body of the collagen helix.

The above studies have led to attempts to achieve a more precise localization of antigenic determinants on the collagen molecule utilizing inhibition of antigen-antibody reactions with well characterized collagen chains and fragments. For this purpose we have developed a sensitive radioimmunoassay. This assay does not require adsorption of antigen to erythrocytes or other particles and is able to detect univalent antigens by subsequent immune precipitation of antigen-antibody complexes. Correlations between the chemical and immunologic properties of collagen using the hemagglutination inhibition technique have also been reported (Michaeli et al., 1969; Timpl et al., 1970; Pontz et al., 1970).

Our results indicate that antigenic determinants exist on both α1 and α2 chains of rat skin collagen and that no cross reactivity exists between antibodies directed toward these determinants. The main antigenic determinant of the α1 chain is located on α1-CB6 (Table 1). α1-CB6 was also identified as the major antigenic determinant in rat and calf skin collagen using hemagglutination inhibition (Timpl et al., 1970; Pontz et al., 1970).

The predominant antigenic determinant of the α2 chain is located on CNBr fragment α2-CB5 (Table 2). Antibodies to α2-CB1 were also demonstrated but only by direct assay with ^{125}I-α2-CB1 (Table 2). No inhibition was observed with large quantities of α2-CB1 when antisera were tested with ^{125}I-α2. The most likely explanation for these findings is the presence, in the antisera tested, of a large proportion of antibodies to α2-CB5 in contrast to a relatively small proportion of antibodies to α2-CB1. Thus labelled α2 chains would be precipitated with antibodies to α2-CB5 precluding significant inhibition by α2-CB1. While these findings are in accord with studies of calf and rat skin collagen from laboratories using the hemagglutination techniques (Timpl et al., 1970; Pontz et al., 1970), α2-CB1 was identified as the major antigenic determinant in guinea pig collagen (Michaeli et al., 1969).

It is clear that any studies of the antigenicity of collagen must take into consideration the structure of collagen in the immunized species (usually the rabbit) and the degree to which the rabbit will produce antibodies to determinants which exist on its own collagen. In addition, the differences in the location of antigenic determinants which have been described may reflect differences in the nature of the immunizing antigen and the method of detection of the antibody response.

Thus the region of the collagen molecule which appears to differ most between species, i.e. the NH_2-terminal end of the α2 chain represented by α2-CB1 (Bornstein, 1968; Bornstein and Kang, 1970) has been identified as the site of an antigenic determinant. Similarly, the absence of an antigenic determinant on rat skin α1-CB1 is consitent with the identity in compositions of rat and rabbit α1-CB1 (Bornstein and Nesse, 1970). Since major antigenic determinants are clustered in the carboxy-terminal regions of the α1 and α2 chains (represented by α1-CB6 and α2-CB5, respectively) considerable interspecies variability may also exist in these regions. There is, in fact, structural evidence for such interspecies variability (Fietzek and Pietz, 1969; Stark et al., 1971; Click and Bornstein, 1970). The concentration of the limited tyrosine content of collagen to the ends of both chains may make existing structural differences immunologically apparent because tyrosyl residues are known to enhance the immunogenicity of collagen and other proteins (Kirrane and Robertson, 1968).

Our experiments, which show that antigenic determinants on α chains are less available in the native protein than in unfolded chains, are in accord with the likelihood that the immunogenicity of certain antigenic determinants will vary with the conformation of the immunizing protein. Similarly, the ability to detect the presence of antibody populations will be affected by the nature of the test antigen. In this regard the consequences of iodination of tyrosyl residues in collagen chains must be considered in the interpretation of these studies. However, no evidence was found for antibody discrimination between iodinated and unlabelled polypeptides.

The experiments reported here, together with parallel studies in other laboratories, indicate that an immunochemical approach will be useful in investigations of the comparative biochemistry of collagen. The ability to produce antibodies to specific regions of the collagen molecule may also assist in the detection of alterations of this protein in human disease.

Acknowledgement. This investigation was supported by grants T1-AM 5602 and AM 11248 from the U.S. Public Health Service. P.B. is a recipient of Research Career Development Award K4-AM 42582 from the U.S. Public Health Service. H.L. was a postdoctoral fellow under U.S. Public Health Service Training Grant AM 1000.

References

Bornstein, P.: Collagen: relatively invariant (helical) and variable (nonhelical) regions. Science 161, 592 (1968).

Bornstein, P., Kang, A.H.: Comparative biochemical studies of collagen. In: Chemistry and Molecular Biology of the Intercellular Matrix (E.A. Balazs, Ed.) Vol. 1, p. 99 - 107. New York: Academic Press 1970.

Bornstein, P., Kang, A., Piez, K.A.: The limited cleavage of native collagen with chymotrypsin, trypsin, and cyanogen bromide. Biochemistry 5, 3803 - 3812 (1966).

Bornstein, P., Nesse, R.: The comparative biochemistry of collagen: The structure of rabbit skin collagen and its relevance to immunochemical studies of collagen. Arch. Biochem. 138, 443 - 450 (1970).

Bornstein, P., Piez, K.A.: The nature of the intramolecular cross-links in collagen. The separation and characterization of peptides from the cross-link region of rat skin collagen. Biochemistry 5, 3460 - 3473 (1966).

Butler, W.T., Piez, K.A., Bornstein, P.: Isolation and characterization of the cyanogen bromide peptides from the α1 chain of rat skin collagen. Biochemistry 6, 3771 - 3780 (1967).

Click, E.M., Bornstein, P.: Isolation and characterization of the cyanogen bromide peptides from the α1 and α2 chains of human skin collagen. Biochemistry 9, 4699 - 4706 (1970).

Davison, P.F., Levine, L., Drake, M.P., Rubin, A., Bump, S.: The serologic specifity of tropocollagen telopeptides. J. exp. Med. 126, 331 - 376 (1967).

Fietzek, P.P., Piez, K.: Isolation and characterization of the cyanogen bromide peptides from the α2 chain of rat skin collagen. Biochemistry 8, 2129 - 2133 (1969).

Helmkamp, R.W., Contreras, M.A., Bale, W.F.: I^{131}-labeling of proteins by the iodine monochloride method. Int. J. appl. radiat. isotopes 18, 737 (1967).

Kang, A.H., Bornstein, P., Piez, K.A.: The amino acid sequence of peptides from the cross-linking region of rat skin collagen. Biochemistry 6, 788 - 795 (1967).

Kirrane, J.A., Glynn, L.E.: Immunology of collagen. In: International Review of Connective Tissue Research (D. Hall, Ed.) Vol. 4, p. 1. New York: Academic Press 1968.

Kirrane, J.A., Robertson, W. Van B.: The antigenicity of native and tyrosylated neutral - salt - soluble rat collagen. Immunology 14, 139 - 148 (1968).

Lindsley, H., Mannik, M., Bornstein, P.: Distribution of antigenic determinants in rat skin collagen. J. exp. Med. 133, 1309 - 1324 (1971).

Michaeli, D., Benjamini, E., Leung, D.Y.K., Martin, G.R.: Immunochemical studies on collagen. II. Antigenic differences between guinea pig skin collagen and gelatin. Immunochemistry 8, 1 - 6 (1971).

Michaeli, D., Martin, G., Kettman, J., Benjamini, E., Leung, D., Blatt, B.: Localization of antigenic determinants in the polypeptide chains of collagen. Science 166, 1522 - 1524 (1969).

Piez, K.A., Eigner, E.A., Lewis, M.S.: The chromatographic separation and amino acid composition of the subunits of several collagens. Biochemistry 2, 58 - 66 (1963).
Piez, K.A., Miller, E.J., Lane, J.M.: The order of the CNBr peptides from the α1 chain of collagen. Biochem. biophys. Res. Commun. 37, 801 - 805 (1969).
Pontz, B., Meigel, W., Rauterberg, J., Kuhn, K.: Localization of two species specific antigenic determinants on the peptide chains of calf skin collagen. Europ. J. Biochem. 16, 50 - 57 (1970).
Schmitt, F.O., Levine, L., Drake, M.P., Rubin, A.L., Pfahl, D., Davison, P.F.: The antigenicity of tropocollagen. Proc. nat. Acad. Sci. (Wash.) 51, 493 - 497 (1964).
Stark, M., Rauterberg, J., Kuhn, K.: Evidence for a non-helical region at the carboxyl terminus of the collagen molecule. FEBS Letters 13, 101 - 104 (1971).
Steffen, C., Timpl, R.: Immunogenicity and specifity of collagen. Z. Immunitätsforsch. 139, 455 - 467 (1970).
Steffen, C.,Timpl, R., Wolff, I.: Immunogenität und Spezifität von Kollagen. Z. Immunitätsforsch. 134, 91 - 100 (1967).
Steffen, C., Timpl, R., Wolff, I.: Immunogenicity and specifity of collagen. Immunology 15, 135 - 194 (1968).
Timpl, R., Fietzek, P., Furthmayr, H., Meigel, W., Kuhn, K.: Evidence for two antigenic determinants in the C-terminal region of rat skin collagen. FEBS Letters 9, 11 - 14 (1970).
Timpl, R., Wolff, I., Furthmayr, H., Steffen, C.: Immunogenicity and specifity of collagen. Immunology 15, 145 - 151 (1968).
Vuust, J., Lane, J.M., Fietzek, P.P., Miller, E.J., Piez, K.A.: The order from the CNBr peptides from the α2 chain of collagen. Biochem. biophys. Res. Commun. 38, 703 - 708 (1970).

Antigenicity of Soluble Collagen Fractions of Different Species

H. Struck

In extensive animal experiments we examined the stimulating effect of native soluble collagen on wound-healing, with visible success. Before using these preparations clinically their immunological behavior was tested also.

By means of a careful extraction (4°C) by 0.45 m NaCl solution the neutral-salt-soluble collagen was isolated from the skin of calves, pigs, rabbits and rats. The next extract with 1/15 m of citrat buffer (pH 3.7) produced the acid-soluble collagen. Both preparations were highly refined by dialysing four times.

Antisera were produced by immunizing rabbits which contain agglutinating as well as precipitating antibodies against the relevant collagen fractions. The precipitating antibodies appeared in the Agar-Gel-Double-Diffusion-Test only after 48 - 96^h (Fig. 1).

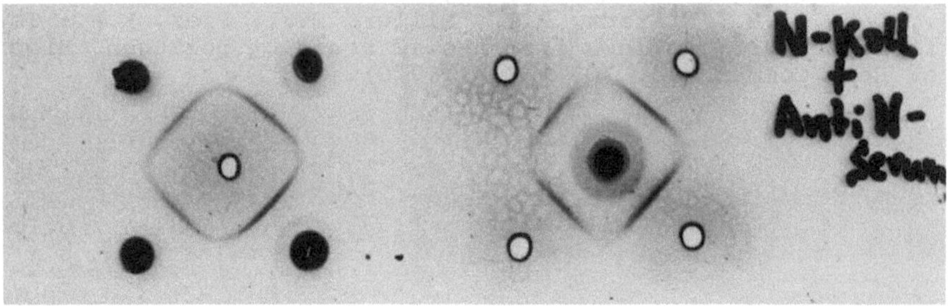

Fig. 1

The antisera developed against neutral-salt-soluble collagen from calf-skin are species-specific and specific of neutral-salt-soluble collagen. The antisera against acid-soluble collagen are only specific for one species. The antisera against rat-skin collagen behave analogously. We reported earlier on experiments indicating specification of these precipitates (Struck and Heinrich, 1967).

Our results were recently confirmed by Needleman and Stefanovic (1969). They produced precipitating antibodies against neutral-salt-soluble collagen as well as antibodies against the acid-soluble one, which was extracted out of tails of rats.

Timpl (1970) determined various agglutinating antibodies against acid-soluble collagen from calf-skin. Our precipitating antiserum against neutral-salt-soluble collagen corresponds to the A-specifity of Timpl, and the one against acid-soluble collagen to the S-specifity. Our earlier examinations showed that with calf-skin collagen, proteolytic enzymes did not influence those specifities (Struck et al., 1966).

The conditions are complicated by the precense of agglutinating as well as precipitating antibodies against soluble fractions of collagen. After burns, agglutinating antibodies against acid-soluble collagen can be found, but not those against the neutral-salt-soluble collagen. Precipitating antibodies were not observed (Struck and Engelhardt, 1970) (Fig. 2). This result showes a complexity of the serology of collagen. It is not surprising that these protein-molecules with the

animal no	weeks of experiments pb 0	1	2	3	4	5
1	ø	8	ø	8	8	8
2	ø	ø	ø	ø	ø	ø
3	ø	ø	ø	ø	ø	ø
4	ø	ø	8	16	16(2)	-
5	ø	4	2	ø	ø	ø
6	ø	16	2	2	-	2
7	ø	ø	16	16	16	16
8	ø	ø	4	4	-	4
9	ø	ø	ø	ø	2	2
10	ø	ø	4	16	16	16

Fig. 2. Titer of hemagglutinin against s-collagen after burns (rats)

great molecular weight of 320,000 (Piez, 1969) possess several serological determinants. The most specific antibodies against n- and s-collagen show that cross-linkage is also important for serological behavior.

In contrast to collagen from calves, rabbits and rats, we did not receive any antiserum from pig-skin collagen.

In the Disc-Electrophorese test we could find no difference between the collagen from pig-skin and that from calf-skin (Fig. 3).

During the active anaphylactic test, we did not see any reaction of the guinea pigs with calf-skin and pig-skin collagen. An epicutaneous test and intracutaneous test on 20 volunteers of both sexes using pig-skin collagen also showed no positive results.

In connection with our research on the stimulation of wound-healing, we would conclude that, from the serological point of view, there is no contraindication to clinical use of soluble collagen.

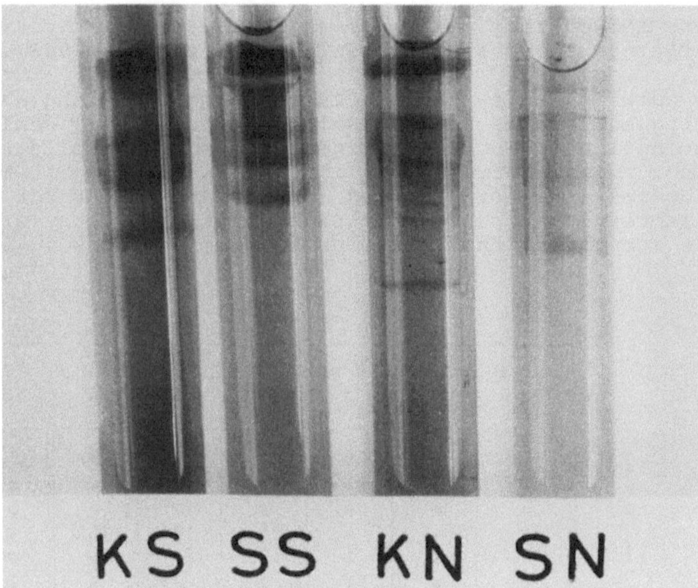

Fig. 3

References

Needleman, S.B., Stefanovic, N.: Production of antibody for rat tail tendon collagen. Experientia (Basel) 25, 1335 - 1337 (1969).
Piez, K.A.: Chemistry of collagen and elastin and biosynthesis of covalent cross-links. In: Thule Internat. Sympos., p. 15, Stockholm: Nordiska Bohandelns Forlag 1969.
Struck, H., Engelhardt,G.H.: Antikorperbildung bei Verbrennungen. Res. Exp. Med. 157, 186 - 188 (1972).
Struck, H., Heinrich, S.: Serologisches Untersuchungen zur Antigenitat und Spezifitat loslicher Kollagenfraktionen. Z. Immun.-Forsch. 133, 376, 466, 479 (1967).
Struck, H., Hernández-Richter, J., Heinrich, S.: Beeinflussung der Wundheilung durch Kollagen, serologische Aspekte und ihre Bedeutung fur die Klinik. Langenbecks Arch. Chir. 316, 577 - 580 (1966).
Timpl, R.: Immunochemistry of collagen. Z. klin. Chem. 9, 83 - 84 (1970).

V. Pathophysiology

Questions of Medicine to Molecular Biology of Connective Tissues

F. Hartmann

Connective tissues fullfill their biological functions by technical properties of the intercellular matrix: stability, plasticity, elasticity, swelling, viscosity, adherence, transparency, diffusibility, ion and water binding capacity, shrinkage at warm temperature and in acids, etc. Of great biological significance is the ability of connective tissues to adapt to changing conditions and demands at any specific point within the organism. This indicates a great variability of the chemical structure of the macromolecular architecture of the matrix.

The physician can gain insight into the biology and pathology of connective tissues from a discussion with engineers and high polymer physicists. This holds true for properties of bone, joints, intervertebral discs, tendons, heart valves, synovial fluid and nucleus pulposus. The cementing substance located between the fibrillar elements of skin, blood vessels and capsules of joints is a tixotropic gel. Its viscosity varies according to mechanical conditions. Metabolic-enzymatic synthesis and degradation of connective tissue cells regulate by mechanical (pressure, traction) chemical (pH, ionic strength) and thermal influences the status of connective tissues designated for certain local (transparency of the lens, permeability of basement membranes etc.) or temporary (callus, wound healing) functions.

Collagen passes several zones of a critical temperature, when gradually heated from 20 - 65°C, in which labile phases of soluble and insoluble collagen can arise. Two critical phases are within the limits of body temperature (TM, TG). Regenerating connective tissue is of inferior quality in respect to the status it is supposed to assume and the function it is expected to perform.

The physician is confronted with two problems in therapeutics:

1. By application of desmotropic pharmaca he is aiming at an abbreviation of the critical phase. The process of fibril formation and of a texture of fibrils-glycoproteins-proteoglycans is supposed to be protected. This we were able to demonstrate in growing tendons treated with phenyl butazones and corticosteroids (x-ray small-angle diagrams, thermo elasticity). Another approach would be the stabilization of soluble collagen by means of a controllable tanning process. Progress was reached in acceleration of bone and wound healing by therapeutic application of collagen and sulfated glycosaminoglycans.

2. The second therapeutic problem is the possibility that the regeneration of damaged connective tissue will lead to scar tissue. The following examples may illustrate this point: shrunken heart valves, shrunken liver, fibrotic lung, turbid lens, crippled joints, hardened and shortened skin in scleroderma. The control-system of connective tissues has failed in certain, still unknown, regulatory mechanisms.

Medical treatment can only become effective with an understanding of the regulatory systems, which control the following connective tissue components: thickness, density and arrangement of fibrils, correlation of soluble to insoluble collagen, quantity of glycoproteins and proteoglycans, the specific proteoglycan pattern of a connective tissue, the grade of sulfation, the nature of conformation of macromolecules, number and kind of cross-links (Hartmann, 1972). So far it can only be . suggested that local mechanical and metabolic conditions of paranchymal adjacent tissues may play a role in the regulation. It should e.g. be known whether uncontrolled synthesis or an enzymatic defect in the constant degradation of connective tissue constituents causes scar tissue formation.

Decomposition of cartilage in degenerative joint diseases differs from destruction of cartilage in inflammation by the fact that in inflammation fibrils and cementing substance will be decomposed at the same time. In osteoarthrosis the quantity and composition of proteoglycans is changed in the beginning. The fibrils become dismanteld and slycken. They become ground by pressure on the joints. They are no longer able to soften the impact of pressure loads, which are then transmitted directly to the cartilage-bone-capsule transition zone. A similar mechanism is found in the destruction of the intervertebral disc.

Basement membranes are of vital importance for connective tissues. They comprise approximately 16% of the body weight, expressed in extracellular water. Basement membranes form a framework, assure mobility of blood vessels, store water, salt, proteins and degradation products, and regulate the transport of nutritive substances into the cell and of degradation products out of the cell (transit route) by means of a molecular sieve effect. Basement membranes vary according to local conditions in the composition of proteoglycans and collagens. Most data available on basement membranes are those of glomerula of the kidney. From experimental results one may conclude that basement membranes may play a role in immunobiology as e.g. absorption and binding of antigen-antibody complexes. In basement membranes in the kidney, it can be demonstrated that with blood circulation antigen-antibody-complexes will be deposited, and/or that antibodies of basement membrane antigens will react with the basement membrane. For basement membranes it is also a fact that at the beginning of a disturbance (inflammation, malnutrition, stress, metabolic irregularity) they will often swell and depolymerize. The distance for diffusion is thus prolonged. Afterwards membranes do not regenerate to the same physiological status, but become more dense and sclerotic. The molecular sieve closes its pores, and transmission changes for the worse. Diffusion and convection of substances into and out of the cell will be hindered and easily precipitable material will be precipitated in the condensed connective tissues of basement membranes. Of clinical importance are alterations in diabetes. In recent papers was reported, besides augmentation of proteoglycans, an augmentation of the hydroxylysin binding and of a'diabetic fibrillosis'.

The pathology and therapy of connective tissues should consider the heterogeneity of their building material at different parts of the body and the function they must fullfill in their composition at a specific location.

References

Hartmann, F.: Ärztliche Fragen an eine Molekularbiologie der Bindegewebe. Z. Rheumaf. 31, 42 - 67 (1972).

Structural Glycoproteins of Connective Tissue: Their Role in Morphogenesis and Immunopathology

L. Robert, and B. Robert

The terminology "structural glycoproteins" refers to glycoproteins extracted from connective tissue by a described procedure (Robert and Dische, 1963; B. Robert et al., 1963; L. Robert et al., 1964b; A.M. Robert et al., 1970). By extension, it may refer to glycoproteins present in tissues, distinct from serum glycoproteins. This distinction between circulating blood serum glycoproteins and tissue glycoproteins refers not only to their different localization but also to their distinct origin. Most if not all of the blood plasma glycoproteins are believed to be synthetized in the liver (except for immunoglobulins): see Schultze and Heremans, 1966; Simkin, 1970). Structural glycoproteins are synthetized in connective tissue by fibroblasts or by other differentiated mesenchymal cells, such as chondrocytes or smooth muscle cells in the vessel wall (L. Robert and Parlebas, 1965a, b, c; L. Robert, 1970). Besides their localization and origin the role they play in the organism is also different from that generally attributed to circulating blood serum glycoproteins. Some serum glycoproteins are believed to function was protease inhibitors, as hemoglobin complexing and sparing agents, or as tissue nutrients (see Jayle et al., 1970). Glycoproteins localized in tissues are supposed to play a structural role during the biosynthesis of the intercellular matrix. We propose to give here a brief summary of our studies and their interpretation in the light of former work from our and other laboratories. Our main objective was to reach a better understanding of the morphogenetic processes at a molecular level as well as of their pathological disturbances.

1. Historical Background

The presence in connective tissue of proteins distinct from collagen and elastin has been known for some time. Early literature on this subject has been given in a former monography (B. Robert, 1959). Some of these *non-collagen proteins* were assumed to be the protein-partners of acid mucopolysaccharides in chondromucoproteins. Others were believed to be present in the interfibrillar substance or ground substance and sometimes called "ground substance proteins". Very frequently such proteins were discarded as "impurities" during the purification of collagen or elastin.

Independently from these studies on non-collagen-proteins, some investigators demonstrated the presence of "*neutral mucopolysaccharides*" in

Abbreviations: TCA: trichloracetic acid
SGP: structural glycoproteins
ME : 2-mercaptoethanol

connective tissue (Glegg et al., 1953, 1954; Dische, 1958; for a historical review see B. Robert, 1959). Leblond and coworkers demonstrated hexose and hexosamine containing "polysaccharides" in reticulin-rich tissues such as the lymph node and the spleen (Glegg et al., 1953, 1954).

It was, however, Dische who investigated in great detail these "neutral mucopolysaccharides", demonstrated their ubiquitous presence in many different connective tissues and showed their great diversity or "microheterogeneity" as far as their carbohydrate composition is concerned (Dische et al., 1958, 1965).

It was during a stay in Dische's laboratory in 1960 that this subject was taken up by us again. Through a fortuitous observation the existence of "non-collagen-proteins" and of "neutral polysaccharides" could be joined together leading to the identification and concept of structural glycoproteins" (L. Robert and Dische, 1963; L. Robert et al., 1964b; A.M. Robert et al., 1970). The above mentioned observation was the following, (see Fig. 1).

In order to study the sugar components of corneal collagen, calf corneas were first extracted with a 1M $CaCl_2$ solution to remove all diffusible molecules, leaving behind the insoluble polymeric stroma. This was believed to be composed only of polymeric collagen. It was submitted to a mild hydrolysis in a dilute trichloracetic acid (TCA)solution shown by Dische to "solubilize" polymeric collagen. However, even after three consecutive extractions in 0.15M TCA at $90^{\circ}C$ for a total of one hour an appreciable insoluble residue remained (see Fig. 1d). All the hydroxyproline of the insoluble polymeric stroma was found in the TCA extract; the TCA-insoluble residue had none. It did contain, however, a protein with attached carbohydrate (L. Robert and Dische, 1963; Moczar et al., 1967, 1969).

The analysis of the TCA-extracted polymeric collagen revealed a "theoretical" OH-proline content (~12%) together with a relatively high carbohydrate content (~5%). All this carbohydrate could be accounted for as approximately equimolar amounts of galactose and glucose (Dische and Robert, 1962; Robert and Dische, 1963). The TCA-insoluble residue could not be solubilized in any of the aqueous and organic solvents which were essayed and could be dissolved only in 8M urea in the presence of 0.1M mercaptoethanol (ME) (B. Robert et al., 1963). This solution was then dialyzed; after dialysis the protein(s) remained in solution for several days in the cold but eventually reprecipitated. These experiments suggested the presence of disulfur linkages in the glycoprotein, confirmed later by the determination by equilibrium-ultracentrifugation of the melocular weight of the SGP of calf cornea in 4M urea, before (74,000) and after reduction (34,000) in 0.1M ME (A.M. Robert et al., 1970). The rapid insolubilization after dialysis was attributed to an aggregation through SH-SS bridge formation. The precipitate formed on standing could again be dissolved in 8M urea-0.1M mercapthoethanol. The glycoprotein present in this urea-extract of corneal stroma was first designated "Keratoglycosaminoglycan" or KGAG (B. Robert et al., 1963; L. Robert et al., 1964b). Later this denomination was replaced by *Structural Glycoproteins* of cornea or SGP-cornea (A.M.Robert et al., 1970). This preparation could be further purified by alcohol precipitation and gel-filtration and shown to contain, besides galactose and glucose, mannose, glucosamine, fucose, sialic acid and about 1% of ester sulfate (L. Robert and Dische, 1963; E. Moczar et al., 1967; M. Moczar and Robert, 1970; E. Moczar et al., 1969, 1970). The absence of hydroxyproline, the presence of all these carbohydrate components, together with its solubility characteristics, suggested from the beginning that this glycoprotein was a distinct molecular entity, different from collagen or proteoglycans previously

Fig. 1a - d. "Chemical dissection" of calf cornea by the method previously described (see Table 1 and L. Robert and Dische, 1963; B. Robert et al., 1963):
a) two calf corneas soaked in the CTC-buffer ($CaCl_2$ 1M tris-citrate pH 8.0);
b) after homogenization in the same buffer;
c) residue after extraction with CTC-buffer (polymeric stroma);
d) residue after extraction of polymeric stroma (see c) with TCA 2.7% at 90°C: polymeric collagen is in the supernate, structural glycoproteins remain insoluble

demonstrated in cornea. This contention was further confirmed by its immunochemical properties (B. Robert et al., 1963; L. Robert et al., 1964a, b, 1970b).

Table 1. Flow sheet of the preparation of structural glycoproteins (left side) and of glycopeptides (right side) from connective tissue

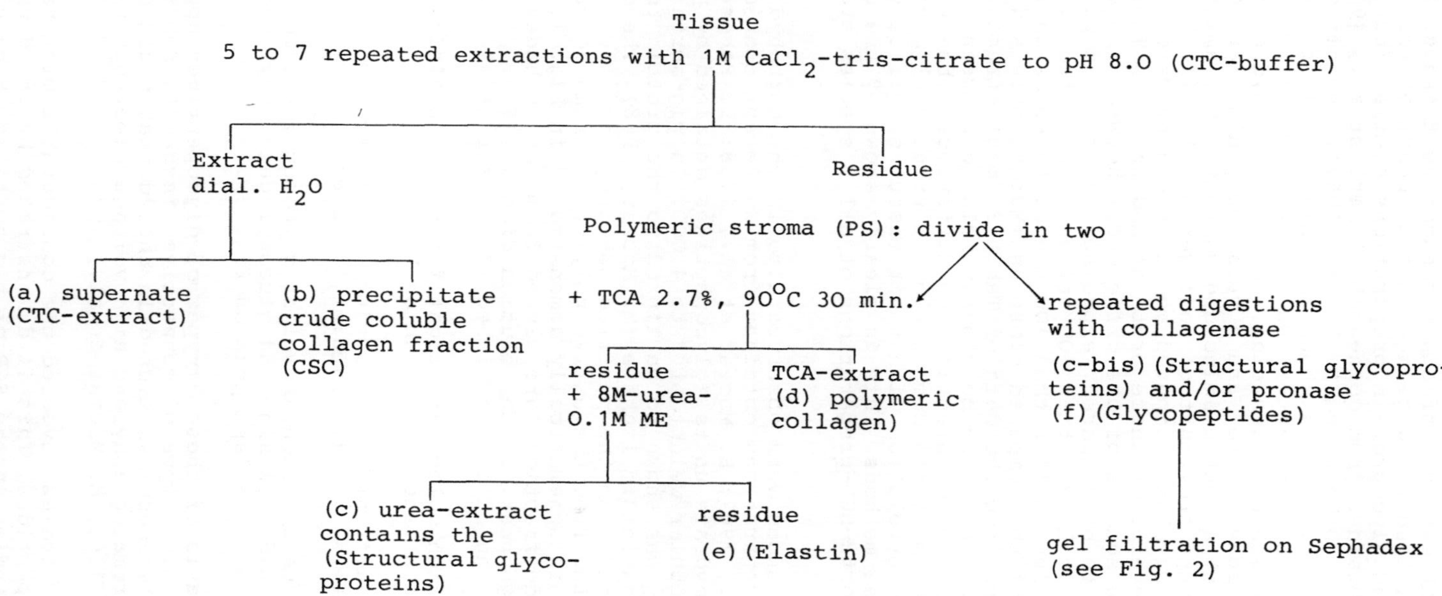

The same extraction procedure (see Table 1) was then used to prepare urea-ME extracts of other connective tissues. All tissues studied (skin, tendon, cartilage, aorta, heart valves, synovial membrane, etc.) yielded glycoprotein preparations similar (but not identical) to the one obtained from cornea (L. Robert et al., 1968; L. Robert and Comte, 1968; L. Robert and B. Robert, 1969). The aminoacid composition of several of them is shown in Table 2. All of these glycoprotein preparations are characterized by a high dicarboxylic aminoacid content, by the presence of cysteine, sometimes of hydroxylysine (in the absence of hydroxyproline) and a relatively high (as compared to collagen or elastin) aromatic aminoacid content. This aminoacid pattern is quite different from that of collagen or elastin. This is demonstrated by the SDQ values computed as described by Marchalonis and Weltman (1971), for several of these proteins and shown in Table 3. SDQ is the sum of the squares of the differences between the aminoacids in two proteins to be compared, expressed as residues per 100 residues. The SDQ value is larger than 100 units if unrelated proteins are compared. An SDQ value smaller than 100 can be taken as an indication of the relatedness of the compared proteins (Marchalonis and Weltman, 1971). As can be seen from the data of Table 3, most structural glycoprotein preparations appear to be closely related between themselves but unrelated to collagen or elastin. The differences observed in the relatedness of the aminoacid composition of structural glycoproteins appear to be in correlation with their tissue specificity observed by immunological methods. This is clearly shown by the close relatedness of The aorta-SGP-preparations of all 5 species studied (Robert et al., 1972).

Differences were observed also in the carbohydrate composition of the SGP-preparations obtained from different connective tissues (Moczar et al., 1967; E. Moczar et al., 1968; L. Robert et al., 1968). The carbohydrate units (glycopeptides obtained by pronase digestion) have a molecular weight of 2 to 4,000. In pig-aorta-SGP the carbohydrate moiety was shown to be linked to the protein through an aspartamido-glycosaminidic linkage (M. Moczar, 1968). The carbohydrate units isolated from the purified SGP-preparation were compared to those isolated from the native polymeric stroma according to a procedure which is schematically demonstrated in Fig. 1. Both types of glycopeptide fractions are eluted from the Sephadex columns after the breakthrough peak. On the elution diagram (Fig. 2) three types of glycopeptides can be distinguished (E. Moczar et al., 1968, 1969).

1. A high molecular weight fraction containing the acidic polysaccharides (see first peak on Fig. 2).

2. An intermediary fraction containing the glycopeptides derived from the SGP-s (see intermediary peak on Fig. 2).

3. A small molecular weight glycopeptide fraction containing the galactosyl-hydroxylysine and glucosyl-galactosyl hydroxylysine derived from collagen (see last peak on Fig. 2).

The ratio of these 3 types of glycopeptides appeared to be specific for a given type of connective tissue (L. Robert and B. Robert, 1970). The glycopeptides eluted in the 2d peak could be further fractionated by chromatography and high voltage electrophoresis (E. Moczar et al., 1968, 1969; M. Moczar et al., 1970).

These studies revealed the considerable microheterogeneity of the glycopeptides. This is demonstrated by the data of Table 4 taken from our studies on the SGP of pig aorta (M. Moczar et al., 1970). The seven

Table 2. Aminoacid composition of several structural glycoprotein preparations (from L. Robert and Comte, 1968 and M. Moczar and Robert, 1970) corrected for OH-pro content and given as residues for 100 residues. The composition of calf skin collagen (Grassman et al., 1960) and calf ligamentum nuchae elastin (Comte and L. Robert, 1968) is also seen

	SGP pig aorta	SGP calf aorta	SGP calf skin	SGP calf-ear cartilage	SGP calf tendon	SGP calf cornea	Collagen calf skin	Elastin calf.lig. nuchae
Aspartic acid	9.5	9.0	10.7	9.1	8.8	10.2	4.6	0.8
Glutamic acid	10.7	13.0	10.6	10.2	10.0	11.3	7.4	1.6
Threonine	5.8	6.3	4.9	5.4	4.1	5.4	1.7	0.8
Serine	6.0	6.0	5.7	5.6	5.5	5.9	2.2	0.8
Proline	5.6	5.0	9.2	4.3	9.6	5.6	13.0	11.6
OH-proline	-	-	-	-	-	-	8.0	-
Glycine	9.5	8.0	8.2	12.5	10.0	10.5	32.5	34.0
Alanine	8.7	9.1	7.3	10.1	7.5	7.0	10.6	23.5
Valine	8.0	6.6	6.7	6.4	6.2	8.6	2.1	14.3
Cysteine	2.8	1.5	1.0	0.8	2.6	1.5	0	0
Methionine	0.1	1.0	1.7	1.1	1.2	1.1	4.3	0
Iso-leucine	4.7	5.4	4.8	4.3	4.1	4.7	1.1	2.5
Leucine	8.2	9.6	9.2	9.5	7.3	9.6	2.1	6.3
Tyrosine	3.0	1.7	2.4	2.1	2.3	1.7	0.3	0.6

Table 3. Degree of homology of the aminoacid composition of structural glycoprotein, elastin, collagen and some other cell-membrane glycoproteins. SDQ values were calculated as described by Marchalonis and Weltman (1971). An SDQ value below 100 is taken as an indication of relatively close relatedness of the aminoacid compositions. The first column of the Table lists the proteins which were taken as reference and the other columns list the proteins which were compared to the one listed in the first column. For the details of aminoacid analysis see L. Robert and Comte (1968), Robert et al. (1972) and Table 2

	SDQ VALUE *compared proteins*								
Reference protein	calf skin collagen	calf lig. elastin	SGP pig aorta	SGP calf aorta	SGP sheep aorta	SGP chick aorta	SGP horse aorta	SGP calf skin	SGP calf-ear cartilage
SGP-pig aorta	969.3	1,183.4	0	17.9	29.3	21.5	78.2	30.4	26.2
SGP-calf aorta	1,075.5	1,333.5	17.9	0				33.9	33.8
SGP-calf skin	930.7	1,295.8	30.4	33.9				0	56.8
SGP-calf tendon	857.7	1,189.3	37.3	55.6				18.8	56.3
SGP-calf cornea	957.6	1,203.6	14.1	23.1				27.7	30.1

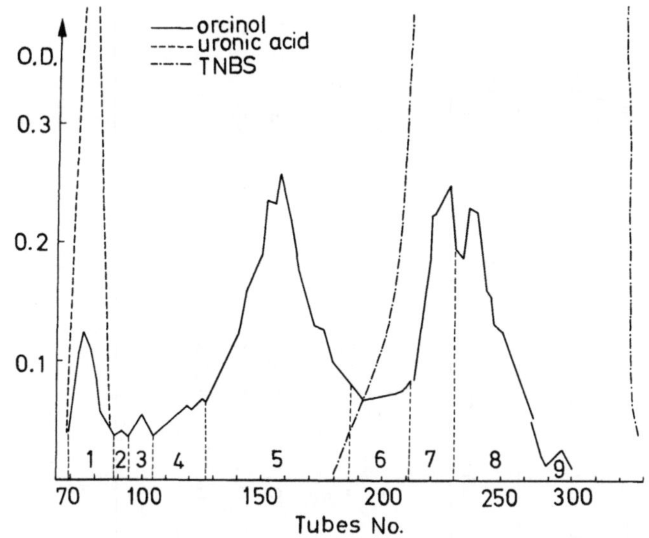

Fig. 2. A typical glycopeptide pattern of the polymeric stroma (PS) of a connective tissue (pig aorta), prepared as described in Fig. 1. A collagenase-pronase digest of aorta-PS was loaded on Sephadex G-50-column and eluted with 0.1N acetic acid. First peak (fraction No. 1) high molecular weight polysaccharide chains from glycosamino-glycuronanes (containing uronic acid and SO_4); second peak (fraction No.4-5): medium molecular weight glycopeptides derived from structural glycoproteins (containing Gal, Glc, Man, GlcNaC, GalNaCl, Fuc, Sial); last peak (fraction No. 7-8): low molecular weight glycopeptides derived from collagen (hydroxylysyl-galactoside and glucosidogalactoside), (from M. Moczar et al., 1970)

glycopeptides separated by high voltage electrophoresis have all different sialic acid/fucose ratios which explains their different migration rates. The similarity of the aminoacid composition, together with the differences in carbohydrate composition and the considerable heterogeneity of the carbohydrate chains, tend to accredit the suggestion that the carbohydrate moiety of structural glycoproteins may play a role in the determination of their tissue and organ specificity.

2. Metabolic Activity

When connective tissues such as cornea, sclera, tendon or aorta were incubated in vitro with labelled aminoacids or sugars, appreciable incorporation has been obtained in the structural glycoproteins isolated from the insoluble polymeric stroma (L. Robert and Parlebas, 1965a, b and c). The specific activity of the urea-extractable glycoprotein fraction was significantly higher than that of the polymeric collagen or elastin. However, when soluble collagen was compared to the glycoprotein fraction accompanying it in the so called "crude soluble collagen" fraction (CSC-fraction), their specific activity was of the same magnitude when either proline or glycine were used as tracers. This observation suggested a similar rate of synthesis for these two proteins: soluble or extractable collagen and the structural

Table 4. Microheterogeneity of the glycopeptides obtained by pronase digestion from structural glycoprotein preparation of pig aorta (from M. Moczar et al., 1970).
Sugar composition of the electrophoretically separated subfractions of the glycopeptide fraction (peak No. 5, Fig. 2) obtained from the insoluble stroma of the media of porcine aorta hydrolyzed successively by collagenase and by pronase. The yields are given in mg of lyophilized material obtained from 42 mg starting material

Fraction number[a]	Yield mg	Molar ratio of sugar (Gal = 1)							Hex/hexosamine ratio	Sial/fucose ratio
		Gal	Glc	Man	Fuc	Sial	GlcN	GalN		
1	2.20	1	1	1	-	-	-	-		
2	3.98	1	0.6	1.1	0.09	0.2	1.56	-	1.75	2.2
3	3.82	1	0.1	0.8	0.1	0.3	1.25	-	1.52	3.0
4	5.61	1	0.07	1	0.1	0.38	1.43	-	1.5	3.8
5	1.31	1	0.2	1.7	0.15	0.5	1.56	0.26	1.69	3.34
6	4.66	1	0.05	1.2	0.13	0.6	1.39	0.07	1.54	4.62
7	3.92	1	0.6	1	0.09	0.6	1.15	0.23	1.88	6.7

[a] see Fig. 1 of M. Moczar, E. Moczar, L. Robert (1970).

glycoprotein fraction "attached" to it. A glycoprotein fraction could be precipitated from the dialyzed supernate of the CTC-extract (see Table 1), with a specific immunserum to SGP extracted from the polymeric stroma (urea-extract, see Table 1) (L. Robert et al., 1964b). When this experiment was done on a water-soluble fraction of a calf cornea incubated previously with ^{14}C-glucose, the immune-precipitate had a very high specific activity. The cross-reacting glycoprotein was eluted from a DEAE-column together with one of the major proteoglycan peaks (L. Robert and Parlebas, 1965c). These results were taken as an indication for the existence of three physically and metabolically distinct forms of SGP-s (L. Robert et al., 1964b):

a) a "free" form, present in the collagen-free supernatant of dialyzed $CaCl_2$-extracts.

b) a form "combined" with the extractable (soluble) collagen fraction and

c) a form "combined" with the fibrous proteins of the insoluble polymeric stroma.

All three forms cross-react immunologically. Their specific activity in radioactive labelling experiments decreased in the order a b c, with the form b) being about equally labelled as extractable (or soluble) collagen.

3. Morphology of Structural Glycoproteins

Recent experiments provided evidence in favor of the identity of "micro-fibrillar" structures described by several authors in connective tissue and considered by Low (1962) as an ubiquitous component, with the structural glycoproteins. This evidence came from the biochemical and morphological studies of the microfibrillar components of elastic tissue (Ross and Bornstein, 1970; L. Robert et al., 1970a) and also from the biochemical and morphological studies of pathological corneas (keratoconus and dystrophies) (Pouliquen et al., 1970; L. Robert et al., 1970c).

Besides the amorphous structures, one can observe in purified elastin preparations Uranyl-acetate and Pb-stainable "microfibrils" inside and around the elastic lamellae. The microfibrils appear to be involved in the morphogenesis of elastic fibers and lamellae.

Fig. 3 shows the mechanism of formation of elastic fibers through the condensation of tropo-elastin molecules on the SGP-fibrils (L. Robert et al., 1970a; B. Robert et al., 1971).

The microfibrillar aggregates appear first during histogenesis (Ross and Bornstein, 1970). As these structural glycoprotein-fibrils are rich in dicarboxylic aminoacids (B. Robert et al., 1971) they can attach the positively charged tropoelastin molecules rich in lysine residues and position them for the action of the lysinoxydase. The structural glycoprotein-microfibrils would thus play a role in the positioning and orienting of the tropoelastin aggregates and render much more efficient the cross-linking process (L. Robert et al., 1970a; B. Robert et al., 1971).

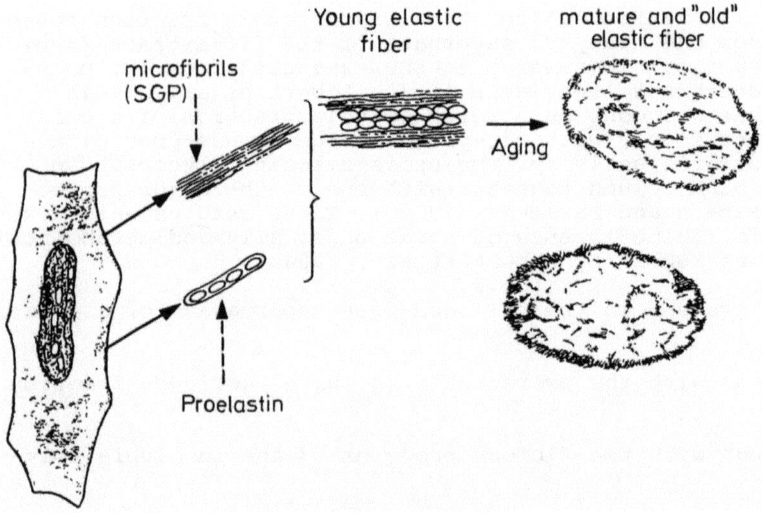

Fig. 3. Schematic drawing showing the successive steps of fibrillogenesis during elastic fiber formation. The cell first extrudes the microfibrils or structural glycoproteins which form the scaffolding onto which the tropoelastin molecules aggregate to be cross-linked. The ratio of SGP (microfibril) to tropoelastin changes during fibrillogenesis and with age (from L. Robert et al., 1970a)

Similar microfibrillar components could also be demonstrated in pathological corneas (Pouliquen et al., 1970). Keratoconus corneas show an increase in glycoprotein components (L. Robert et al., 1970c), and are rich in microfibrillar aggregates (Pouliquen et al., 1970). The overproduction of the structural glycoprotein-microfibrils could result, in this case, from a genetically determined shift of the rate of their production with respect to collagen synthesis. We could provide evidence for the tissue specific regulation of the relative rates of synthesis of SGP-s and collagen (L. Robert and Parlebas, 1965b). Pathological alteration with an abnormal morphology of connective tissue may thus result from a shift of their relative rate of production.

4. Correlation between Cell Membrane Glycoproteins and Structural Glycoproteins

We demonstrated recently a pronounced homology between the aminoacid composition of SGP-preparations from several connective tissues and glycoproteins isolated from cell membranes ("transplantation antigens") (Robert et al., 1972).

These observations lend support to our hypothesis attributing the origin of SGP-s to the cell membrane glycoproteins. This relationship also explains the cornea graft rejections obtained in rabbits pre-immunized with donor-cornea-SGP-preparation (L. Robert et al., 1970b).

Also worth noting is the similarity of the aminoacid and carbohydrate composition of structural glycoproteins and glycoproteins isolated from basement membrane (Kefalides, 1970) as well as to the Cl_q component of complement (Yonemasu et al., 1971).

5. Role of Structural Glycoproteins in Morphogenesis

Differentiation of connective tissues is characterized by the oriented synthesis (*vectorial synthesis*) of fibrillar macromolecules. These macromolecules form a fibrous pattern with organ - or tissue - specific morphological parameters, such as fiber density, fiber diameter, orientation and disposition of the fibers, interfibrillar distance, ratio of collagen to elastin fibers. These morphological parameters can be correlated to chemical parameters such as the quality and quantity of acidic mucopolysaccharides or glycosaminoglycans, the ratio of structural glycoproteins to collagen or elastin, the ratio of extractable (soluble) components to insoluble polymeric components (degree of cross-linking). These quantitative, chemical parameters could, in their turn, be correlated with the rate of incorporation of labelled aminoacids or sugars in the macromolecular components such as collagen and structural glycoproteins reflecting their rate of biosynthesis (L. Robert and Parlebas, 1965a, b; L. Robert and B. Robert, 1967, 1970). Although a definitive, clearcut interpretation of these results is not yet possible, they inspire several suggestions:

a) A peptide (glycopeptide) component of the cell membrane could "accompany" the freshly synthetized collagen molecules during their extrusion from the cell. An oriented invagination of the cell wall could explain the oriented synthesis of collagen as well as the presence of cell wall glycopeptides in the polymeric collagen and elastin preparations. A similar mechanism was recently suggested for the participation of structural glycoproteins (or microfibrils) in the biosynthesis and deposition of elastic fibers or lamellae (see chapter 3).
b) A peptide (glycopeptide) could be synthetized along with the chains of collagen, being coded by the same structural gene. These peptides would be cleaved later by a specific peptidase, after the folding of a triple helix. This hypothesis is supported by the recent findings of Layman et al. (1971). Such a mechanism would explain the formation of the disulfur bridges in the SGP-preparations. The SS-bridges would stabilize the three α-chains of tropocollagen during the folding of the triple helix.
c) Both mechanisms could coexist, suggesting the presence of several types of glycopeptides associated with fibrous proteins. The varying ratio of "microfibrils" to elastin during elastogenesis, as well as the varying ratio of SGP to collagen in several connective tissues, are in favor of the first hypothesis.

The second hypothesis, suggesting a mechanism for the biosynthesis of collagen similar to that of insulin or zymogens requires the presence of a constant ratio of peptide (glycopeptides) to freshly synthetized tropocollagen. It implies also that the trimerization of the freshly synthesized collagen α-chains has to be preceeded by disulfur bridge formation before the assemblage of the triple helix. Such a mechanism would also explain the inhibitory effect of SGP-s on the lateral aggregation of collagen fibrils as we suggested before (L. Robert and B. Robert, 1967, 1970). Further experiments are required before these hypothetical mechanisms can be definitely accepted.

6. Immunopathological Importance of Structural Glycoproteins

In spite of the denaturing and hydrolytic methods used for its preparation (heating in 0.15M TCA, 8M urea-mercaptoethanol extraction) structural glycoproteins were shown to be quite strong immunogens (B. Robert et al., 1963; L. Robert et al., 1964a, b, 1970b). The purified preparations elicited the formation of precipitating antibodies in rabbits and chickens. Delayed hypersensitivity could be induced in rats and guinea pigs as well as a strongly accelerated corneal graft rejection (L. Robert et al., 1966; Payrau, 1968; L. Robert et al., 1965, 1970b; L. Robert and B. Robert, 1968). The serum of a rabbit immunized with SGP-preparations from cornea when injected in the skin of another rabbit produced a strong infiltrated, papulous skin reaction. Some human sera were shown to contain antibodies to structural glycoproteins as well as to elastin (Crouzet and Robert, 1970).

As SGP-s are significantly more immunogenic than collagen, most of the anti-collagen sera tested could be shown to contain anti-SGP antibodies. Traces of SGP are difficult to eliminate from tropocollagen preparations for reasons which are discussed above (see chapter 5). For the same reason anti-collagen antibody titers obtained with even highly purified tropocollagen preparations cannot be accepted as specific for collagen until the absence of SGP-s in the tropocollagen preparations used for the titration is definitively proven. The use of purified α-chains obtained from tropocollagen preparations appears to be the best method to overcome this difficulty. Another precaution could be the absorption of the sera on purified SGP-preparation before titration with collagen. As structural glycoproteins posses tissue and species specificity they might be involved in graft rejection phenomena. This claim was substantiated by producing accelerated rejection of lamellar corneal grafts in rabbits sensitised with corneal SGP-s (L. Robert et al., 1965, 1970b). It was claimed that the soluble, diffusible form of SGP (see form a and b in chapter 2) could act as a sensitising antigen diffusing from the transplant into the recipient organism. The "bound" form (see form c chapter 2) could then act as a target antigen for the humoral antibodies as well as for the competent cells and thus localise the immune-reaction. This claim is substantiated by the tissue specificity of the SGP-s. Similar or identical mechanisms might be involved in any organ transplant containing connective tissue, such as skin, cartilage, bone, heart valve, vessel wall or other organs.

As an extension of the above mechanism, cross sensitisation between SGP and related glycoproteins of bacterial or other origin might be responsible for some of the autoimmune mechanisms such as observed in several connective tissue diseases (Zabrisky, 1967; Kunkel, 1970). Cross-reactions were observed between SGP-s of heart valve or aorta and streptococcus A polysaccharide (Goldstein et al., 1967) and streptococcus A cell membrane (L. Robert and Zabrisky, 1971). The stuctural glycoproteins of heart muscles, heart valves and synovial membranes might thus be the main target antigens involved in cross sensitisation with streptococcus A.

Rabbits immunised for prolonged periods with SGP of aorta were shown to develop frequently typical sclerotic lesions of their aorta similar to those obtained in animals immunised with elastin (A.M. Robert et al., 1971; L. Robert et al., 1970a). Electron microscopic observation showed that calcium deposition often occurs on the "microfibrillar" aggregates near or on elastic lamellae (L. Robert et al., 1971). Their high dicarboxylic acid content renders these structures apt to fix calcium by coordination to COO^- groups.

Acknowledgements. The experiments described or cited were carried out from 1960 to 1962 in collaboration with Dr. Z. Dische, Department of Biochemistry and Ophtalmology Research, Coll. Phys. Surg. Columbia University New York; from 1962 to 1967 in the Biochemistry Laboratory of the Institut of Immunobiology de l'INSERM, in collaboration with Professor B.N. Halpern and from 1967 at the Laboratoire de Biochimie de Tissue Conjonctif (Equipe de Recherche du CNRS No. 53), Fac. Méd., Univ. Paris-Val de Marne, 6 r du Gl Sarrail, 94000 Créteil.

The skillful and devoted collaboration of the following persons is thankfully acknowledged:
Dr. A.M. Robert, Dr. E. Moczar, Dr. M. Moczar, Dr. J. Parlebas, Mrs. S. Junqua, Miss F. Chavarot, Mr. and Mrs. J.C. Derouette, Miss D. Brechemier, Mr. J. Ouzilou, Mr. J. Fayolle and Mr. W. Hornebeck.

Grants from the CNRS (ER No. 53), Délégation Générale à la Recherche Scientifique et Technique (No. 68 01 476) from the Direction des Recherches et des Moyens d'Essay (70 34 220) from the Fondation pour la Recherche Médicale Francaise and from NIH (Nat. Eye Inst., from 1964 to 1967) are thankfully acknowledged.

This manuscript was completed in 1970 and slightly revised in 1971.

References

Comte, P., Robert, L.: Composition en aminoacides de la kappa-élastine. Bull. Soc. Chim. biol. (Paris) 50, 1349 (1968).

Crouzet, J., Camus, J.P., Robert, L.: Antigénicité du tissu conjonctif. Considérations théoriques et recherches cliniques en rhumatologie. Presse med. 78, 1185 (1970).

Dische, Z.: An exercise in "Organismic" biochemistry by an outsider. Exp. Eye Res. 4, 265 - 282 (1965).

Dische, Z., Danilczenko, A., Zelmenis, G.: The neutral heteropolysaccharides in connective tissue. In: Chemistry and Biology of Mucopolysaccharides. Ciba Found. Symp. (Eds. G.E.W. Wolstenholme, Maeve O'Connor), p. 116 - 139. London: Chruchill 1958.

Dische, Z., Robert, L.: Protein linked carbohydrate of corneal stroma. Fed. Proc. 21, 172 (1962).

Glegg, R.E., Eidinger, D., Leblond, C.P.: Some carbohydrate components of reticular fibers. Science 118, 614 - 616 (1953).

Glegg, R.E., Eidinger, E., Leblond, C.P.: Presence of carbohydrates, distinct from acid mucopolysaccharides in connective tissue. Science 120, 839 - 840 (1954).

Goldstein, L., Halpern, B., Robert, L.: Immunological relationship between streptococcus A polysaccharide and the structural glycoproteins if heart valve. Nature 213, 5071 (1967).

Grassmann, A.W., Hannig, K., Schleyer, M.: Zur Aminosäuresequenz des Kollagens, II. Hoppe-Seylers Z. physiol. Chem. 322, 71 (1960).

Jayle, M.F., Engler, R., Degrelle, H.: Les sialoglycoprotéines plasmatiques dans le syndrome d'agression et la croissance tissulaire. Expos. ann. Biochim. med. 30, 149 - 172 (1970).

Kefalides, N.A.: Comparative biochemistry of mammalian basement membranes. In: Chemistry and Molecular Biology of the Intercellular Matrix (Ed. E.A. Balazs) Vol. 1, p. 535 - 573. London, New York: Academic Press 1970.

Kunkel, H.G., Tan, E.M.: Autoantibodies and disease. In: Dixon, F.J., jr., Humphrey, J.H. (eds.): Advances in Immunology, Vol. 4, p. 351. New York: Academic Press 1964.
Layman, Don L., McGoodwin, E.B., Martin, G.R.: The nature of the collagen synthesized by cultured human fibroblasts. Proc. nat. Acad. Sci. (Wash.) 68, 454 - 458 (1971).
Low, F.N.: Microfibrils: fine filamentous components of the tissue space. Anat. Rec. 142, 131 - 138 (1962).
Marchalonis, J.J., Weltman, J.K.: Relatedness among proteins: a new method of estimation and its application to immunoglobulins. Comp. Biochem. Physiol. 38, 609 - 625 (1971).
Moczar, M.: Mise en évidence d'une liaison aspartamidoglycosaminidique dans les glycoprotéines de structure de l'aorte. FEBS Letters 1, 169 - 170 (1968).
Moczar, M., Moczar, E., Robert, L.: Composition of the glycopeptides isolated from the structural glycoproteins of aorta of different species. Biochem. biophys. Res. Commun. 28, 380 (1967).
Moczar, E., Moczar, M., Robert, L.: Isolation and characterization of the glycopeptides of the structural glycoproteins of corneal stroma. Life Sci. 8, Part II, 757 (1969).
Moczar, M., Moczar, E., Robert, L.: Composition of glycopeptides obtained by proteolytic digestion of the media of porcine aorta. Artherosclerosis 12, 31 - 40 (1970).
Moczar, M., Robert, L.: Extraction and fractionation of the media of the thoracic aorta. Atherosclerosis 11, 7 - 25 (1970).
Moczar, E., Robert, L., Moczar, M.: Etudes sur les fractions glucidiques de la trame fibreuse insoluble du stroma cornéen. Europ. J. Biochem. 6, 213 (1968).
Payrau, P.: Biochemistry and immunochemistry of the cornea. In: Biochemistry of the Eye. XXth Intern. Congress Ophthal. Munich 1966 (Eds. M.U. Dardenne, B. and J. Nordmann), p. 1. Basel, New York: S. Karger 1968.
Pouliquen, Y., Graf, B., de Kozak, Y., Bisson, J., Faure, J.P.: Etude morphologique du kératocône. Arch. Ophtal. 30, 497 - 532 (1970).
Robert, A.M., Grosgogeat, Y., Reverdy, V., Robert, B., Robert, L.: Lésions artérielles produites chez le lapin par immunisation avec l'élastine et les glycoprotéines de structure de l'aorte. Atherosclerosis 13, 427 - 449 (1971).
Robert, A.M., Robert, B., Robert, L.: Chemical and physical properties of structural glycoproteins. In: Chemistry and Molecular Biology of the Intercellular Matrix. (Ed. E.A. Balazs), Vol. 1, p. 237 - 242. London, New York: Academic Press 1970.
Robert, B.: Etudes sur les glycoprotéines sériques et tissulaires. Thèse de Médecine, Fac. Méd. Paris 1959.
Robert, B., Parlebas, J., Robert, L.: Etude immunochimique d'une glycoprotéines de la cornée, la Kératoglycosaminoglycane. C. R. Acad. Sci. (Paris) 256, 323 (1963).
Robert, B., Szigeti, M., Derouette, J.C., Bouissou, H., Fabre, M.T., Robert, L.: Europ. J. Biochem. 21, 507 (1971).
Robert, L., Comte, P.: Aminoacid composition of the structural glycoproteins. Life Sci. 7, 493 (1968).
Robert, L., Darrell, R.W., Robert, B.: Immunological properties of connective tissue glycoproteins. In: Tristram G.R., Balazs, E.A. (Eds.): Nato Symposium on the Chemistry and Molecular Biology of the Intercellular Matrix, Vol. 3, p. 1991. New York, London: Academic Press 1970b.
Robert, L., Dische, Z.: Analysis of a sulfated sialofucoglucosaminogalactomannosidoglycan from corneal stroma. Biochem. biophys. Res. Commun 10, 209 (1963).

Robert, L., Fayolle, J., Derouette, S., Zabriskie, J.: Homology of aminoacid composition of structural glycoproteins, transplantation antigens, cell wall glycoproteins and Streptococcus A cell membrane. Transplant. Proc. 4, 415 - 418 (1972).
Robert, L., Grosgogeat, Y., Robert, A.M., Robert, B.: Mechanism of calcification of elastic tissue. Israel J. med. Sci. 7, 431 (1971).
Robert, L., Moczar, E., Robert, B.: Biochemical and immunological properties of the structural glycoproteins of connective tissue. In: 4th Int. Colloque on Cystic Fibrosis of the Pancreas, Part II, p. 309. Basel, New York: S. Karger 1968.
Robert, L., Parlebas, J.: Rapport entre la nature de la trame fibreuse et le taux d'incorporation d'aminoacides marqués dans différentes types de tissu conjonctif. C. R. Acad. Sci. (Paris) 261, 842 (1965a).
Robert, L., Parlebas, J.: Biosynthèse in vitro des glycoprotéines de la cornée. Bull. Soc. Chim. biol. (Paris) 47, 1853 (1965b).
Robert, L., Parlebas, J.: Etudes sur la biosynthèse des glycoprotéines tissulaires. In: Colloque sur la Biochimie et la Physiologie du Tissu Conjonctif (Ed. P. Comte), p. 81. Lyon: 1965c.
Robert, L., Parlebas, J., Poullain, N., Robert, B.: Données nouvelles sur l'immunochimie des protéines fibreuses du tissu conjonctif. In: Protides of Biological Fluids, Vol. 11, p. 109. Amsterdam: Elsevier 1963.
Robert, L., Payrau, P., Pouliquen, Y., Parlebas, J., Robert, B.: Role of a structural glycoprotein of corneal stroma in transplantation immunity. Nature 207, 383 (1965).
Robert, L., Reyss-Brion, M., Junqua, S., Salaun, J.: Composition chimique de la cornée de l'embryon de poulet et du poulet adult. C. R. Acad. Sci. (Paris) 269, 491 (1969).
Robert, L., Robert, B.: Structural glycoproteins of membranes and connective tissue: biochemical and immunopathological properties. In: Protides of Biological Fluids, Vol. 15, p. 143. Amsterdam: Elsevier 1967.
Robert, L., Robert, B.: Biochemical and immunochemical properties of a structural glycoprotein of corneal stroma, keratoglycosaminoglycan (KGAG). In: Biochemistry of the Eye, XXth Int. Congress Ophtal. Munich 1966 (Ed. M.U. Dardenne, B., J. Nordmann), p. 20. Basel, New York: S. Karger 1968.
Robert, L., Robert, B.: Structural glycoproteins, their metabolism in normal and pathological connective tissue. In: Nutritional Aspects of the Development of Bone and Connective Tissue. Symposium of the Group of European Nutritionists, Cambridge, 1968. Bibl. "Nutritio et Dieta", No. 13, p. 123. Basel, New York: S. Karger 1969.
Robert, L., Robert, B., Parlebas, J., Zweibaum, A., Oudea, P.: Immunochemistry of structural proteins and glycoproteins. In: Structure and Function of Connective and Skeletal Tissue, NATO Symposium, St. Andrews/Scotland (Tristram, G.R., Ed.), p. 406. London: Butterworth 1964b.
Robert, L., Robert, B., Robert, A.M.: Molecular biology of elastin as related to aging and atherosclerosis. Exp. Geront. 5, 339 - 356 (1970a).
Robert, L., Robert, B., Zweibaum, A., Oudea, P., Parlebas, J.: Etudes sur la spécificité d'espèce d'une glycoprotéines du tissue conjonctif la Kératoglycosaminoglycane. In: Protides of Biological Fluids, Vol. 12, p. 110. Amsterdam: Elsevier 1964a.
Robert, L., Schillinger, G., Moczar, M., Junqua, S., Moczar, E.: Etude biochimique du kératocône. Arch. Ophtal. 30, 589 - 608 (1970c).
Ross, R., Bornstein, P.: Studies of the components of the elastic fiber. In: Chemistry and Molecular Biology of the Intercellular Matrix (Ed, E.A. Balazs) Vol. 1, p. 641 - 655. London, New York: Academic Press 1970.

Schultze, H.E., Heremans, J.F.: Molecular biology of human proteins. Vol. 1: Nature and Metabolism of Extracellular Proteins. Amsterdam: Elsevier 1966.
Simkin, J.L.: Biosynthesis of plasma glycoproteins. Expos. ann. biochim. méd. 30, 19 - 38 (1970).
Yonemasu, K., Stroud, R.M., Niedermeier, W., Butler, W.T.: Chemical studies on Cl_q; a modulator of immunoglobulin biology. Biochem. biophys. Res. Commun. 43, 1388 - 1394 (1971).
Zabrisky, J.B.: Mimetic relationship between group A streptococci and mammalian tissues. In: Advances in Immunology (Ed. F.J. Dixon, jr., H.G. Kunkel), Vol. 7, p. 147. New York: Academic Press 1967.

Experimental Chronic Arthritis in Pigs after a Single Intraarticular Injection of Heterologous Proteoglycans

R. Fricke, W. Bollwahn, and C. Messow

The main interest in animal models of chronic arthritis originates in the still unknown pathogenesis of chronic arthritis in man. Various inorganic and organic substances were tested for their arthritis causing potency. Models employing organic materials are more realistic in imitating rheumatoid arthritis, than those using inorganic material. After systemic application of Freund's complete adjuvans in rats, polyarthritis develops in many ways resembling rheumatoid arthritis in man (Pearson, 1956). Intraarticular application of protein, e.g. of heterologous horse serum protein, in rabbits causes, after immunization with the same antigen, a chronic arthritis (Klinge, 1930). Employing autologous and heterologous fibrin, intraarticular chronic arthritis can be produced in rabbits after immunization (Dumonde and Glynn, 1962). The same results can be achieved by applying ovalbumin (Glynn, 1968). Split products of immunoglobulin also provoked chronic arthritis after repeated application in joints of rabbits (Fehr, 1971, 1972). Lysosomal enzymes repeatedly injected in rabbits also produced chronic arthritis, similar to that of the so-called protein arthritis (Weissmann et al., 1969). It is generally accepted that chronic arthritis is sustained by an autoimmunological process (Dumonde and Glynn, 1962). Connective tissue constituents were therefore also considered to be involved in the pathogenesis of rheumatoid arthritis. Collagen autoantibodies were demonstrated in joint effusions in rheumatoid arthritis (Steffen et al., 1968). Glycosaminoglycans so far have not been successfully employed in production of chronic arthritis. It is not yet known whether proteoglycans, to which glycosaminoglycans are connected in nature, are involved in chronic arthritis. An attempt therefore was made to investigate the role of proteoglycans in chronic arthritis by employing them in an experimental model in pigs (Fricke et al., 1972).

Proteoglycan fractions, prepared by precipitation in ethanol and $MgCl_2$ from intervertebral discs of human, bovine and porcine origin, were injected once into talocrural joints of pigs, two to three months old. The substances applied were sterile. In the Ouchterlony technique antigenicity of the proteoglycans, keratansulfate and chontroitinsulfate was established according to chemical data. Serum proteins were not detected. For controls saline solutions were injected into contralateral joints.

Six weeks after intraarticular injection, pigs having received human or bovine proteoglycans started to go lame. X-ray analysis of the joints injected revealed pocket erosions at the articular surface of the talocrural joints. There were clinical signs of inflammation such as effusions, periarticular joint swelling and rise of temperature.

At sacrifice villous thickening of the synovial tissue and erythrocyte imbibation was seen. The articular surface revealed cartilage erosions reaching down to the bone (Fig. 1). Histology of the synovial tissue revealed thickening of the synovial lining cell layer in all joints treated. There was round cell infiltration, predominantly lymphocytes (Fig. 2). In the subsynovial tissue region cell augmentation was also noted. Granulation tissue was undermining the cartilage (Fig. 3). As a result destruction of cartilage and bone developed.

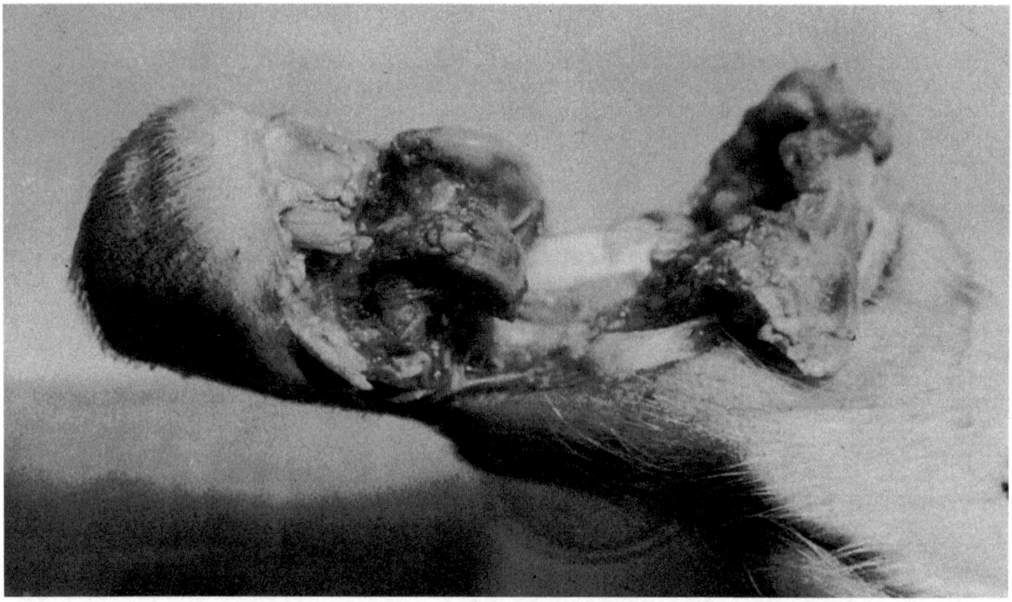

Fig. 1. Talocrural joint of a pig eight weeks after intraarticular application of human proteoglycans

It should be pointed out that, whereas synovitis was seen after injection of each proteoglycan preparation as well as after ovalbumin, erosive destruction of cartilage and bone was only seen after intraarticular application of human and bovine, i.e. heterologous, proteoglycans. The same substances did not cause chronic arthritis, when injected into knee joints of rabbits only once.

The results of these experiments show that intraarticular application of preparations of heterologous proteoglycans can produce chronic arthritis after a single injection. In rabbits on the other hand the same substances did not provoke chronic arthritis after one application. In this experiment ovalbumin was also applied, which was reported to cause arthritis only after immunization (Glynn, 1968).

In rodents therefore chronic arthritis can only be produced after immunization, in contrast to our experiments in pigs.

This comparison demonstrated the difference of the results of the present experiment from those of other groups. The fact that a single injection of heterologous antigenic proteoglycans provokes a pertenuating chronic arthritis in pigs but not in rodents may indicate a species specific difference in response to antigenic substances.

From the observation that only heterologous proteoglycans provoked chronic arthritis, one may conclude that these connective tissue

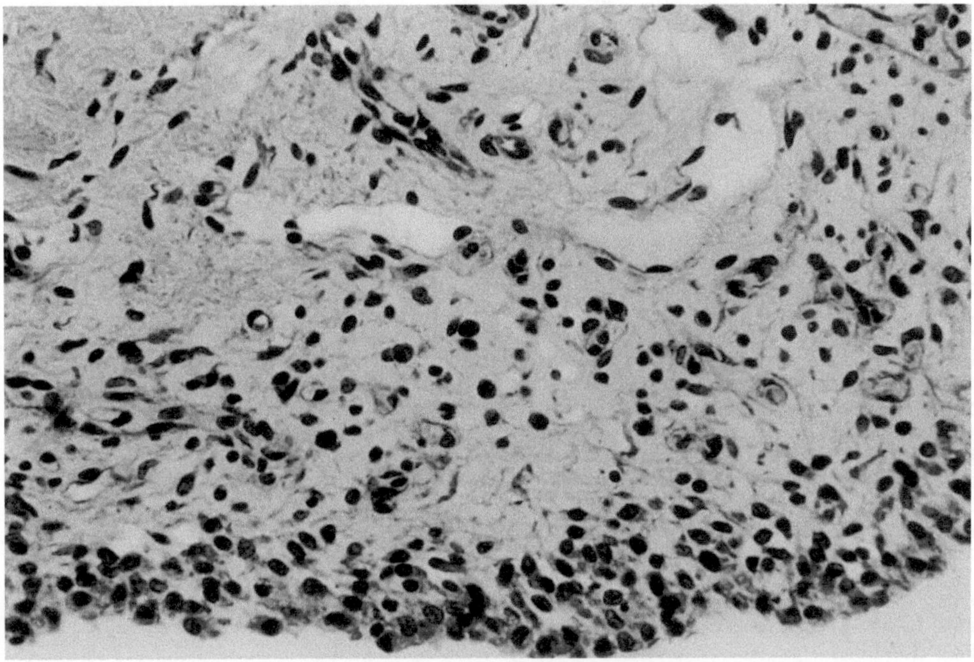

Fig. 2. Synovial tissue eight weeks after intraarticular application of human proteoglycans

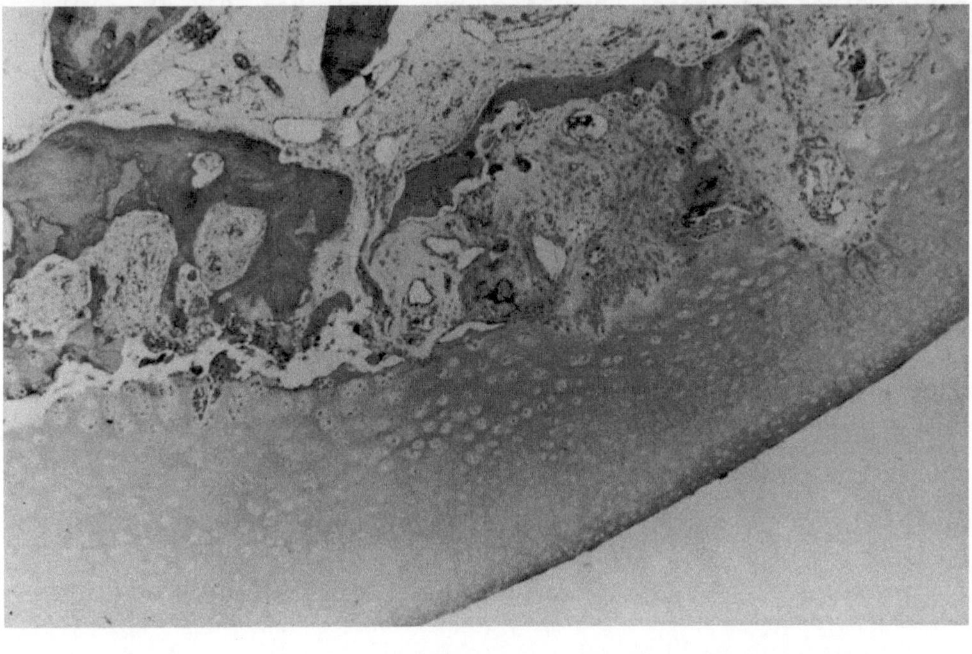

Fig. 3. Granulation tissue undermining cartilage eight weeks after application of human proteoglycans in a talocrural joint of a pig

constituents may also play a role in the pathogenesis of chronic arthritis.

This work was supported by the Deutsche Forschungsgemeinschaft.

References

Dumonde, D.C., Glynn, L.E.: The production of arthritis in rabbits by an immunological reaction to fibrin. Brit, J. exp. Path. 43, 373 - 383 (1962).

Fehr, K.: Gammaglobulin G fragment antigens produced by digestion of IgG with intracellular proteases and their possible relation to inflammation and rheumatoid factor formation in rheumatoid arthritis. In: Rheumatoid Arthritis Colloquia Geigy, p. 383 - 390 (Ed. W. Müller, H.G. Harwerth, K. Fehr). London, New York: Academic Press 1971.

Fehr, K.: Pathogenese der progredient chronischen Polyarthritis (PcP). Bern: Huber Verlag 1972.

Fricke, R., Bollwahn, W., Schulze, W., Messow, C.: Chronic arthritis produced in pigs by proteoglycans. Scand. J. Clin. Lab. Invest. 29. Suppl. 123, 12 (1972).

Glynn, L.E.: Heberden Oration 1967. The chemistry of inflammation and its significance in rheumatoid arthritis. Ann. rheum. Dis. 27, 105 - 121 (1968).

Klinge, F.: Die Eiweißüberempfindlichkeit (Gewebsanaphylaxie) der Gelenke. Experimentelle pathologisch-anatomische Studie zur Pathogenese des Gelenkrheumatismus. Beitr. Path. Anat. 83, 186 - 216 (1930).

Pearson, C.M.: Development of arthritis, periarthritis and periostitis in rats given adjuvants. Proc. Soc. exp. Biol. Med. 91, 95 - 101 (1956).

Steffen, C., Schuster, R., Tausch, G., Timpl, R., Pecker, I.: Klin. Wschr. 46, 976 - 981 (1968).

Weissmann, G., Spilberg, I., Krakauer, K.: Arthritis induced in rabbits by lysate of granulocyte lysosomes. Arthr. Rheum. 12, 103 - 116 (1969).

Hyaluronidase Activity in Lysosomes and Extracellular Fluid of Tumor Cells

B. Fiszer-Szafarz and D. Szafarz

During tumor cell invasion the peritumoral connective tissue is destroyed. One of the enzymes probably involved in this phenomenon is hyaluronidase, which depolymerizes hyaluronic acid of the ground substance of connective tissue (Fiszer-Szafarz, 1968).

Working with Walker carcinoma in the rat, we have shown that in tumors hyaluronidase activity is present in the extracellular fluid at a higher level than in normal tissue. This fluid also contains hyaluronic acid, but in the tumor an important fraction of it has been degraded and carried away into the blood stream. In the blood serum of tumor-bearing animals we have demonstrated the presence of a competitive hyaluronidase inhibitor, which very probably consists of the degradation products of hyaluronic acid (Fiszer-Szafarz and Gullino, 1970a, 1970b).

As hyaluronidase is stored in lysosomes, we compared its activity in the lysosomes of chemically induced hepatomas with normal rat liver. We found that, whereas normal liver lysosomes exhibit an important hyaluronidase activity, hepatoma lysosomes only display a scarcely detectable activity (Fiszer-Szafarz and Szafarz, 1973).

Our results also showed that the hepatoma lysosomes contain, in fact, the enzyme in an inactive form, which can be activated by a dialyzable factor present in normal lysosomal extracts.

If these conclusions might be extended to other lysosomal hydrolases, they could explain the apparent absence of these enzymes in various tumors as shown by histochemical techniques.

References

Fiszer-Szafarz, B.: Demonstration of a new hyaluronidase inhibitor in serum of cancer patients. Proc. Soc. exp. Biol. Med. 129, 300 - 302 (1968).

Fiszer-Szafarz, B., Gullino, P.M.: Hyaluronic acid content of the interstitial fluid of Walker carcinoma 256. Proc. Soc. exp. Biol. Med. 133, 597 - 600 (1970a).

Fiszer-Szafarz, B., Gullino, P.M.: Hyaluronidase activity of normal and neoplastic interstitial fluid. Proc. Soc. exp. Biol. Med. 133, 805 - 807 (1970b).

Fiszer-Szafarz, B., Szafarz, D.: Lysosomal hyaluronidase activity in normal rat liver and in chemically induced hepatomas. Cancer Res. 33, 1104 - 1108 (1973).

Clinical and Experimental Investigations of the Metabolism of Collagen, Ground Substance, and Mesenchymal Cells

U. Gerlach and W. H. Hauss

This paper deals with changes in the metabolism of the connective tissue and of mesenchymal cells, which we consider to be of primary and essential importance in the pathogenesis of certain chronic diseases.

The term "mesenchymal cells" describes cells that produce the ground substance and collagen fibers of connective tissues. The mesenchymal cells include the parent cells of the mesenchymal system.

The *methods* applied in our investigations include the incorporation of radioactive precursors into chondroitin sulfate, collagen and other proteins. They also include the measurement of the activity of the sulfate activating enzyme system (activation of sulfate to phosphoadenosine phosphosulfate = PAPS), and the measurement of the reduplication rate of cells with the aid of ^{3}H-thymidine. In addition, histo-chemical and electron microscopic investigations were carried out. The results were tested for statistical significance.

1. In the first section of this paper, consisting of three chapters, experiments will be shown, illustrating what Hauss and his colleagues have called "non-specific mesenchymal reactions". These reactions will first be demonstrated in animal experiments and then in human diseases.

Fig. 1 shows that various factors produce marked changes in the mesenchymal metabolism. In detail, it was shown that defined stimuli such as the action of toxins, mechanical, chemical or nervous stimuli as well as complex stimuli like infection, hypertension, excess of muscular activity, specialized diet and hypoxemia alter the metabolism of the mesenchyma. In response to the stimuli mentioned the rate of incorporation of sulfate into chondroitin sulfate increases in every case.

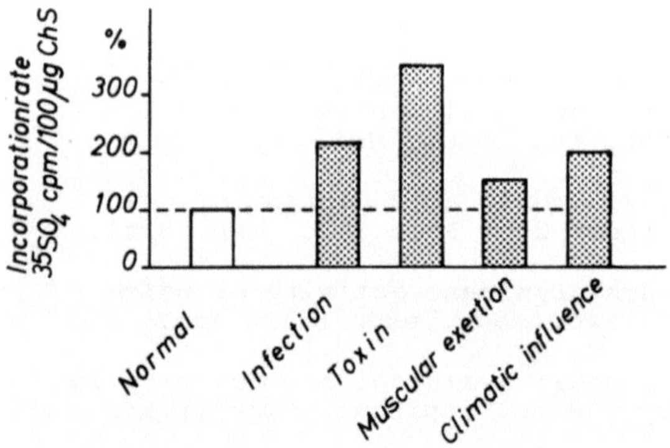

Fig. 1. Acceleration of mesenchymal metabolism by different stimuli (Hauss et al.)

As shown in Fig. 2, the incorporation of sulfate into chondroitin sulfate increases rapidly after a toxic stimulus within a few hours: even the incorporation of proline into collagen, which is reduced in the beginning, increases above the normal rate. This increase affects both the neutral salt soluble and the collagen insoluble in neutral salt solution.

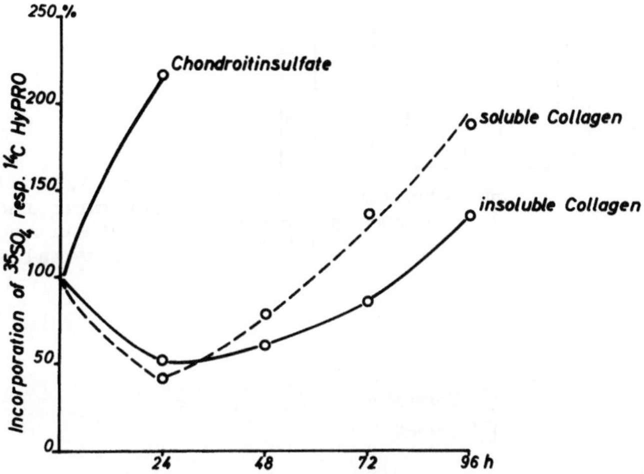

Fig. 2. Incorporation of $^{35}SO_4$ resp. ^{14}C PRO in connective tissue of rats after toxin stimulus

Changes of the mesenchymal metabolism will generally disappear without leaving any pathological tissue damage. If, however, the individual stimuli are too strong and long lasting, or repetitive, then the mesenchymal reaction will increase in such a way that structural changes in the tissue will result. This form of pathological change occurs primarily when various stimuli coincide. The typical reaction resulting from this is called an additive phenomenon (Fig. 3).

The experimental data described so far show that the regulation of the mesenchymal metabolism is readily and rapidly altered by effective stimuli.

As various agents can produce this reaction it is called "non-specific mesenchymal reaction".

2. In the second section, the importance of this "non-specific mesenchymal reaction" in the pathogenesis of human diseases will be demonstrated.

At first, the reaction of the mesenchyma during the course of a classical connective tissue disease, rheumatoid arthritis, will be shown. In Fig. 4 the intensity of the mesenchymal metabolism in the skin of a healthy person is compared with the intensity of metabolism in a patient with rheumatoid arthritis.

One can see the distinct rise in the incorporation rate of radioactive precursors into chondroitin sulfate and collagen. It should be

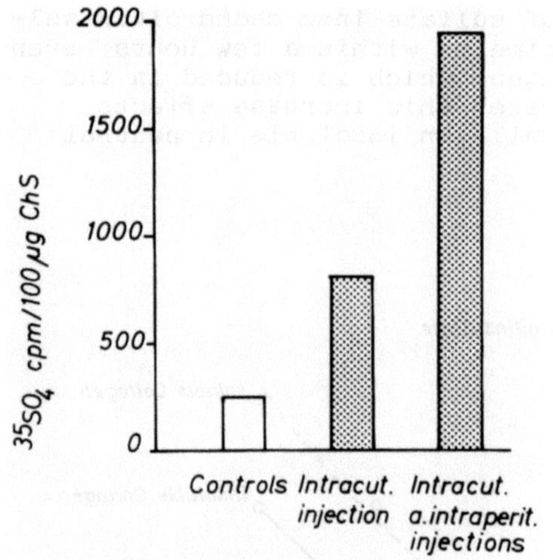

Fig. 3. Addition phenomenon. Significant increase of mesenchymal reaction induced by coincidence of several stimuli (Hauss et al.)

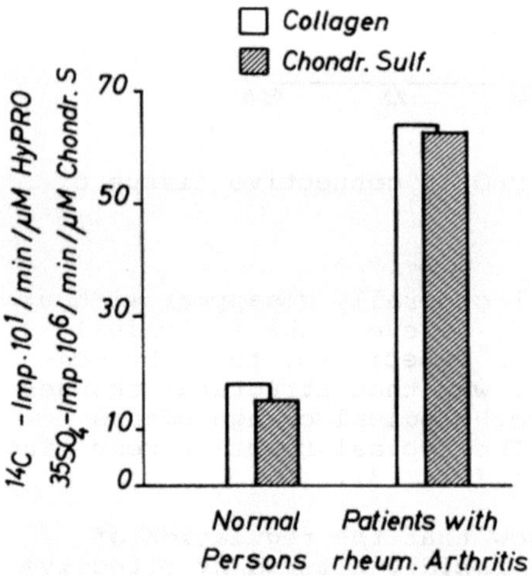

Fig. 4. Mesenchymal metabolism in the skin of patients with rheumatoid arthritis (incubated excision material; Hauss et al.)

emphasized that skin samples of patients with rheumatoid arthritis were taken from an area beyond the affected joint.

More detailed studies showed that the degree of the mesenchymal reaction was greater the more active the rheumatic process was.

It was possible to reduce the pathologically increased mesenchymal metabolism by anti-rheumatic drugs such as Cortisol, Phenylbutazone and anti-metabolites (Fig. 5).

The changes in mesenchymal metabolism can also be determined by investigating the excretion of hydroxyproline in patients with rheumatoid arthritis. The excretion of this aminoacid, which is representative of collagen, is increased in patients with rheumatoid arthritis. The excretion of chondroitin sulfate in this group of patients is similarly raised (Fig. 6).

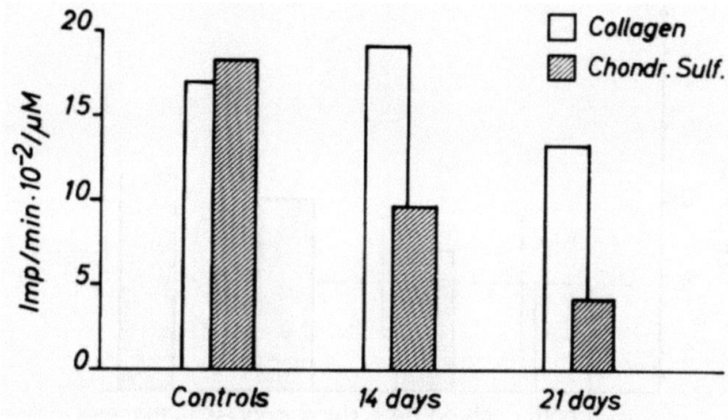

Fig. 5. Suppression of increased mesenchymal metabolism in skin by antirheumatic drugs (Hauss et al.)

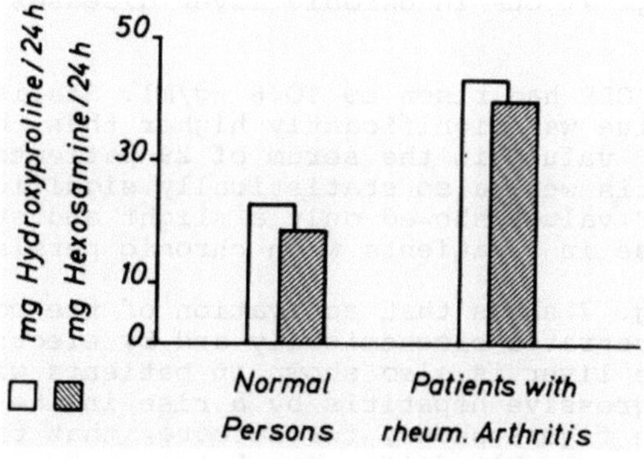

Fig. 6. Mesenchymal metabolites in the urine of patients with rheumatoid arthritis (Hauss et al.)

Acceleration of the mesenchymal metabolism is a general reaction of the body in the pathogenesis of *various* chronic inflammatory diseases. One example of this is the change in metabolism in *chronic liver diseases*. Preceding investigations showed that progress of a chronic liver disease was related to a measurable increase of activity of the mesenchymal metabolism of the liver. Thus, it could be shown in liver biopsies that the incorporation of radioactive sulfate into chondroitin sulfate was greater when inflammation in the liver was more active. Furthermore, electron microscope findings, together with enzymological studies, showed that the increase of collagen in a cirrhotic liver is correlated to the activity of the chronic disease. Following this line of thought, we shall report on the attempt to draw conclusions on the activity of the mesenchymal processes in the liver from the more readily available blood serum samples. We investigated whether determination of the "collagen-like protein" (CLP) in blood serum permits any assessment of the activity of the mesenchymal metabolism in liver. CLP means a plasma fraction whose aminoacid composition resembles that of collagen. One may assume that it represents a collagen fraction. The hypothesis that a pathological increase in collagen metabolism in liver could lead to an increase of CLP in the serum during liver cirrhosis was studied in 84 patients. The results are shown in Fig. 7.

The mean level of CLP in normal serum is 7.8 µg/ml. In 37 patients with clinically and histologically proved liver cirrhosis the level

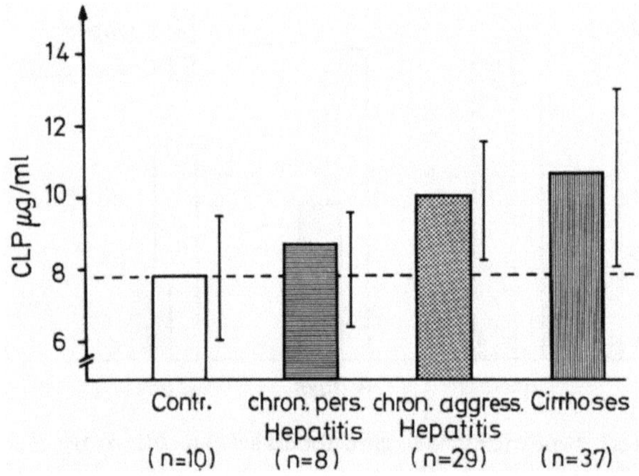

Fig. 7. CLP in chronic liver diseases

of CLP had risen to 10.6 µg/ml. Statistical analysis showed that this value was significantly higher than the control value ($P<0.002$). The CLP values in the serum of 29 patients with chronic aggressive hepatitis were also statistically significantly raised. ($P<0.002$). The CLP values showed only a slight and statistically non-significant rise in 8 patients with chronic persistent hepatitis.

Fig. 7 shows that activation of the collagen metabolism that was demonstrated biochemically and by electron microscopy in cylinders of the liver is also shown in patients with liver cirrhosis and chronic aggressive hepatitis by a rise in the CLP level in the blood serum. The figure shows, furthermore, that there is possibly a biochemically determinable difference between chronic persistent hepatitis and chronic aggressive hepatitis.

In the group of patients with clinically and histologically confirmed liver cirrhosis, the cholinesterase activity of the serum was also assessed in addition to the CLP level. Serum cholinesterase is an enzyme of which reduction of the level is an indication of a reduced ability of synthesis of the liver. Comparison of cholinesterase activity and the CLP level in serum showed that the higher the CLP level in the serum is, the lower is the cholinesterase activity (Fig. 8).

The regression coefficient $r = -0.709$ and is statistically significant ($P<0.005$). This finding also means that the more advanced the cirrhosis is, the higher is the CLP value.

If one wants to know the clinical value of the measurement of the CLP level, one can say that by the method applied it is possible to determine a further parameter of liver function, namely of collagen metabolism. This may improve the possibilities of diagnosis and aid the therapeutic assessment.

In the following section the importance of the mesenchymal reaction for the development and progress of arteriosclerosis is described. In an investigation of the metabolism in the mesenchyma of postmortal human aortas, Hauss and Junge-Hülsing (1961) showed that the incorporation of sulfate into chondroitin sulfate was reduced with increasing age. These results will not be discussed in detail here. In addition

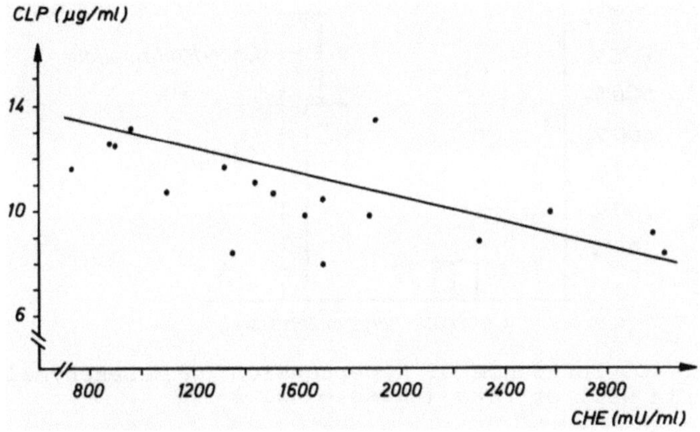

Fig. 8. The higher the collagen-like-protein (CLP) level in serum, the lower is the cholinesterase (CHE) activity

the authors found that the mesenchymal metabolism in an arteriosclerotic human aorta was raised considerably in comparison to the age-dependant normal level (Fig. 9). In order to confirm the importance of this finding, the effect of one of the strongest sclerogenic factors, hypertension, on the mesenchymal metabolism of the aorta was investigated. For this investigation 100 hypertensive rats were compared to 50 normotensive animals.

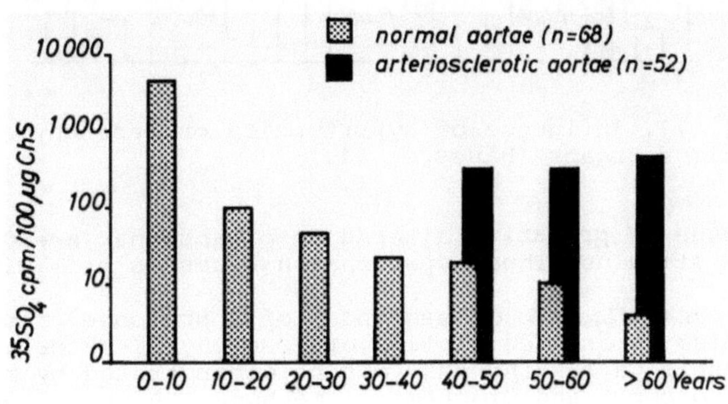

Fig. 9. Influence of arteriosclerosis on the incorporation of $^{35}SO_4$ into chondroitin sulfate of human aortae (incubated material; Hauss et al.)

Fig. 10 shows that the result of this investigation was a marked rise in the mesenchymal metabolism of chondroitin sulfate and collagen in the aortae of rats 6 weeks after continuous hypertension. The rate of incorporation of sulfate into chondroitin sulfate and of proline into collagen was raised 7-fold and 5-fold, respectively.

It is noteworthy that the connective tissue metabolism in aorta and heart is altered by hypertension, but not in other tissues (Fig. 11). This means that the regulation of the metabolism of the connective

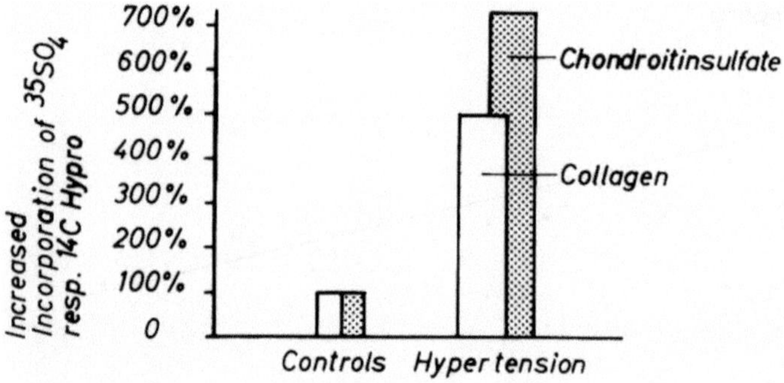

Fig. 10. Influence of hypertension on mesenchymal metabolism in the aortic wall of rats (Hauss et al.)

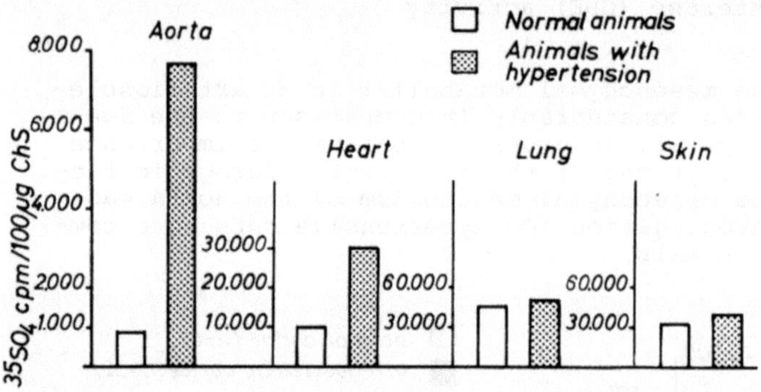

Fig. 11. Influence of hypertension on mesenchymal metabolism in various organs (Hauss et al.)

tissue is primarily altered in organs that are directly affected by the stimulus. Thus hypertension stresses primarily the heart and aorta.

The *morphological* consequences of a prolonged rise in mesenchymal metabolism is an accumulation of mesenchymal material in the wall of blood vessels, a finding that can be demonstrated by microscopic, electron-microscopic and biochemical methods, and which can easily be performed. If one looks at the electron-microscopic magnification of a normal aorta (Fig. 12a), one can see only a thin layer of mesenchymal material in the sub-endothelial space.

Several weeks later the structure of the aorta has changed considerably by hypertension (Fig. 12b), the sub-endothelial zone has become broadened and condensed. The mesenchymal substances have increased. At higher magnification the broadened sub-endothelial zone shows numerous collagen fibers embedded in the ground substance. This is the morphologically visible expression of biochemical changes in the metabolism of connective tissue described in the previous paragraphs.

As a result of mesenchymal reactions in the aorta, two further morphological changes take place (Fig. 13). The elastic fibers appear to be

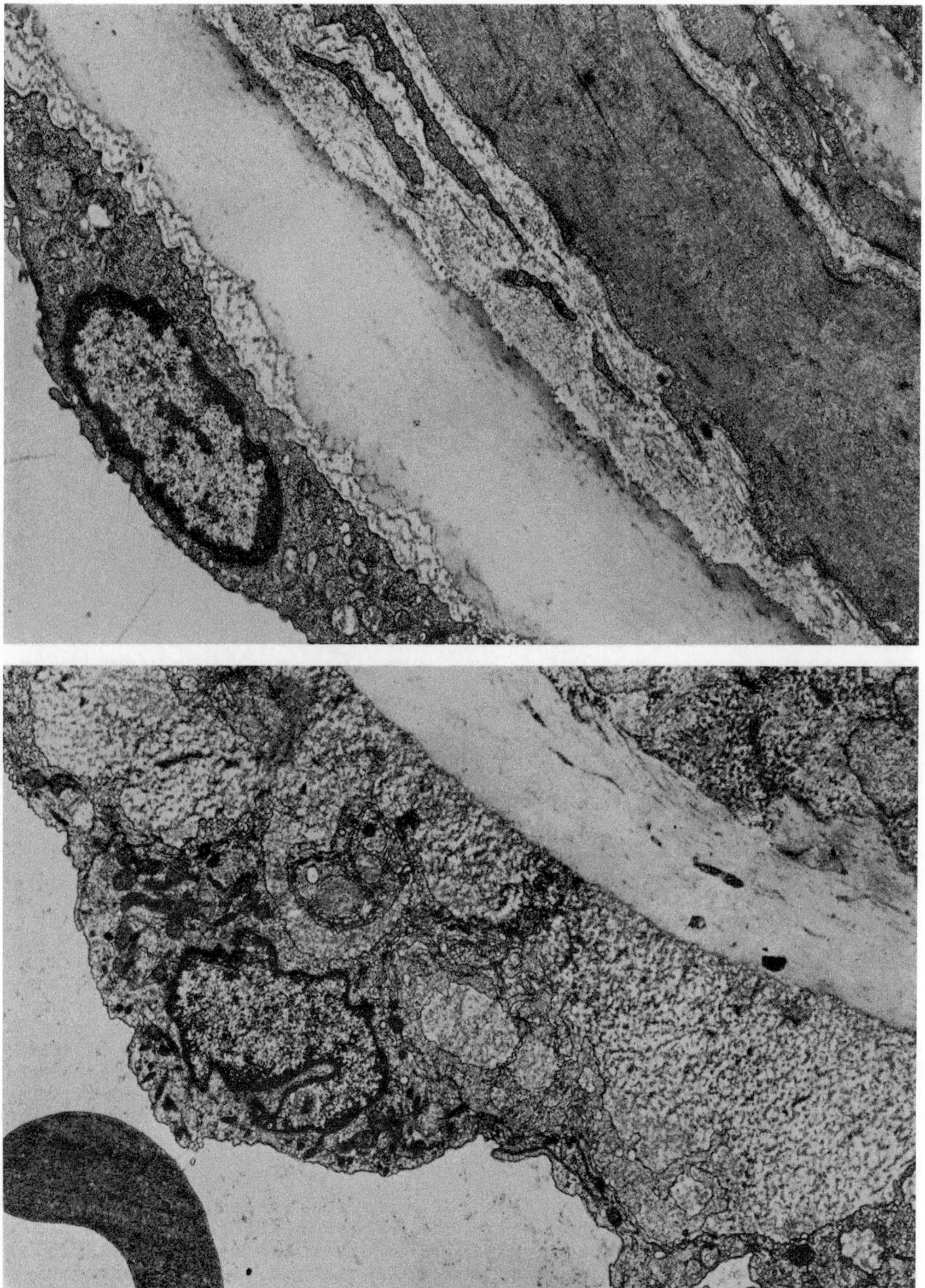

Fig. 12a and b. Blood vessel wall (electronmicroscopy), a) normal aorta; b) aorta several weeks after start of hypertension thinner and become destroyed after several weeks of the action of hypertension. Transformation is seen in smooth muscle cells. The cells transformed penetrate the sub-endothelial space, in which they seem to be capable of forming mesenchymal substances.

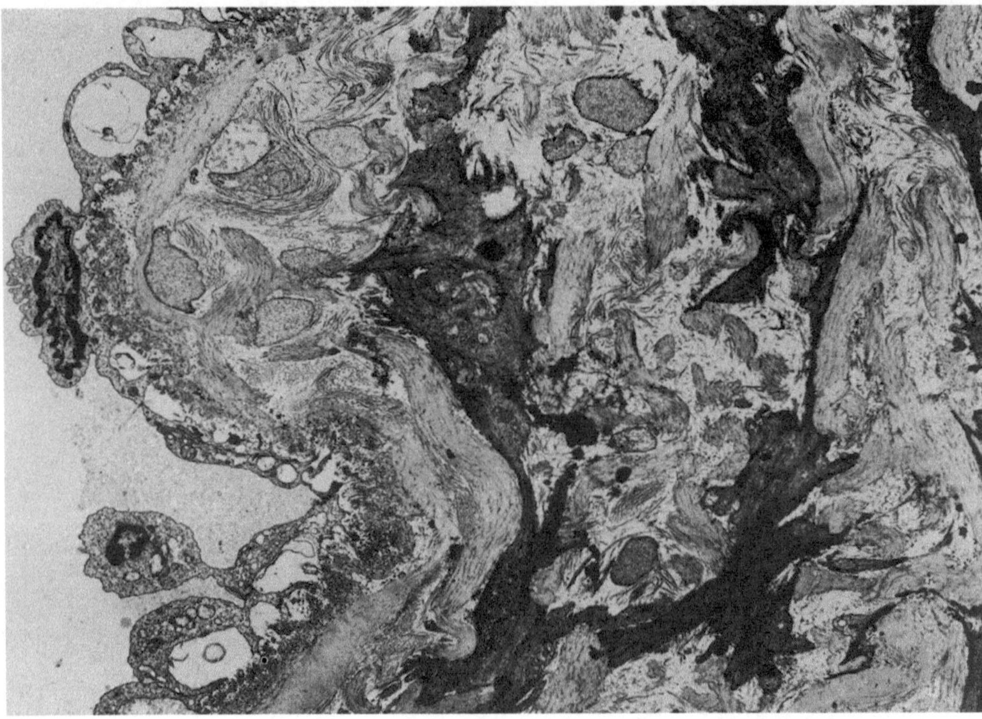

Fig. 13. Aorta several weeks after increase of blood pressure

The *pathophysiological* consequences of the mesenchymal reaction are a change in those functions that are associated with the amount and structure of the mesenchyma. These functions affect particularly water retention and the transport of metabolites. The change in the transport function is particularly critical in the tissue "transit zone" (Hauss, 1963). This transit zone consists of the endothelium of the vessel, the basal membrane and the variable layer of chondroitin sulfate, protein and collagen. All transportation from the blood stream into the tissues and from the tissues into the blood must pass through this transit zone. Under this assumption any malfunction that alters the transit zone will be of general importance.

3. Based on the results reported, the question arises whether the biochemical and structural changes in the walls of arteries are caused by increased metabolic activity of *individual* local mesenchymal cells, or whether the number of mesenchymal cells in the walls of arteries is increased within the course of the mesenchymal reaction. The electron microscopic finding of numerous cells in the sub-endothelial zone following the action of hypertension suggests that the sclerogenic factors not only cause an increased production of mesenchymal sub-

stances but can also alter the number of mesenchymal cells in the tissues affected.

In order to pursue this question, autoradiographic experiments with ^{3}H-thymidine were carried out. Fig. 14 shows that one hour after injection of ^{3}H-thymidine, only a small number of cells in the normal rat aorta are labelled, which means that only a few cells in the normal arterial wall were in the DNA reduplication phase. From this it can be deduced that cell proliferation in the normal rat aorta is small.

	Normal	Acute hypertension (1 hour)	Chronic hypertension (weeks)
Intima	3	14	59
Media	12	136	235
Adventitia	4	218	255
Total	19	368	549

Fig. 14. Cell reduplication in the aortic wall of rats. Influence of hypertension on the number of ^{3}H-thymidine labelled cells (Hauss et al.)

Fig. 14 also shows that the reduplication rate in the aorta during acute hypertension is considerably increased. Apparently hypertension is a stimulus that can increase the number of cells in the blood vessels. The increase in cell proliferation is even greater if the hypertension lasts for several weeks (Fig. 14). It is noteworthy that all layers of the aorta (intima, media and adventitia) are involved in cell proliferation.

If one labels the cells in the aorta with thymidine at various intervals, before and after stimulation, one can decide whether the cells labelled later are reduplicated in loco, i.e. within the wall of the aorta, or whether the reduplication has taken place elsewhere. By applying this method it could be demonstrated that the proliferation that is caused by hypertension takes place in loco, i.e. within the wall of the aorta, (Schmitt and Hauss, 1972).

It should, however, be emphasized that not all pathogenic stimuli produce a cell proliferation in the same way. Thus, the vessel wall is infiltrated after the action of bacterial toxins by mononuclear cells that are not reduplicated within the arterial wall, but probably in the spleen, lymph glands and thymus. Apparently various ways lead to a cell proliferation within the mesenchyma.

References

Gerlach, U.: Untersuchungen über den Bindegewebsstoffwechsel unter klinischen und experimentellen Bedingungen. Z. ges. exp. Med. 139, 542 - 554 (1965).

Hauss, W.H.: Pathogenese der Coronarsklerose und des Herzinfarktes. Verh. dtsch. Ges. inn. Med. 69. Kongress, Wiesbaden 1963.

Hauss, W.H., Gerlach, U., Junge-Hülsing, G.: Über die Reaktion des Mesenchymstoffwechsels auf endogene und exogene Reize. Méd. et Hyg. (Genève) 23, 208 (1965).

Hauss, W.H., Junge-Hülsing, G.: Über die universelle unspezifische Mesenchymreaktion. Dtsch. med. Wschr. 86, 763 (1961).

Hauss, W.H., Junge-Hülsing, G., Gerlach, U.: Die unspezifische Mesenchymreaktion. Stuttgart: Thieme 1968.

Hauss, W.H., Junge-Hülsing, G., Gerlach, U., Wirth, W.: Über Veränderungen des Mesenchymstoffwechsels durch Umweltfaktoren durch Hormone und bei rheumatischen Erkrankungen. Rheumatismus 36, 40 (1965).

Hauss, W.H., Junge-Hülsing, G., Matthes, J.K., Wirth, W.: Über den Einfluß von Schock und Hyperlipidämie auf den Lipidgehalt, die Lipidsynthese und die Mucopolysaccharidsynthese der Gefäßwand. J. Atheroscler. Res. 5, 451 - 465 (1965).

Hauss, W.H., Junge-Hülsing, G., Wirth, W., Albrecht, H.J.: Über die Mesenchymreaktion bei allergischen und parallergischen Erscheinungen. Z. ges. exp. Med. 135, 384 (1962).

Husen, N. van, Fegeler, K., Eberhardt, G., Gerlach, U.: „Collagen-like Protein" im Blutserum von Patienten mit chronischen Leberkrankheiten. Med. Welt 22 (N.F.), 1877 - 1878 (1971).

Junge-Hülsing, G.: Über den Stoffwechsel von Sulfomukopolysacchariden in Granulationsgewebe. Z. Rheumaforsch. 18, 9 (1959).

Junge-Hülsing, G.: Untersuchungen zur Pathophysiologie des Bindegewebes. Heidelberg: Hüthig 1965.

Junge-Hülsing, G., Hauss, W.H.: Über den Schwefeleinbau in normales und pathologisches Bindegewebe. In: W.H. Hauss, H.Losse: Struktur und Stoffwechsel des Bindegewebes. Stuttgart: Thieme 1960.

Matthes, K.J., Junge-Hülsing, G., Schmitt, G., Wagner, H., Oberwittler, W., Hauss, W.H.: Über die Beziehungen zwischen gestörtem Mesenchymstoffwechsel und Veränderungen der Lipidkonzentration in der Gefäßwand bei arterieller Hypertension. J. Atheroscler. Res. 9, 305 - 318 (1969).

Schmitt, G., Hauss, W.H.: Über die Reaktion der Gefäßwand auf mechanische, nervöse und toxische Reize. In: Die venöse Insuffizienz. Baden-Baden: G. Witzstrock 1972.

Schmitt, G., Knoche, H., Hauss, W.H.: Über den Einfluß zentralnervöser Reize auf die Aortenwand. Verh. dtsch. Ges. Kreislauf.-Forsch. 36, 336 - 341 (1970).

Schmitt, G., Knoche, H., Junge-Hülsing, G., Koch, R., Hauss, W.H.: Über die Reduplikation von Aortenwandzellen bei arterieller Hypertonie. Z. Kreisl.-Forsch. 59, 481 - 487 (1970).

Schmitt, G., Müller, U.St., Hauss, W.H.: Zur Pathogenese der Koronarsklerose. In: Vascular Smooth Muscle/ Der Gefäßmuskel (Ed. Betz, E.). Berlin, Heidelberg, New York: Springer 1972.

Collagen Maturation in Granulation Tissue

A. Delaunay and S. Bazin

Collagen is synthetized in granulation tissue. It reaches a high amount, more or less rapidly, according to the type of inflammation. The collagen locally formed will be either progressively removed (acute inflammation) or will remain (subacute inflammation) (Bazin and Delaunay, 1964 - 1967). The question remains whether collagen is formed in a general manner in a granulation tissue and whether maturation may favor its degradation or stabilization in an inflammatory process, which may develop either towards healing or towards chronicity.

The present paper deals with preliminary results relevant to this problem. First we developed a method to estimate the cross-linkage of collagen. In the second step we examined collagen maturation in a slowly developing inflammation.

1. Material and Methods

a) Method of Estimation of Collagen Maturation

Normal subcutaneous tissue was taken from the back of ten young female Wistar rats, and minced. It was homogenized and extracted stepwise with the following degrading agents, consecutively:

- first extract in phosphate buffer 0.066 M, pH 7.8, at 0°C,
- second extract in citrate buffer 0.1 M, pH 3.5, at 0°C,
- third extract in phosphate buffer 0.05 M, NaCl 0.15 M, pH 6.6, at 60°C,
- fourth extract in trichloracetic acid 5%, at 90°C.

Collagen of the two first extracts was purified by NaCl precipitation and redissolution in 0.1 M acetic acid. Collagen of the two last extracts was isolated by dialysis followed by lyophilization.

The collagen of these four extracts was estimated by hydroxyproline titration (Stegemann, 1958), after hydrolysis and chromatography (Partridge and Elsden, 1961).

These four fractions of collagen were designated as: neutral salt soluble collagen (first extract); acid soluble collagen (second extract); 60°C labile collagen (third extract); insoluble collagen (fourth extract).

The collagen of each fraction was examined by polyacrylamide gel electrophoresis and the proportions of αchains β and γ components determined (Nagai et al., 1964).

b) Tissues Examined

Subacute slowly developing inflammation was induced in several groups of ten rats each by subcutaneous implantation of a polyvinyl sponge at the back of 2 months old female Wistar rats (Bazin and Delaunay, 1964). The inflamed tissues were harvested from 10 rats of each group, 7 days to 8 months after sponge implantation. The tissues taken from the rats of each group were minced and pooled, homogenized and extracted as indicated above.

As the animals were between 2 and 10 months old during the experiments, the effect of aging was examined in normal subcutaneous tissue from groups of two-, three- and ten-months old rats. The tissues were taken from the same back site where sponges had been implanted in the animals of the experiment. The normal tissues were homogenized and extracted as indicated above.

2. Results

a) Validity of our Method of Estimation of Collagen Cross-Linkage

The proportions of αchains, β and γ components were determined in the collagen of the four fractions extracted from normal subcutaneous tissue. The results obtained with the first three fractions are shown in Table 1.

Table 1. Cross-linkage of collagen fractions from subcutaneous tissue

Fractions of collagen	%		
	α	β	γ
Neutral salt soluble	64.7	34.7	0.6
Acid soluble	46.3	47.8	5.9
$60^{\circ}C$ labile	28.9	34.8	36.3

It can be seen that in neutral salt soluble collagen, the proportion of α chains is higher and the proportion of β components lower than in acid soluble collagen. In the latter collagen, the proportions of α chains and of β components are higher while the proportion of γ components is much lower than in the $60^{\circ}C$ labile collagen.

As for the collagen of the fourth extracted fraction, there were in polyacrylamide gel components smaller than α chains, than those observed in gelatin which was obtained by breaking peptide bonds of the collagen molecule. Thus it appears that collagen of this fraction has been irreversibly denatured during its extraction. As it could not be dissolved otherwise (this is the reason why it was designated "insoluble collagen"), it is obvious that this fraction corresponds to the most cross-linked collagen.

Altogether these results show that collagen of the four extracted fractions is increasingly cross-linked in this order: neutral salt soluble, acid soluble, 60°C labile, and insoluble. Therefore, these results demonstrate that our method of step-wise extractions is valuable and that the relative proportions of these four collagen fractions can be taken as an estimation of the collagen maturation.

b) Inflammatory Tissues

The proportions of the four extracted fractions of collagen are given in Table 2.

Table 2. Repartition of collagen fractions in the total collagen of inframmatory tissue

Time inflammation of tissues	Total collagen content in the tissues[a]	% of total collagen			
		neutral salt soluble	acid soluble	60°C labile	insoluble
7 days	4.95	4	13	53	30
14 days	10.84	3.4	11.2	49	36.4
21 days	13.62	3.1	10.9	40.2	45.8
30 days	13.95	3.2	7.5	39.5	49.8
2 months	14.95	3	3.7	48.5	44.8
3 months	19.87	3.1	7.8	55.9	34.2
4 months	22.78	2	6.9	59.8	31.2
5 months	22.28	2.1	10.9	36.8	50.2
8 months	22.81	1.9	11.6	34.5	52

[a] in mg hydroxyproline/g dry tissue.

It can be seen, from the proportion of insoluble collagen (the most cross-linked fraction), that the collagen synthetized in an inflammatory tissue does not develop regularly with increasing age of the granuloma. In fact, the maturation was more rapid during the first two months, then slowed down during the fourth and fifth month.

However, the proportion of neutral salt soluble collagen decreased progressively up to the fourth month indicating a decreasing biosynthesis during this period. Thus, the process of maturation does not appear directly related to the degree of biosynthesis.

Moreover, the proportions of acid soluble collagen and of 60°C labile collagen were found to develop quite similarly during the first month, but differently from then on. The increased proportion of acid soluble collagen observed when the collagen content increased no further (4th to 8th month) probably indicated a remodeling of the structure of collagen fibrils in the tissue. However, the irregular changes of the 60°C labile collagen appear at this point difficult to understand. This may have something to do with aging of the animals, especially during the last months of development of the granulation tissue.

c) Normal Subcutaneous Tissue

The proportions of the four fractions of collagen, extracted from subcutaneous tissue of two-, three and ten-months-old rats are given in Table 3.

Table 3. Repartition of collagen fractions in the total collagen of normal subcutaneous tissue

Age of animals	Total collagen content in the tissue[a]	% of total collagen			
		neutral salt soluble	acid soluble	60°C labile	insoluble
2 months	15.54	1.8	24.7	58.5	15.2
3 months	18.78	1.3	18.6	63	17.1
10 months	16.97	1.2	14.7	31	53.1

[a] in mg hydroxyproline/g dry, not defatted tissue.

It can be seen that in normal tissues of two- and three-months-old animals, the collagen maturation, estimated through the proportion of insoluble collagen, is less pronounced than in a granulation tissue at any time of its development. Moreover, the proportion of acid soluble collagen in the normal tissue appears to be higher than that found in the granulation tissue, as compared with the proportion of 60°C labile collagen (Tables 2 and 3).

However, when the animals grow older (10-months-old rats) the collagen maturation appears rather similar to that in granulation tissue. The only apparent difference is a slightly smaller proportion of acid soluble collagen and a slightly higher proportion of 60°C labile collagen in the granulation tissue (Tables 2 and 3).

3. Discussion

Our results show that collagen synthesized in a granulation tissue develops rather rapidly. These observations raise several questions: which factors may induce a very rapid maturation of collagen during the first phase of its synthesis in an inflammatory tissue? Are there some unknown "tanning" substances in the inflammatory tissues? Is the environment of the secreted collagen molecules able to enhance the normal process of cross-linkage? An influence of glycoproteins may have to be discussed, since collagen maturation appears to be slow in the later phase of development of the granuloma, when a higher content of insoluble glycoprotein is present in the tissue (Bazin and Delaunay, 1967). But up to now the possible role of glycoproteins in collagen maturation is not clear.

Moreover, what is the biochemical significance of the 60°C labile collagen fraction? A "labile collagen" has been defined by Verzar (1961) as an immature collagen, solubilized at temperature of hydrothermal shrinkage of normal skin. It has been found in a high proportion in dermal wounds, even long after healing (Verzar and Willenegger, 1961).

As for our experiments, we did not find that the proportion of 60°C labile collagen in inflammatory tissues differs from normal ones. Furthermore, we observed independent fluctuations of acid soluble collagen and of 60°C labile collagen, e.g. two fractions which had not been separated in Verzar's "labile collagen". Therefore, the biochemical mechanism of the collagen cross-linkage in these two fractions has to be investigated further. Some experiments about the role of aldehydes in this process are in progress; the data will be published later on.

References

Bazin, S., Delaunay, A.: Biochimie de l'inflammation. VI. Fluctuations du taux du collagène et des protéines non fibrillaires dans différents types de foyers inflammatoires. Etudes comparatives. Ann. Inst. Pasteur 107, 163 - 172 (1964).

Bazin, S., Delaunay, A.: Biochimie de l'inflammation. VII. Variations du taux des constituants de la trame conjonctive dans des foyers inflammatoires d'évolution lente. Ann. Inst. Pasteur 112, 316 - 328 (1967).

Nagai, Y., Gross, J., Piez, K.A.: Disc electrophoresis of collagen components. Ann. N.Y. Acad. Sci. 121, 494 - 500 (1964).

Partridge, S.M., Elsden, D.F.: Rapid methods for the determination of glucosamine, galactosamine and hydroxyproline. Bioch. J. 80, 34P (1961).

Stegemann, H.: Mikrobestimmung von Hydroxyprolin mit Chloramin T und p. Dimethylaminobenzaldehyd. Z. physiol. Chem. 311, 41 (1958).

Verzar, F.: Increase in the binding of hydroxyproline in collagen of the skin with age. Gerontologia (Basel) 4, 104 - 111 (1961).

Verzar, F., Willenegger, H.: Das Altern des Kollagens in der Haut und in Narben. Schweiz. med. Wschr. 91, 1234 - 1236 (1961).

A Negative Role for Collagenase in Observed Differences between Human Normal and Rheumatoid Polymeric Collagens

Jacqueline B. Weiss, and S. J. Leibovich

Studies on the action of mammalian collagenases have tended to use as substrate either tropocollagen monomers in solution or "native fibrils" assembled *in vitro* from such monomers. Both these forms of collagen contain only intramolecular cross-links, that is cross-links between chains within the same tropocollagen molecule, and no stable cross-links between one tropocollagen molecule and another occur. The very lack of such intermolecular cross-links is, of course, the essential reason for their solubility. Rat tail tendon and guinea pig skin have also been used as substrates for mammalian collagenase, and in both instances the assumption has been made that the collagen fibers of these tissues were intermolecularly cross-linked and, therefore, insoluble. However, it has been shown that rat tail tendon collagen and guinea pig skin collagen can, with a little patience, be totally dissolved in dilute acetic acid (Hamerman and Schubert, 1968; Forest and Jackson, in press). It has also been observed that the denatured products of these solubilized collagen fibers contain no intermolecular crosslinked forms (Forest and Jackson). This would suggest that the forces holding together the fibers of rat tendon and guinea pig skin collagen are not stable covalent bonds but hydrophobic and hydrogen bonds. In the human situation, on the other hand, the collagen fibers visible in the intact tissue in the light microscope are quite insoluble in dilute acetic acid and are known to contain extensive intermolecular cross-links. It seems clear, therefore, that caution must be exercised in comparing the results from soluble collagen, or from collagens from other species, with the effect of collagenase on human intact fibrils. Steven and Jackson (1967) have shown that insoluble intermolecularly crosslinked fibrils can be extracted in very pure form from the intact fiber and have termed their preparation polymeric collagen.

We have recently demonstrated a difference in the susceptibility of polymeric collagen extracted from normal aged human synovia and comparable rheumatoid arthritic synovial polymeric collagen to attack by pepsin. Human polymeric collagen appears resistant to attack by this enzyme and is not depolymerized, whereas between 20 and 70% of the polymeric collagen from rheumatoid synovia is depolymerized, giving rise to a slightly degraded tropocollagen molecule. This degraded molecule exhibits the same defects of fibrillogenesis, viz. formation of polarized and symmetric fibrillar and tactoidal forms, which we have shown in calf skin tropocollagen, to be caused by removal of specific structural sites at both ends of the molecule (Leibovich and Weiss, 1970a, b).

As the activity of collagenase in rheumatoid synovia is known to be increased relative to normal synovia (Evanson et al., 1967) it seemed worth investigating whether the susceptibility we had observed of rheumatoid polymeric collagen to pepsin was due to preliminary partial degradation of the collagen by such specific collagenases[1].

1 See footnote next page.

Normal human synovial polymeric collagen, after treatment with partially purified preparations of collagenases from rheumatoid synovia, either at 27°C or 37°C for 48 hours, did not reveal any degradation products attributable to the action of collagenase, nor did such treated polymeric collagen exhibit any subsequent susceptibility to pepsin attack. In fact, we were unable to demonstrate any difference in polymeric collagen after treatment with collagenase. When solutions of tropocollagen were incubated together with polymeric collagen and collagenase at 27°C, the tropocollagen was characteristically degraded to 1/4 and 3/4 segments, showing that no inhibitors of collagenase were present in the polymeric collagen preparations. The results of our experiments are summarized in Table 1.

Table 1. Distribution of soluble to insoluble hydroxyproline for various collagen preparations after treatment with rheumatoid synovia collagenase for 48 hours at either 27°C or 37°C

Substrate	27°C Digest		37°C Digest	
	Insol. hypro.	Sol. hypro.	Insol. hypro.	Sol. hypro.
Reconstituted tropocollagen fibrils			2%	98%
Polymeric collagen	98.5%	1.5%	96%	4%
Tropocollagen and polymeric collagen together in 1:1 ratio (50% tropocollagen)	53%	47%	49%	51%

These results seem to favor the argument that human rheumatoid collagenase may act only on newly synthesized tropocollagen or on breakdown products of polymeric collagen resulting from a previous attack by some more specific enzyme system, perhaps cathepsins. They also lend force to our contention that it may be dangerous to relate the action of collagenases on skin and tendon of laboratory animals to human skin, tendon or synovial membrane. It is also, obviously, better to study the action of this enzyme on polymeric collagen in the synovia as we find that this form constitutes over 80% of the total collagen in this tissue.

We have investigated the possibility that the observed lability of rheumatoid polymeric collagen to pepsin is due to an increased rate of turnover of collagen in rheumatoid arthritis, which would result in a younger form of polymeric collagen, lacking its full complement of intermolecular cross-links. However, we have found that foetal and baby polymeric collagens do not resemble rheumatoid polymeric collagen in their susceptibility to pepsin. We are, therefore, further investigating the nature of the abnormality in rheumatoid polymeric collagen.

---

1 The collagenase preparations used in this study were kindly provided by Professor J. Evanson.

References

Evanson, J.M., Jeffrey, J.J., Krane, S.M.: Human collagenase, identification and characterisation of an enzyme from rheumatoid synovium in culture. Science 158, 499 (1967).

Forest, L., Jackson, D.S.: Personal communication.

Hamerman, D., Schubert, M.: A primer on Connective Tissue, p. 11 and 20. Philadelphia: Lea and Febiger 1968.

Leibovich, S.J., Weiss, J.B.: Evidence for the role of terminal regions of tropocollagen in fibrillogenesis: Specific enzymic degradation studies. Biochem. J. 117, 22 p (1970).

Leibovich, S.J., Weiss, J.B.: Electron microscope studies of the effects of endo- and exopeptidase digestion on tropocollagen: A novel concept of the role of terminal regions in fibrillogenesis. Biochim. biophys. Acta (Amst.) 214, 445 (1970).

Steven, F.S., Jackson, D.S.: Purification and amino acid composition of monomeric and polymeric collagens. Biochem. J. 104, 534 (1967).

The Possible Role of Cross-Linking of Collagen in Rheumatoid Arthritis

M. Adam, Z. Deyl, and J. Rosmus

Klemperer et al. (1942) introduced the term "collagen diseases", although the creation of the term was greatly influenced by Neumann's incorrect assumption that fibrinoid damage of connective tissue reflects an alteration of collagen fibers. On the other hand, studies during the last years concerning the collagen metabolism in inflammatory rheumatic diseases seem to be of more than academic interest and seem to support the original concept of Klemperer, i.e. that collagen is damaged in these diseases and that it may even play an important role in their aetiopathogenesis. An alteration of collagen metabolism in rheumatoid arthritis has been reported by Steven (1965), who found a reduced cross-linking of the collagen structure in patients suffering from this disease.

It should be mentioned that it is not quite clear whether all phases of collagen turnover are responsible for the decrease of cross-linking found by Steven. Both an increased catabolism and an increased collagen formation undoubtedly contribute to the final appearance of the collagen structure in rheumatoid arthritis.

The above mentioned findings demonstrate the effect of collagen in rheumatoid arthritis. They give, however, no firm support to the concept of its active role in this disease. The active role of collagen in the aetiopathogenesis of rheumatoid arthritis seems to be supported by findings of Steffen et al. (1965). These authors determined collagen antibodies in 66% of their patients with rheumatoid arthritis. It can, therefore, be anticipated that the decreased cross-linking of the collagen structure in patients with rheumatoid arthritis may further enable its break-down. The break-down products may then act as antigens and cause the antibody formation.

The beneficial effect of corticosteroids in rheumatoid arthritis and the smaller incidence of this disease in males are well known. From this point of view the hormonal control of collagen metabolism is most interesting.

Isotope studies have shown the decreased collagen synthesis in hypophysectomized animals (Daughaday and Mariz, 1962). On the other hand, it has been demonstrated that the mechanical strength of tendon fibers is decreased in hypophysectomized rats (Giles and Everitt, 1967). These findings were further related to the biological age of the animal. This, however, is a strictly mechanical point of view. It can hardly help to decide what actual processes are causing mechanochemical changes of collagen fibers.

For this reason we investigated the distribution of collagen in different organs and its solubility. Further, we tried to screen the possible influence of some other pituitary dependent glands and their hormones.

1. Methods

Investigations were carried out using male Wistar rats with an initial body weight of 100 g which corresponds to the age of 7 weeks. The animals were fed with standard pelleted Larsen chow and tap water *ad libitum.*

Hypophysectomy. The pituitary was removed parapharyngeally and 7 rats surviving 63 days after operation were used as subjects.

Adrenalectomy and *gonadectomy* were done by the usual techniques. Adrenalectomized rats were killed 18 days and gonadectomized animals 74 days after operation.

Adrenalectomized animals were maintained on 1% NaCl drinking water. Hydroxyproline content was estimated according to Stegemann (1958).

Neutral salt-soluble, acid soluble and urea soluble collagen were prepared as described elsewhere (Deyl et al., 1971). The rate of contraction - relaxation of tendon fibers was investigated according to Chvapil and Zahradnik (1960).

2. Results

With respect to the hormonal control of collagen polymerization two effects occur: 1. the proportion of connective tissue proteins is changed in relation to the total weight of organs; 2. the quality of collagen present, i.e. its solubility, is changed as well. The data of solubility, summarized in Table 1 indicate the effect of hypophysectomy, adrenalectomy and gonadectomy in collagen formation and maturation. Hypophysectomy influences negatively the collagen formation, which results in apparent aging of the collagen structur. The same is true for the effect of gonadectomy in male rats. Adrenalectomy has an opposite effect. This effect, however, is not as distinct as in the experiments mentioned before because of the limited survival of adrenalectomized animals. For this reason the animals were killed at the eighteenth day after adenalectomy. These results correspond to those of Günter and Carsten (1961) and may be explained by the removal of an inhibitory effect of cortisone on collagen formation.

Hypophysectomized animals were substantially retarded in growth, as shown by the weights at the end of the experiment 136.6 ± 11.5 g, as compared to the weight of the controls, 337.4 ± 15.6 g. The concentration of collagen in the organs investigated in kidney, heart, liver, testes and spleen, was increased (Table 2). The differences from the control group were highly significant ($p<0.001$). The greatest difference was observed in testes, where the hydroxyproline concentration after hypophysectomy was more than tripled. While in most organs of the animals the hydroxyproline concentration was increased, the hydroxyproline concentration of aorta was lower by aprox. 20%. However, this difference was not significant.

These data represent a concentration of collagenous stroma in various organs only. The total amount of collagen in hypophysectomized animals, however, distinctly decreased in comparison to controls (Table 3), which stands for all organs investigated, except for aorta and lungs in which the discrepancy is presumably due to infectious affections which were quite frequent in our rats. The collagen formation is

Table 1. Solubility of skin collagen in hypophysectomized and control rats

	Body weight g	Total hydroxyproline content mg/100 mg dry skin	Neutral salt soluble collagen		Acid soluble collagen	Insoluble collagen	
			Soluble in 0.15 M NaCl	Soluble in 0.45 M NAcl		6 M urea extractable collagen	Residue
			Percentage of total collagen				
Hypophysectomized (63 days)	136.6	6.98	0.15	0.86	0.53	34.80	54.76
Controls	337.4	8.00	0.36	4.20	2.34	35.30	51.50
Adrenalectomized (18 days)	142	5.3	1.11	6.76	9.55	29.34	48.56
Controls	155	5.8	0.83	5.53	8.60	28.2	49.83
Gonadectomized (74 days)	258	4.76	0.39	2.09	3.4	22.91	67.29
Controls	239	7.58	0.80	4.50	6.04	34.59	51.41

Table 2. Collagen proportion in different organs in hypophysectomized rats

	Hypophysectomized			Controls			t-Test	
Organ	Organ weight mg	HYP[a] g/wet weight	Dry weight %	Organ weight mg	HYP[a] g/wet weight	Dry weight %	Weight	HYP[a]
Kidney								
$\bar{x}$	755.0	197.4	23.0	2410.0	104.6	22.1	5.98	12.98
s	44.8	3.93		273.0	6.08		p<0.001	p<0.001
Heart								
$\bar{x}$	319.0	225.3	25.1	948.0	85.9	24.0	10.5	15.4
s	14.4	7.26		58.2	5.32		p<0.001	p<0.001
Liver								
$\bar{x}$	4500.0	126.7	32.1	11900.0	76.5	29.2	6.23	10.23
s	400.0	3.9		1025.0	3.02		p<0.001	p<0.001
Spleen								
$\bar{x}$	310.0	136.0	24.4	1040.0	103.6	22.8	5.8	4.15
s	9.42	7.48		124.2	2.4		p<0.001	p<0.001
Aorta								
$\bar{x}$	51.0	506.9	48.0	106.5	617.7	23.3	0.87	1.23
s	46.7	94.1		13.35	17.3		p<0.05	p<0.05
Epididymal fat $\bar{x}$	687.0	82.4	70.0	3440.0	55.4	58.8	1.04	0.60
s	137.0	36.2		1005.0	30.9		p<0.05	p<0.05
Lung								
$\bar{x}$	1490.0	345.0	24.4	1520.0	221.0	21.5	0.1	0.93
s	164.0	129.1		244.0	28.9		p<0.05	p<0.05
Testes								
$\bar{x}$	211.0	495.0	22.9	2620.0	130.9	11.1	13.5	4.30
s	35.9	81.7		174.0	23.1		p<0.001	p<0.001

[a] HYP = Hydroxyproline

also influenced by hypophysectomy, but less then the formation of other components.

As indicated in Table 4 there is a distinct difference between the contraction-relaxation profile of tail tendons of hypophysectomized and of control animals. This may be ascribed to the decreased food intake and lower over-all body weight of experimental animals. However, the contraction-relaxation curve of control animals, of the same weight as those hypophysectomized, shows that the effect observed should be ascribed to pituitary removal rather than to the decreased food intake.

Table 3. Total amount of collagen per organ in hypophysectomized and control rats

Organ	Hypophysectomy (Hypro per organ)	Control (Hypro per organ)
Kidney	149.037	252.086
Heart	71.870	71.433
Liver	570.150	910.350
Spleen	42.160	107.744
Aorta	25.852	65.785
Epididymal fat	56.609	190.576
Lung	514.050	335.920
Testes	104.445	342.958

In light of the above results, one can see that the overall picture of the effect of pituitary on the properties of connective tissue - mainly on the aggregational state of collagen - is rather difficult to draw, and apparently some partial results reported in the past might be misleading.

The first fact to be discussed is the decreased mechanical strength as indicated by the breakage time of tail tendons of hypophysectomized animals in urea. This phenomenon is thought to be due to blocked cross-linking formation, and the effect of pituitary has to be considered to paralell or to overlap the well-known decreased food intake in these animals. However, while in long-term food depriviation decreased collagen cross-linking was unequivocally established, the data related to a decreased level of cross-linking in hypophysectomized animals were not always verified. According to the experience of different authors the density of collagen cross-linking may vary greatly (Steinmetz et al., 1966; Deyl et al., 1967; Giles and Everitt, 1967). Data presented in the current paper indicate that the overall amount of fractions of soluble collagen in hypophysectomized animals is quite low, and that at least in skin the main portion of collagen present in hypophysectomized animals is the insoluble fraction. This statement holds, however, for animals surviving for a sufficiently long period after pituitary removal. In earlier stages a relatively high proportion of soluble collagen may appear due not only to the general hormonal disbalance of the organism but also presumably due to the abrupt decrease in food intake. It should be kept in mind that growth hormone increases collagen formation to a higher extent than other proteins, and it increases specifically collagen degradation (Valavaara and Kulonen, 1969; Valavaara et al., 1968). Thus from data of collagen solubility after hypophysectomy and adrenalectomy, resp., it can be concluded that the removal of pituitary influences the collagen synthesis more via absence of somatotropic hormone than via absence of the hormone corticotropin.

If we accept the autoimmune theory of the aetiopathogenesis of rheumatoid arthritis, the increased stabilization of collagen structure, i.e. the resulting decreased solubility and increased resistance to proteolytic enzymes, could definitely lead to a decreased rate of denaturation and breakdown products, which would participate in the formation of autoantigens. From the point of the autoimmune theory

Table 4. Contraction-relaxation of rat tail tendon fibers (5 cm long, 50 mg load)

Animal		Fiber length after min. (mm) 1 min[a]	30	60	90	120
	n	15	15	15	14	13
Hypophysectomized	$\bar{x}$	19.9	44.3	49.4	52.0	53.9
	s	0.33	0.77	0.68	0.66	0.45
	n	8	8	8	8	7
	$\bar{x}$	15.6	33.9	44.6	48.3	49.6
Controls	s	0.33	1.51	1.38	0.10	0.91
(same age)	t	9.18	6.12	3.14	3.08	4.26
	p	0.001	0.001	0.01	0.01	0.001
	n	10	10	10	10	10
	$\bar{x}$	21.3	34.5	41.6	51.4	56.7
Controls	s	0.094	1.06	0.64	0.78	0.56
(same weight)	t	1.4	7.44	8.34	0.59	3.9
	p	0.05	0.001	0.001	0.05	0.001

[a] 1 min = minimal length practically one minute after submersing in 2 M $KClO_4$.
$\bar{x}$ = mean, s = standard deviation

the findings presented may probably help to elucidate the low incidence of rheumatoid arthritis in males and may help to explain the mode of corticosteroid action in rheumatoid arthritis.

References

Chvapil, M., Zahradnîk,R.: A study of the chemical shrinkage and relaxation of collagen fibers. Biochim. biophys. Acta (Amst.) 40, 329 (1960).
Daughaday, W.H., Mariz, I.V.: Conversion of proline-U-^{14}C to labeled hydroxyproline by rat cartilage in vitro: effects of hypophysectomy, growth hormone, and cortisol. J. Lab. clin. Med. 59, 741 (1962).
Deyl, Z., Everitt, A.V., Rosmus, J.: Exp. Geront. 2, 249 (1967).
Deyl, Z., Juřicová, M., Rosmus, J., Adam, M.: The effect of food deprivation on collagen accumulation. Exp. Geront. 6, 383 (1971).
Giles, J.S., Everitt, A.V.: The role of the thyroid and food intake in the ageing of collagen fibers. I. In the young rat. Gerontologia (Basel) 13, 65 (1967).
Günther, T., Carsten, P.M.: Über die Löslichkeit von Kollagen in Abhängigkeit von Nebennierenrindenhormonen. Naturwissenschaften 48, 699 (1961).
Klemperer, P., Pollack, A.D., Baehr, G.: Diffuse collagen disease. J. Amer. med. Ass. 119, 331 (1942).
Steffen, C.: Praktische Bedeutung neuerer serologischer Untersuchungsmethoden. Dtsch. med. J. 16, 465 (1965).
Stevens, F.S.: Evidence for the chemical modification of collagen in rheumatoid arthritis. Ann. rheum. Dis. 24, 473 (1965).
Stegemann, H.: Microbestimmung von Hydroxyprolin mit Chlaromin-T und p-Dimethylaminobenzaldehyd. Z. physiol. Chem. 311, 41 (1958).
Steinetz, B.G., Beach, V.L., Elden, H.R.: Some effects of hormones on contractile properties of rat tail tendon collagen. Endocrinology 79, 1047 (1966).
Valavaara, M., Heikkinen, E., Kulonen, E.: Effect of hypophysectomy on the incorporation of proline in the collagen fractions. Experientia (Basel) 24, 779 (1968).
Valavaara, M., Kulonen, E.: The pituitary control of granulation tissue. Scand. J. clin. Lab. Invest. Suppl. 108, 23 (1969).

Studies on Collagen Biosynthesis in Rheumatoid Synovial Tissue

J. Uitto, S. Lindy, H. Turto, and K. Vainio

Aspects of collagen biosynthesis were investigated in human synovial tissue taken from patients with rheumatoid arthritis and from control subjects. These studies employed determinations of the rate of ^{14}C-hydroxyproline formation in vitro as well as assays of protocollagen proline hydroxylase activity.

The rate of collagen formation was measured by determining the radioactivity of ^{14}C-hydroxyproline in non-dialyzable proteins, after the tissue specimen had been incubated in a medium (Uitto, 1970) containing radioactive proline. The formation of radioactive hydroxyproline (dpm/μg hydroxyproline) in synovial tissue from 12 patients with rheumatoid arthritis was 47.6 ± 12.1 (mean ± SEM). The corresponding value in synovial tissue specimen taken from 9 control subjects was 10.9 ± 2.1.

The activity of protocollagen proline hydroxylase was assayed by incubating aliquots of the 15,000 × g supernatant of the tissue homogenate with biological, ^{14}C-proline-labelled protocollagen substrate (Uitto et al., 1970). Protocollagen proline hydroxylase activity (dpm ^{14}C-hydroxyproline synthesized with 50,000 dpm radioactive protocollagen substrate per mg supernatant protein) in the synovial tissue of 18 patients with rheumatoid arthritis was 3,130 ± 595 (mean ±SEM). The corresponding value in the synovial tissue of 10 control subjects was 1,500 ± 115.

Hydroxyproline is an imino acid almost exclusively found in collagen, and consequently its formation can be taken as an index of the rate of collagen biosynthesis. Protocollagen proline hydroxylase is an enzyme that participates in the biosynthesis of collagen by catalyzing the hydroxylation of proline in protocollagen, a hydroxyproline-deficient polypeptide precursor of collagen. The activity of this enzyme has been detected in several animal tissues (Halme, 1970) and also in human liver (Takeuchi and Prockop, 1969) and skin (Uitto et al., 1969, 1970). In granulomatous tissue, the activity of protocollagen proline hydroxylase has been shown to parallel the rate of collagen formation (Juva, 1968). Consequently, the present results seem to suggest that increased collagen formation in the synovial tissue may form an essential part of the disease process in rheumatoid arthritis.

References

Halme, J.: Protocollagen Proline Hydroxylase. Thesis, University of Helsinki, Helsinki 1970.

Juva, K.: Hydroxylation of proline in the biosynthesis of collagen. An experimental study with chick embryo and granulation tissue of rat. Acta physiol. scand., suppl. 308 (1968).

Takeuchi, T., Prockop, D.J.: Protocollagen proline hydroxylase in normal liver and in hepatic fibrosis. Gastroenterology 56, 744 (1969).
Uitto, J.: A method of studying collagen biosynthesis in human skin biopsies *in vitro*. Biochim. biophys. Acta (Amst.) 201, 438 (1970).
Uitto, J., Halme, J., Hannuksela, M., Peltokallio, P., Kivirikko, K.I.: Protocollagen proline hydroxylase activity in the skin of normal human subjects and of patients with scleroderma. Scand. J. clin. Lab. Invest. 23, 241 (1969).
Uitto, J., Hannuksela, M., Rasmussen, O.: Protocollagen proline hydroxylase activity in scleroderma and other connective tissue disorders. Ann. clin. Res. 2, 235 (1970).
Uitto, J., Lindy, S., Turto, H., Rokkanen, P., Vainio, K.: Increased protocollagen proline hydroxylase activity in synovial tissue in rheumatoid arthritis. Clin. chim. Acta 30, 741 (1970).

Changes in Skin Collagen in Osteoporosis

F. B. Reed and D. A. Hall

A number of workers have investigated changes in skin collagen with age and disease. Shuster (1967) has shown that the collagen content of skin declines with age. He demonstrated that this trend is more marked in women than in men, and for the arm than for the thigh. Females were found to have less collagen per unit area skin surface than males of a similar age. Wright and Joplin (1969) have shown that skinfold thickness decreases with age.

Everitt et al. (1970), Bakerman (1965) and a number of other workers have shown that the quantity of soluble collagen declines during growth, and subsequently remains at low values. Harris and Sjoerdsma (1966) have confirmed this observation, measuring acetic acid soluble collagen in skin from live subjects.

Keech (1955), in an electron microscope study, found qualitatively more digestion of collagen from the skin of young subjects, compared with that from old subjects, on incubation with bacterial collagenase. Kohn and Rollerson (1960), using collagen from the central tendon of the human diaphragm, have also found increased collagenase susceptibility in collagen obtained from a young group, compared with post-mortem samples from an old group of subjects.

The decreased solubility and increased stability of collagen with aging has been related by Jackson and Steven (1969) to an increase in the formation of covalent bonds between adjacent collagen molecules.

Deviations from the normal type of skin collagen have been found in a number of conditions:

Scarborough and Shuster (1960) have shown that both in corticosteroid induced purpura and senile purpura, there is sparse abnormally distributed collagen in the skin. C.J. Stevenson et al. (1960) have shown that in osteogenesis imperfecta there is decreased collagen content in skin, indicating a widespread defect of connective tissue. Wright and Joplin (1969) have shown that subjects with Cushing's syndrome tend to have thinner skin than normal subjects of a similar age. Shuster et al. (1967) have found that a group of rheumatoid patients treated with corticosteroids have decreased collagen in their skin when compared with a similar group of rheumatoid patients not receiving corticosteroid therapy. McConkey et al. (1967) have found a significantly higher incidence of transparent skin in female osteoporotic subjects as compared with normal controls. They suggested that transparency of the skin and osteoporosis may both be the consequence of the same underlying disorder of connective tissue. Similarly, Shuster has raised the question of whether the enhanced collagenolytic effect of corticosteroids on skin might also explain the corticosteroid-induced osteoporosis that has been observed. The prevalence of both osteoporosis and thin skin in subjects with Cushing's syndrome, lends further support to the hypothesis that the pathological changes in bone in osteoporosis might be accompanied by changes in the metabolism of collagen in the skin.

The object of the present study has been to follow up the results of these and other workers by obtaining measurements, which indicate the constitution of skin collagen in subjects with and without a high degree of osteoporosis. For our purposes osteoporosis of this type was only diagnosed in those patients with crush fractures and rarefaction and biconcavity of the vertebrae. The majority of such subjects were elderly, but at all ages they demonstrated a degree of osteoporosis in excess of that normally associated with their age group. They were, therefore, suffering from a syndrome which differed markedly from normal "aging osteoporosis", which has been described by Newton-John and Morgan (1968) as being a natural concommittant of the aging process. These subjects are referred to below as "pathological osteoporotics".

1. Material and Methods

The technique of Shuster, in which the collagen in a standard area of skin surface was measured, was followed throughout. Thus changes in total collagen in full skin thickness present could be followed, as well as alterations in the percentages of the different collagen fractions.

A surgical biopsy punch, diameter 6mm, driven at 20,000 rpm by a power operated hand-drill, was used to obtain biopsy specimens under local anaesthesia. Samples were obtained from the mid-line of the lateral aspect of the thigh from 82 subjects, both normal and osteoporotic, male and female, ranging in age from 22 to 80 years, and 69 similar post mortem skin samples were also obtained within 24 hours of death.

The skin samples thus obtained were trimmed free of excess fat, and suspended in 4ml 0.15 *M*. NaCl for 72 hours at 4°C. The solution so obtained (A) was retained for analysis, and the residual tissue frozen in liquid nitrogen, crushed in a steel percussion mortar, and transferred quantitatively to 4 ml 2.0 *M* NaCl at 4°C. After 24 hours, the residue was separated by centrifugation, and the supernatant (B) stored for analysis. The residue was then transferred to 4 ml pH 7.4 Tris buffer, 0.01 *M* with respect to Ca Cl_2 containing 500 µg bacterial collagenase (Sigma type 1). This was incubated at 37°C for 24 hours; the residue (D) and the supernatant (C) separated by centrifugation, and both preserved for analysis. A, B, C and D were then hydrolyzed, and analyzed for hydroxyproline by the method of Grant (1964). From the hydroxyproline content of A + B the salt-soluble collagen can be calculated, from that of C the collagenase susceptible collagen, and from D the insusceptible residue. From these the total collagen extracted, (= A+B+C+D), can then be expressed as µg collagen present per mm^2 skin surface for the full skin thickness.

Collagenase was assayed by a method adapted from Grant and Alburn (1959). Collagen obtained from bovine achilles tendon (as supplied by Koch-Light, Colnbrook, Bucks., England) of 20 - 40 mesh was used as substrate. This was incubated, with shaking at 37°C, in the presence of collagenase in pH 7.4 Tris buffer, containing 0.01 *M* Ca Cl_2. The reaction was terminated after 1 hour, and the released hydroxyproline determined. This method provided a reliable test of collagenase activity, enabling one (a) to determine when a collagenase preparation had deteriorated, and (b) to compare the activity of different batches of collagenase. Thus it was possible to maintain standard enzymic conditions in the extraction of collagenase susceptible collagen from skin.

Quantitative bone radiology was carried out on all patients by Mr. A. Horsman of the MRC Mineral Metabolism Unit, Leeds. Metacarpal Density was measured by the method of Anderson et al. (1966), using a conversion figure from aluminium to bone of 1.74 mg ash/cm^2 for each cm of the aluminium step wedge.

2. Results

Fig. 1 shows the mean values of total skin collagen for 151 males and females. No significant differences were found between the skin samples

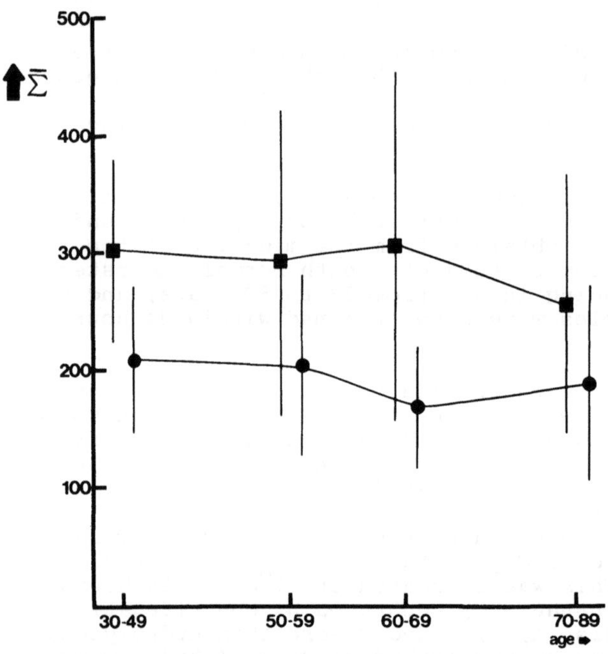

Age groups	30 - 49 years	50 - 59 years	60 - 69 years	70 - 89 years
Males ■	302 ± 78	291 ± 131	305 ± 149	256 ± 111
Females ●	209 ± 63	204 ± 77	168 ± 52	188 ± 83
p	<0.0025	<0.025	<0.0025	<0.025

Fig. 1. Age change in mean total skin collagen - males and females. $\bar{\Sigma}$ (mean µg collagen mm^2 skin surface) plotted against age groups

obtained from live subjects, and those obtained from post-mortem. Similarly, no significant differences were found between values for normal and osteoporotic subjects. The graph thus shows the means and standard deviations for all males and all females. From this it can be seen that males have significantly more collagen in their skin for each of the four age groups shown. Over the age range 30 to 90 years, although there appears to be a slight fall with age, there is no significant change in the total collagen content. The histograms in Fig. 2

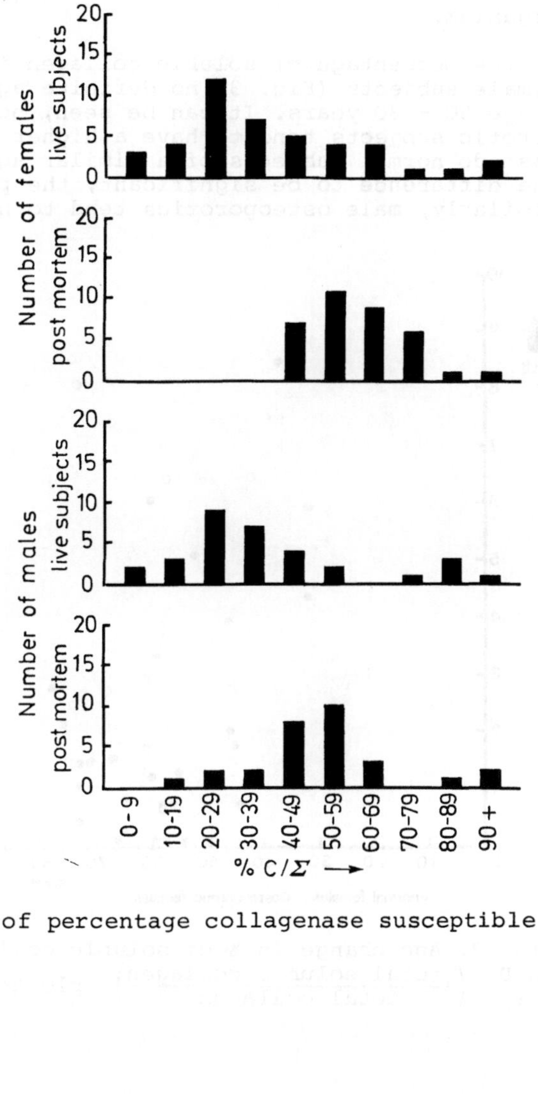

Fig. 2. Distributions of values of percentage collagenase susceptible collagen in total skin sample

show the differences in distribution of the percentage of collagenase-susceptible collagen in the skin. As can be seen for both males and females, there is a dramatic increase in susceptibility in those samples obtained post-mortem. This suggests that the process of collagen breakdown is markedly increased after death. Hence we cannot consider results obtained from the fractionation of collagen from post-mortem skin together with results obtained from live subjects. A second point arising from this result is that a study of some of the degradative changes that occur in the first few hours after death may yield valuable information about changes that occur in the living organism.

If the percentage of soluble collagen is plotted against age for live female subjects (Fig. 3) no definite age change emerges for the age range 50 - 80 years. It can be seen, however, that pathological osteoporotic subjects tend to have a higher proportion of soluble collagen than do normal subjects of a similar age. Analysis of the data shows the difference to be significant, the p value being less than 0.01. Similarly, male osteoporotics tend to have more soluble collagen than

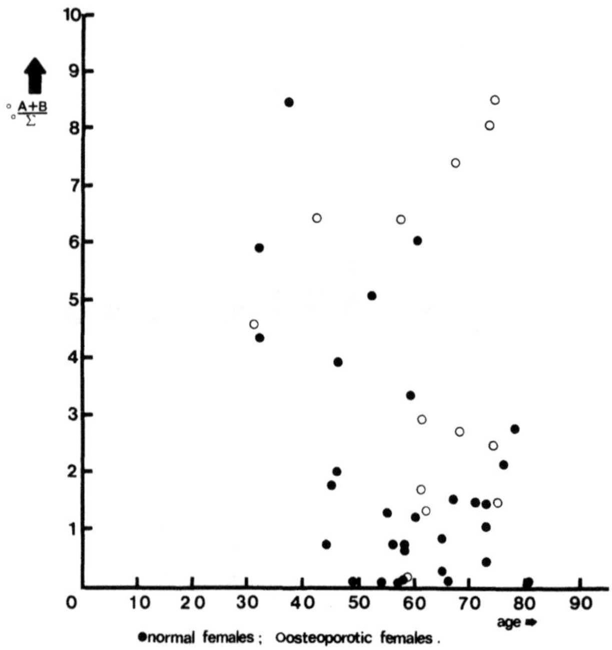

Fig. 3. Age change in % of soluble collagen in skin biopsies-females $\%\frac{A+B}{\Sigma}$ $\left(\%\frac{\text{total soluble collagen}}{\text{total collagen}}\right)$ plotted against age in years

normal male subjects, but in this case, due to the smaller population, the difference is not significant.

Fig. 4 shows collagenase susceptible collagen plotted against age for female subjects. Pathological osteoporotic patients tend to have a higher percentage of susceptible collagen than normal subjects of the same age and sex, but this difference is not significant for either males or females.

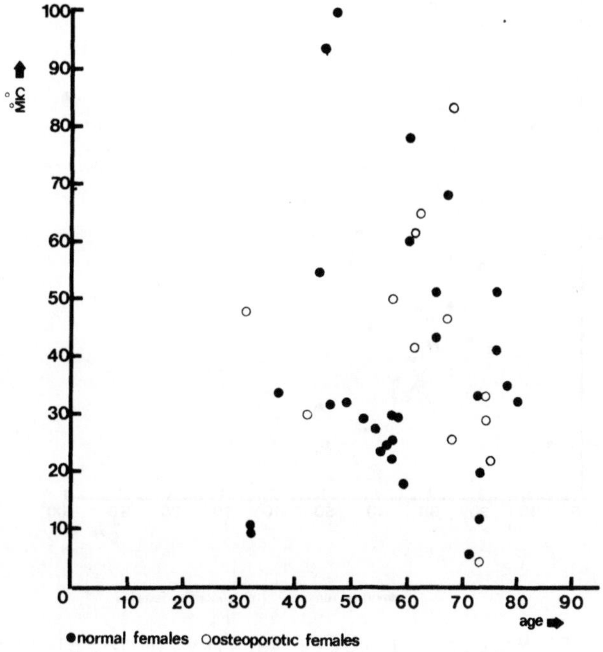

Fig. 4. Age change in the percentage of collagenase susceptible collagen in skin biopsies-females
$\% \frac{C}{\Sigma}$ $\left(\% \frac{\text{total collagenase susceptible collagen}}{\text{total collagen}}\right)$ plotted against age in years

In Fig. 5 this last skin parameter is plotted against Metacarpal Total Density (MTD). This shows that females with excessive osteoporosis have (a) a low bone density and (b) a higher percentage of susceptible collagen.

This trend is again seen in plotting MTD against percentage soluble collagen for female subjects (Fig. 6). It appears that subjects with low bone density and high soluble collagen in skin are more likely to sustain crush fractures of the spine than is the normal population.

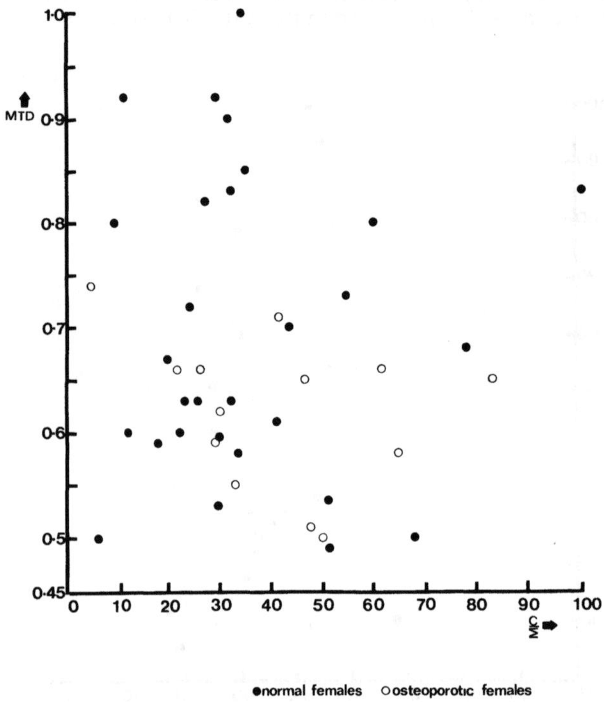

Fig. 5. Metacarpal total density (MTD) plotted against the percentage of collagenase susceptible collagen-females

It has recently been suggested that osteoporosis is not a disease, but a consequence of the generalized loss of bone that occurs during aging. These results confirm that crush fractures do occur in subjects with low bone density; however, as Newton-John and Morgan (1968) have pointed out, there are other unknown factors which select those people who will suffer a fracture from a group of persons with thin bones. Our findings suggest that these other factors include changes in the metabolism of collagen in the body.

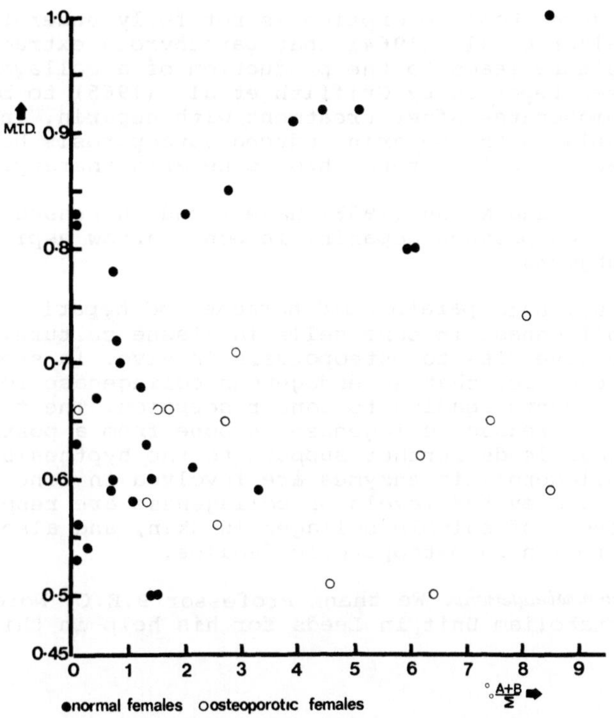

Fig. 6. Metacarpal total density (MTD) plotted against the percentage of total soluble collagen-females

3. Discussion

Albright and Reifenstein (1948) have described osteoporosis as a disease of tissue metabolism as distinct from calcium metabolism. He based this statement on his observation that treatment of post-monopausal osteoporotic women with oestrogen resulted in a decrease in pain, increase in weight, and also increased skin thickness. Other workers, notably Nordin (1961), claim to have shown that a negative calcium balance, which can be the consequence of a low calcium diet, leads eventually to the production of osteoporosis. Recent work on the mechanism of bone resorption has led Houssain et al. (1970) to suggest that these hypotheses are not mutually exclusive. Jowsey (1968) has shown that cats fed a low calcium diet develop osteoporosis, whereas cats fed a low calcium diet, do not. It thus appears that PTH is necessary for the production of osteoporosis in cats. Zanelli (1969) has shown that PTH added to mouse calvaria in tissue culture results in increased bone resorption. She has also confirmed the observation of Nisbit (personal communication) that this effect can be blocked by the addition of stilboesterol to the culture medium. Oestrogen can thus be seen to protect bone from PTH induced resorption. This protective action of oestrogen is lost in postmenopausal women, and hence their increased incidence of osteoporosis. The mechanism by which PTH induces bone resorption is not fully understood. It has been shown by Walker et al. (1964) that parathyroid extract added to bone in tissue culture leads to the production of a collagenase. Collagenase has also been reported by Griffith et al. (1965) to be increased in rat bone homogenates after treatment with heparin. In Griffith's view this explains the heparin induced osteoporosis he has noted in patients receiving long-term chronic heparin therapy.

Frame and Nixon (1968) have found increased numbers of mast cells (which produce heparin) in bone marrow aspirates from osteoporotic subjects.

Thus, both parathyroid hormone and heparin stimulate production of collagenase in bone cells in tissue culture, and both have been shown to give rise to osteoporosis in vivo. It seems reasonable to suppose, therefore, that an endogenous collagenase forms one link in the chain of events leading to bone resorption. The finding of Nichols (1966) of increased collagenase in bone from a postmenopausal osteoporotic woman lends further support to the hypthesis of Vaes (1968) that collagenolytic enzymes are involved in bone resorption. It is possible that elevated levels of collagenase are responsible for both increased levels of soluble collagen in skin, and also for increased bone resorption in osteoporotic females.

Acknowledgement. We thank Professor B.E.C. Nordin of the M.R.C. Mineral Metabolism Unit in Leeds for his help in this project.

References

Albright, F., Reifenstein, E.C.: The Parathyroid Glands and Metabolic Bone Disease, p. 145. Baltimore: Williams and Wilkins 1948.
Anderson, J.B., Shimmins, J., Smith, D.A.: A new technique for the measurement of metacarpal density. Brit. J. Radiol. 39, 443 - 450 (1966).
Bakerman, S.: Alteration in the distribution of the α, β and γ components of human skin collagen with disease. Nature 206, 634 - 635 (1965).
Everitt, A.V., Gal, A., Steele, M.G.: Age changes in the solubility of tail tendon collagen throughout the lifespan of the rat. Gerontologia (Basel) 16, 30 - 40 (1970).
Frame, B., Nixon, R.K.: Bone marrow mast cells in osteoporosis. New Engl. J. Med. 279, 626 - 630 (1968).
Grant, N.H., Alburn, H.E.: Studies on the collagenases of clostridium histolyticum. Arch.Biochem. 82, 245 - 255 (1959).
Grant, R.A.: Estimation of hydroxyproline by the autoanalyser. J. clin. Path. 17, 685 - 686 (1964).
Griffith, G.C., Nichols, G., Asher, J.D., Flanagan, B.: Heparin osteoporosis. J. Amer. med. Ass. 193, 91 - 94 (1965).
Harris, E.D., Sjoerdsma, A.: Collagen profile in various clinical conditions. Lancet 1966II, 707 - 711.
Houssain, M., Smith, D.A., Nordin, B.E.C.: Parathyroid activity and postmenopausal osteoporosis. Lancet 1970I, 809 - 811.
Jackson, D.S., Steven, F.S.: Age changes in the crosslinking of human collagen. Gerontologia (Basel) 15, 77 - 84 (1969).
Jowsey, J., Raisy, L.: Experimental osteoporosis and parathyroid activity. Endocrinology 82, 384 - 396 (1968).
Keech, M.: Electron microscopy of human skin collagen. Ann. rheum. Dis. 14, 19 - 49 (1955).

Kohn, R.R., Rollerson, E.: Ageing of human collagen in relation to its susceptibility to collagenase. J. Geront. 15, 10 - 14 (1960).
McConkey, B., Walton, K.W., Carney, S.A., Lawrence, J.C., Ricketts, C.R.: Significance of occurence of transparent skin: a study of histologic characteristics and biosynthesis of dermal collagen. Ann. rheum. Dis. 26, 219 - 225 (1967).
Newton-John, H.F., Morgan, D.B.: Osteoporosis: Disease or senescence? Lancet 1968I, 232 - 233.
Nichols, G., jr.: The Significance of Collagenase Activity in Human Bone. Fourth Europ. Symp. Calc. Tiss. 1966. No. 60. Abridged proc. pub. Excerpta Medica, Amsterdam.
Nordin, B.E.C.: The pathogenesis of osteoporosis. Lancet 1961I, 1011 - 1014.
Scarborough, H., Shuster, S.: Corticosteroid purpura. Lancet 1960I, 93 - 94.
Shuster, S., Raffle, E.J., Bottoms, E.: Skin collagen in rheumatoid arthritis and the effect of corticosteroids. Lancet 1967I, 525 - 527.
Stevenson, C.J., Bottoms, E., Shuster, S.: Skin collagen in osteogenesis imperfecta. Lancet 1960I, 860 - 861.
Vaes, G.: On the mechanism of bone resorption. J. Cell Biol. 39, 676 to 697 (1968).
Walker, D.G., Lapiere, C.M., Gross, J.: A collagenolytic factor in rat bone promoted by parathyroid extract. Biochem. biophys. Res. Commun. 15, 397 - 402 (1964).
Wright, A.D., Joplin, C.F.: Skin fold thickness in normal subjects and in patients with acromegaly and Cushing's syndrome. Acta Endocrinologica (Kbhg.) 60, 705 - 711 (1969).
Zanelli, J.M., Lea, D.J., Nisbet, J.A.: A bioassay method in vitro for parathyroid hormone. J. Endocr. 43, 33 - 46 (1969).

Collagen and Elastin Metabolism in Scurvy: Studies on the Nature of the Collagen Lesion

M. J. Barnes

The participation of ascorbic acid in collagen metabolism has long been recognized (see Gould, 1968; Chvapil and Hurych, 1968; Barnes, 1969) but the precise nature of this involvement is still a matter of some speculation. Despite the known ability, established in recent years, of ascorbic acid to participate, at least in isolated cell-free systems, in the hydroxylation of protocollagen (see Prockop, 1970; Udenfriend, 1970), the accumulation *in vivo* of a proline-enriched hydroxyproline-deficient collagen in scorbutic guinea pigs could not be demonstrated in earlier studies (Gould et al., 1960; Gross, 1959; Robertson et al., 1959). The role of ascorbic acid in the hydroxylation of peptidyl proline and lysine remains undefined, for the role of co-substrate in the mixed-function oxidase reaction has been clearly ascribed to α-ketoglutarate. Further, the requirement for ascorbate is not absolute and in this respect it is natural to consider that other substances (e.g. tetrahydrofolate; see Gould, 1970) might function in the hydroxylation *in vivo* either as an alternative to or even to the exclusion of ascorbic acid. The impaired formation of collagen in scurvy might then be seen as implying some other role of ascorbic acid in collagen metabolism distinct form that of hydroxylation.

1. Isotope Incorporation Studies

We have sought evidence for the occurrence of protocollagen in scurvy from isotope incorporation studies. A comparison was made in scorbutic guinea pigs between the metabolism of collagen and elastin, since the latter also contains some hydroxyproline, although much less than in collagen. The results (Barnes et al., 1969a, 1970) indicated that in the case of elastin, the effect of ascorbic acid deprivation could, in fact, be interpreted simply as an elimination of the hydroxylation of elastin proline with the formation of a polymer increasingly deficient in hydroxyproline, a finding entirely in accord with the concept of the role of ascorbic acid in connective tissue protein metabolism as one solely in the hydroxylation of peptidyl proline or lysine. It seems probable that the hydroxylase involved in the hydroxylation of prolyl residues in elastin is, in fact, collagen proline hydroxylase (Rhoads and Udenfriend, 1969).

The effect of ascorbic acid deprivation on collagen metabolism, however, is not so easily defined in these terms (Barnes et al., 1969a, 1970). There was no accumulation of unhydroxylated material; incorporation of radioactivity into both collagen hydroxyproline *and* proline showed a rapid decline. The proline: hydroxyproline specific activity ratio showed only a slight but nevertheless consistent increase to a value around 1.2. These results have been interpreted as indicative of some impairment in hydroxylation with the formation, *but in rapidly diminishing amounts*, of a collagen only slightly under-hydroxylated. The slight reduction in the level of hydroxylation did not appear to interfere with the aggregation and cross-linking of this collagen (Barnes and Kodicek, 1972).

The question that arises is why collagen is not simply replaced by protocollagen, a question that could have a bearing, of course, on the role of hydroxylation in collagen metabolism. The failure to detect (in dialyzed skin collagen extracts) at any period of ascorbic acid deficiency, even when examining the tissue as shortly as 2 hours after administration of the isotope, a collagen deficient in hydroxyproline beyond the small reduction already noted might imply that such a collagen is very rapidly degraded or alternatively not synthesized. Conceivably under-hydroxylated peptides may be released from the ribosome before they attain non-diffusible proportions or possibly an inhibition of ribosomal amino acid incorporation may occur.

2. Urinary Excretion Studies

Despite the rapid decline in the incorporation of radioactivity into skin collagen of scorbutic guinea pigs, the incorporation into the diffusible hydroxyproline of skin, which may be regarded as arising by the degradation of collagen, remained high. This is in agreement with the observation that the urinary hydroxyproline excretion in these animals remains undiminished for some considerable period after collagen metabolism is first affected by the avitaminosis and only falls when inanition becomes a complicating factor (Barnes et al., 1969b). In humans with scurvy, an increased hydroxyproline excretion has been reported (Burkley, 1968; Efron et al., 1968).

Although at first sight these observations are not compatible with impaired hydroxylation, we regard the normal or increased hydroxyproline excretion in scurvy as reflecting extensive degradation of partially-hydroxylated material.

Preliminary fractionation studies have indicated that approximately 1/3 of the total urinary hydroxyproline excretion in guinea pigs is non-diffusible (through Visking 18/32 tubing). Isotope incorporation data are consistent with the derivation of this non-diffusible fraction in scurvy from, in addition to the degradation of preformed insoluble collagen, the degradation of the slightly-under-hydroxylated collagen, the presence of which in skin has already been described. The incorporation data indicated that the latter collagen is degraded at a relatively constant rate despite a diminishing concentration in the tissues.

Collagen-derived diffusible peptides in urine are now being examined for evidence of an impairment in hydroxylation beyond the slight reduction detectable in collagen of skin extracts or the non-diffusible hydroxyproline-containing peptides of urine.

3. Impaired Amino Acid Incorporation

There remains the likelihood that much of the reduction in the incorporation of radioactivity into collagen in scurvy can be attributed to an inhibition of ribosomal collagen protein synthesis, a suggestion that receives some support from electron microscopal studies (Ross, 1968). If such inhibition should be regarded as independent of impaired hydroxylation, then it is reasonable to suppose that under the appropriate conditions it should be possible to demonstrate *in vivo* impaired hydroxylation unaccompanied by impaired amino acid incorporation (i.e. the formation of protocollagen). We decided to examine the situation in chronic hypovitaminosis C on the assumption that a low intake of the vitamin might prove adequate for collagen protein synthesis, while insufficient for hydroxylation. However, we could only detect the formation of the same slightly under-hydroxylated collagen already observed in the acute condition (Barnes and Kodicek, 1972). As judged from the level of incorporated radioactivity, this collagen was formed at a relatively constant rate that was about half that of normal collagen synthesis in control animals. Hydroxyproline excretion, as in the acute condition, remained unchanged. Thus it was not possible in this instance to demonstrate impaired hydroxylation without reduced incorporation.

From consideration of the results with elastin, it might seem reasonable to conclude, by analogy, that the primary lesion in collagen synthesis is impaired hydroxylation and that the failure of protocollagen to accumulate in scurvy should be regarded as a consequence of this impairment. Recently evidence has been presented which implies that hydroxylation of both prolyl and lysyl residues is necessary for the normal extrusion of collagen from the cell (Ramaley and Rosenbloom, 1971; Rosenbloom and Prockop, 1971; Rosenbloom and Christner, 1971). It is possible that with a failure in hydroxylation, material either unhydroxylated or substantially low in hydroxylated residues will accumulate within the cell. This may lead to its lysosomal degradation and also to a feed-back inhibition of further synthesis.

The features of impaired collagen metabolism observed *in vivo* in scurvy, namely the appearance in diminishing amounts of a collagen only slightly under-hydroxylated together with the occurrence of a normal or increased level of degradation products, bear a similarity to the situation described in 3T6 mouse fibroblast cultures, when hydroxylation is inhibited with α, α' dipyridyl (Ramaley and Rosenbloom, 1971).

References

Barnes, M.J.: Ascorbic acid and the biosynthesis of collagen and elastin. Bibl. "Nurtr. et Dieta" (Basel) 13, 86 - 98 (1969).

Barnes, M.J., Constable, B.J., Kodicek, E.: Studies *in vivo* on the biosynthesis of collagen and elastin in ascorbic acid-deficient guinea pigs. Biochem. J. 113, 387 - 397 (1969a).

Barnes, M.J., Constable, B.J., Kodicek, E.: Excretion of hydroxyproline and other amino acids in scorbutic guinea pigs. Biochim. biophys. Acta (Amst.) 184, 358 - 365 (1969b).

Barnes, M.J., Constable, B.J., Morton, L.F., Kodicek, E.: Studies in vivo on the biosynthesis of collagen and elastin in ascorbic acid-deficient guinea pigs: Evidence for the formation and degradation of a partially hydroxylated collagen. Biochem. J. 119, 575 - 585 (1970).

Barnes, M.J., Kodicek, E.: Biological hydroxylations and ascorbic acid with special regard to collagen metabolism. Vitamins and Hormones 30, 1 - 43 (1972).

Burkley, K.: Urinary hydroxyproline excretion in scurvy. M. Sc. Thesis, University of Iowa 1968.
Chvapil, M., Hurych, J.: Control of collagen biosynthesis. Int. Rev. Connect. Tissue Res. 4, 67 - 196 (1968).
Efron, M.L., Bixby, E.M., Hockaday, T.D.R., Smith, L.S., Meshorer, E.: Hydroxyprolinemia: The origin of free hydroxyproline in hydroxyprolinemia. Collagen turnover. Evidence for a biosynthetic pathway in man. Biochim. biophys. Acta (Amst.) 165, 238 - 250 (1968).
Gould, B.S.: In: Treatise on Collagen (Ed. Gould, B.S.), Vol. 2A, p. 323. London, New York: Academic Press 1968.
Gould, B.S.: In: Chemistry and Molecular Biology of the Intercellular Matrix (Ed. Balazs, E.A.) , Vol. 1, p. 431. London, New York: Academic Press 1970.
Gould, B.S., Manner, G., Goldman, H.M., Stolman, J.: Some aspects of collagen formation. Ann. N.Y. Acad. Sci. 85, 385 - 398 (1960).
Gross, J.: Effect of vitamin C deficiency on the neutral salt-extractible collagen of skin. J. exp. Med. 109, 557 - 569 (1959).
Prockop, D.J.: In: Chemistry and Molecular Biology of the Intercellular Matrix (Ed. Balazs, E.A.), Vol. 1, p. 335. London, New York: Academic Press 1970.
Ramaley, P.B., Rosenbloom, J.: Inhibition of proline and lysine hydroxylation prevents normal extrusion of collagen by 3T6 fibroblasts in culture. FEBS Letters 15, 59 - 64 (1971).
Rhoads, R.E., Udenfriend, S.: Substrate specificity of collagen proline hydroxylase: Hydroxylation of a specific proline residue in bradykinin. Arch. Biochem. 133, 108 - 111 (1969).
Robertson, W. van B., Hewitt, J., Herman, C.: The relation of ascorbic acid to the conversion of proline to hydroxyproline in the synthesis of collagen in the carrageenan granuloma. J. biol. Chem. 234, 105 - 108 (1959).
Rosenbloom, J., Christner, P.: Effects of 4,5-dehydrolysine on collagen biosynthesis in embryonic chick tibiae. Fed. Proc. 30, 1195 Abs. (1971).
Rosenbloom, J., Prockop, D.J.: Incorporation of cis-hydroxyproline into protocollagen and collagen. J. biol. Chem. 246, 1549 - 1555 (1971).
Ross, R.: The fibroblast and wound repair. Biol. Rev. 43, 51 - 96 (1968).
Udenfriend, S.: In: Chemistry and Molecular Biology of the Intercellular Matrix (Ed. Balazs, E.A.), Vol. 1, p. 371. London, New York: Academic Press 1970.

Subject Index

MIX
Papier aus verantwortungsvollen Quellen
Paper from responsible sources
FSC® C105338

If you have any concerns about our products,
you can contact us on
ProductSafety@springernature.com

In case Publisher is established outside the EU,
the EU authorized representative is:
Springer Nature Customer Service Center GmbH
Europaplatz 3, 69115 Heidelberg, Germany

Printed by Libri Plureos GmbH
in Hamburg, Germany